Lignocellulosic Biomass to Liquid Biofuels

Lignocellulosic Biomass to Liquid Biofuels

Edited by

ABU YOUSUF
Department of Chemical Engineering & Polymer Science, Shahjalal University of Science and Technology, Sylhet, Bangladesh

DOMENICO PIROZZI
Department of Chemical, Materials and Production Engineering, University of Naples "Federico II", Naples, Italy

FILOMENA SANNINO
Department of Agricultural Sciences, University of Naples "Federico II", Naples, Italy

ELSEVIER

ACADEMIC PRESS
An imprint of Elsevier

Academic Press is an imprint of Elsevier
125 London Wall, London EC2Y 5AS, United Kingdom
525 B Street, Suite 1650, San Diego, CA 92101, United States
50 Hampshire Street, 5th Floor, Cambridge, MA 02139, United States
The Boulevard, Langford Lane, Kidlington, Oxford OX5 1GB, United Kingdom

British Library Cataloguing-in-Publication Data
A catalogue record for this book is available from the British Library

Library of Congress Cataloging-in-Publication Data
A catalog record for this book is available from the Library of Congress

ISBN: 978-0-12-815936-1

For Information on all Academic Press publications
visit our website at https://www.elsevier.com/books-and-journals

Publisher: Katie Hammon
Acquisition Editor: Raquel Zanol
Editorial Project Manager: Mariana Pizzolatto Henriques
Production Project Manager: Anitha Sivaraj
Cover Designer: Mark Rogers

Typeset by MPS Limited, Chennai, India

Dedicated

To the future biofuel researchers, who will work for a less polluted earth.

Contents

List of contributors

Naveenji Arun
Catalysis and Chemical Reaction Engineering Laboratories, Department of Chemical and Biological Engineering, University of Saskatchewan, Saskatoon, SK, Canada

Sundaram Arvindnarayan
Department of Mechanical Engineering, Rajas Engineering College, Vadakkankulam, India

A.E. Atabani
Alternative Fuels Research Laboratory (AFRL), Energy Division, Department of Mechanical Engineering, Faculty of Engineering, Erciyes University, Kayseri, Turkey

Azrina Abd Aziz
Faculty of Engineering Technology, Universiti Malaysia Pahang, Gambang, Malaysia

M. Balajii
Department of Biotechnology, PSG College of Technology, Coimbatore, India

Alessandro Blasi
ENEA—Italian National Agency for New Technologies, Energy and Sustainable Economic Development, Energy Technologies Department, Research Centre of Trisaia, Rotondella, Italy

Vincenza Calabrò
Department of Computer Engineering, Modeling, Electronics and Systems Engineering (DIMES), University of Calabria, Rende, Italy

Patrizia Casella
ENEA, National Agency for New Technologies, Energy and Sustainable Economic Development, Department of Sustainability, Portici, Italy

Sudip Chakraborty
Department of Computer Engineering, Modeling, Electronics and Systems Engineering (DIMES), University of Calabria, Rende, Italy

Ajay K. Dalai
Catalysis and Chemical Reaction Engineering Laboratories, Department of Chemical and Biological Engineering, University of Saskatchewan, Saskatoon, SK, Canada

Jeyaprakash Dharmaraja
Division of Chemistry, Faculty of Science and Humanities, Sree Sowdambika College of Engineering, Virudhunagar, India

Alberto Figoli
CNR, Institute on Membrane Technology, National Research Council, University of Calabria, Rende, Italy

Gabriella Fiorentino
Department of Science and Technology, Parthenope University of Naples, Centro Direzionale – Isola C4, Naples, Italy

Angela Iovine
ENEA, National Agency for New Technologies, Energy and Sustainable Economic Development, Department of Sustainability, Portici, Italy; Department of Engineering, University of Campania "Luigi Vanvitelli", Aversa, Italy

Gopalakrishnan Kumar
Institute of Chemistry, Bioscience and Environmental Engineering, Faculty of Science and Technology, University of Stavanger, Stavanger, Norway

Dehua Liu
Key Laboratory for Industrial Biocatalysis, Ministry of Education of China, Institute of Applied Chemistry, Department of Chemical Engineering, Tsinghua University, Beijing, P.R. China

Catia Giovanna Lopresto
Department of Computer Engineering, Modeling, Electronics and Systems Engineering (DIMES), University of Calabria, Rende, Italy

Tiziana Marino
CNR, Institute on Membrane Technology, National Research Council, University of Calabria, Rende, Italy

Sanjeet Mehariya
ENEA, National Agency for New Technologies, Energy and Sustainable Economic Development, Department of Sustainability, Portici, Italy; Department of Engineering, University of Campania "Luigi Vanvitelli", Aversa, Italy

Antonio Molino
ENEA, National Agency for New Technologies, Energy and Sustainable Economic Development, Department of Sustainability, Portici, Italy

Minhaj Uddin Monir
Faculty of Engineering Technology, Universiti Malaysia Pahang, Gambang, Malaysia; Department of Petroleum and Mining Engineering, Jessore University of Science and Technology, Jessore, Bangladesh

Dino Musmarra
Department of Engineering, University of Campania "Luigi Vanvitelli", Aversa, Italy

S. Niju
Department of Biotechnology, PSG College of Technology, Coimbatore, India

Domenico Pirozzi
Department of Chemical, Materials and Production Engineering, University of Naples "Federico II", Naples, Italy

Arivalagan Pugazhendhi
Innovative Green Product Synthesis and Renewable Environment Development Research Group, Faculty of Environment and Labour Safety, Ton Duc Thang University, Ho Chi Minh City, Vietnam

Filomena Sannino
Department of Agricultural Sciences, University of Naples "Federico II", Naples, Italy

Neeta Sharma
ENEA, National Agency for New Technologies, Energy and Sustainable Economic Development, Department of Sustainability, ENEA Research Centre Trisaia, Rotondella, Italy

Sutha Shobana
Department of Chemistry & Research Centre, Mohamed Sathak Engineering College, Ramanathapuram, India

M. Swathika
Department of Biotechnology, PSG College of Technology, Coimbatore, India

Sergio Ulgiati
Department of Science and Technology, Parthenope University of Naples, Centro Direzionale – Isola C4, Naples, Italy; School of Environment, Beijing Normal University, Beijing, P.R. China

Manokaran Vadivel
Department of Chemistry & Research Centre, Mohamed Sathak Engineering College, Ramanathapuram, India

Alessandra Verardi
Department of Computer Engineering, Modeling, Electronics and Systems Engineering (DIMES), University of Calabria, Rende, Italy

Abu Yousuf
Department of Chemical Engineering and Polymer Science, Shahjalal University of Science and Technology, Sylhet, Bangladesh

Jingzhi Zhang
Key Laboratory for Industrial Biocatalysis, Ministry of Education of China, Institute of Applied Chemistry, Department of Chemical Engineering, Tsinghua University, Beijing, P.R. China

Xuebing Zhao
Key Laboratory for Industrial Biocatalysis, Ministry of Education of China, Institute of Applied Chemistry, Department of Chemical Engineering, Tsinghua University, Beijing, P.R. China

Haifeng Zhou
Key Laboratory of Low Carbon Energy and Chemical Engineering, College of Chemical and Environmental Engineering, Shandong University of Science and Technology, Qingdao, P.R. China

Amalia Zucaro
Department of Science and Technology, Parthenope University of Naples, Centro Direzionale – Isola C4, Naples, Italy

Gaetano Zuccaro
Laboratory of Environmental Biotechnology, National Institute of Agronomic Research, University of Montpellier, Narbonne, France; Department of Chemical, Materials and Production Engineering, University of Naples "Federico II", Naples, Italy

About the editors

Abu Yousuf holds a PhD degree in Chemical Engineering from the University of Naples Federico II, Italy. His primary research interests include biorefinery, bioenergy, bioremediation, and waste-to-energy. He published more than 50 papers in reputed ISI and Scopus indexed journals and 5 book chapters. He has been editing three books, which will be published by Elsevier. He has been serving as an editorial board member of several reputed journals. He won UNESCO Prize on E-learning course of "Energy for sustainable development in Asia," Jakarta, Indonesia, 2011. He attended in the "BIOVISION. Next Fellowship Programme 2013" at Lyon, France, after a selection based on scientific excellence, mobility, involvement in civil society. He also received several awards for his research outcomes from UK, Malaysia and Bangladesh. He successfully accomplished 10 research grants including the grants provided by The World Academy of Science (TWAS), Italy; Ministry of Higher Education, Malaysia; and Ministry of Science and Technology, Bangladesh. He is the member of IChemE, American Chemical Society (ACS), American Association for Science and Technology (AASCIT), and Bangladesh Chemical Society (BCS). He has 12 years' experience of teaching in undergraduate and postgraduate levels having very good remarks from the student. Currently, he is serving as a Professor in Chemical Engineering and Polymer Science, Shahjalal University of Science and Technology, Bangladesh. Previously, he held the position of Senior Lecturer at the Faculty of Engineering Technology, Universiti Malaysia Pahang, Malaysia. He presented his research work in Germany, France, Italy, India, Vietnam, Malaysia and Bangladesh. He can be contacted at ayousufcep@yahoo.com/ayousuf-cep@sust.edu.

Domenico Pirozzi is Professor of Biochemical Engineering Principles at the University Federico II of Naples and member of the Interdepartmental Center for Environmental Research (C.I.R.AM). He received his PhD in Chemical Engineering with a specialization in Biochemical Engineering at the University Federico II and subsequently spent 18 months at University of Strathclyde (Glasgow, United Kingdom) within the TMR Program. His primary research interests include enzymatic reactors, biofuel production from wastes, and bioremediation of polluted waters. He has been involved in several research projects and consultant for several companies and is author of about 100 papers in ISI indexed journals and of several book chapters. He has a long experience of teaching Biochemical Engineering principles, biochemical reactors, transport phenomena at the University Federico II of Naples and has tutored several graduate and PhD theses. He has also taught undergraduate and graduate courses at the Universidad National del Litoral (Santa Fe, Argentina) and the Tsing-Hua University (Bejing, China) as visiting Professor.

Filomena Sannino holds PhD in Agricultural Chemistry from the University of Naples Federico II, Italy. Her primary research interests include bioremediation, heterogeneous catalysis, bioenergy, and valorization of agricultural biomasses. She is the founding partner of Spin-Off GreenAmbioTech s.r.l. She is the author of the patent "Process for producing triacetylhydroxytyrosol from olive oil mill waste waters for use as stabilized antioxidant; De Martino A., Sannino F., Manna C., Gianfreda L., Capasso R. PCT/IT2005/000781."

She is a member of The Interdepartmental Research Centre on Nuclear Magnetic Resonance (NMR) for the Environment, Agroo-food and New Materials (CERMANU) of the University of Naples Federico II.

She is a member of Technical Scientific Committee for the "Food Security and New Technologies" Area, partnership between the Interdepartmental Research Centre "Laboratory of Urban and Regional Planning" (LUPT) of University of Naples "Federico II" and European Parliament.

She published 83 papers in ISI indexed journals and several book chapters. She is a member of Agricultural Chemistry Society and International Humic Substances Society (IHSS).

She has 19 years' experience of teaching for undergraduate and post-graduate students in agricultural biochemistry, inorganic chemistry, use and recycling of agricultural biomasses, plant biotechnology for environmental phytoremediation, and biochemistry of plant hormones and growth regulators.

Currently, she is a researcher in Agricultural Chemistry at Agricultural Sciences Department of University of Naples "Federico II."

Preface

Lignocellulosic biomass is a potential source of biofuels, which is abundant and readily available around the world. Simultaneously, liquid biofuels have great demand in transportation and industrial sectors. Therefore, this book particularly focuses on the processing of lignocellulosic biomass to liquid biofuels. Currently, a growing number of liquid biofuels are evaluated for commercial exploitation, such as bioethanol, biodiesel, methanol, and reformulated gasoline components. Some technological barriers still arise in the production of biofuels from lignocellulosic biomass, but robust research is going on to overcome those obstacles.

At the same time, new routes have been developed to obtain these biofuels from lignocellulosic materials, involving a wide array of biological, chemical, and thermal technologies. This book compiles the state of the art of those processes.

Chapter 1, Fundamentals of lignocellulosic biomass, describes the fundamental structure, chemistry, and biorefinery of lignocellulosic biomass. To design and develop sustainable technology, it is urgent to understand the composition, characters, and their refinery process. This chapter also gives an updated research finding about the source, chemistry, and biorefinery of this valuable material.

Chapter 2, Pretreatment of lignocellulosic biomass for efficient enzymatic saccharification of cellulose, explains the methods used for biomass pretreatment and corresponding action models and mechanism, focusing on physical, chemical, biological, and combined pretreatments. It criticizes merits and drawbacks of each pretreatment process in terms of efficiency, formation of inhibitors to subsequent enzymatic hydrolysis and fermentation, energy consumption as well as operation costs. The chapter also recommends further actions to improve the economic feasibility of the pretreatment processes.

Chapter 3, Bioconversion of lignocellulosic biomass to bioethanol and biobutanol, describes processes involved both in bioethanol and biobutanol production after a brief introduction on suitable strains and their productivity of bioethanol and biobutanol. It emphasizes on simultaneous saccharification and fermentation, effect of inhibitors on fermentation, and strategies for minimizing inhibitor effects.

Chapter 4, Lignocellulosic biomass to biodiesel, provides an overview of the biodiesel production using agro-industrial lignocellulosic biomasses as feedstock. The chapter attends the lipid accumulation capability of oleaginous microorganisms such as microalgae, bacillus, and fungi (molds and yeasts). It also highlights the potentiality of lignocellulosic biomasses as a source of biodiesel, oleaginous strains and their productivity, and catalysts for biodiesel synthesis.

Chapter 5, Biobutanol from lignocellulosic biomass: bioprocess strategies, gives an outlook on the major conversion technologies to produce biobutanol from various lignocellulosic biomass feedstocks by focusing on their typical performances that has established along with their applications. The major issues of the chapter are pretreatment and hydrolysis of different solid lignocellulosic biomass materials, effects of degradation products on growth and biobutanol production by advanced ABE fermenting microbial strains and strategies for improving the production of biobutanol.

Chapter 6, Syngas fermentation to bioethanol, starts by presenting the microbes that can be involved in the fermentation process with their competence of using syngas (CO, H_2, CO_2, and CH_4). The chapter critically reviews the syngas cleaning process as it is the significant factors for fermentation of syngas. Biochemical processes for product enhancement and bioreactor design are also systematically discussed in this chapter. This chapter closes with an important discussion on the opportunities and challenges of the hybrid gasification–fermentation process and future research directions in syngas fermentation.

Chapter 7, Fischer–Tropsch synthesis of syngas to liquid hydrocarbons, provides a detailed summary of Fischer–Tropsch (FT) synthesis of syngas to liquid hydrocarbons, role of different catalysts in FT synthesis, kinetic modeling and process simulation for FT synthesis, role of carbon nanofibers/carbon felt composites, plug the knowledge gaps, and provide new research directions to improve the FT synthesis.

Chapter 8, Constraints, impacts, and benefits of lignocellulose conversion pathways to liquid biofuels and biochemicals, focuses on the investigation of the physical and environmental constraints for the large-scale implementation of lignocellulosic biofuels chains. It explores the potential improvements of biofuel technology, in order to remove the existing infrastructural, energy, and environmental barriers. The chapter also paves the way to the penetration of lignocellulosic biofuels into the energy market as a complement to other renewable and nonrenewable options.

Chapter 9, Environmental and socioeconomic impact assessment of biofuels from lignocellulosic biomass, discusses the trends in the growth of biofuels sector, importance of by-products in life cycle analysis and the immediate need for socioeconomic and environmental assessment to ensure the credibility of novel biofuel production processes.

Chapter 10, Pretreatment of lignocellulosic sugarcane leaves and tops for bioethanol production, summarizes the attributes and agronomy of sugarcane tops, different pretreatment methods employed for their bioconversion and hydrolysis techniques, which can be implemented for the production of bioethanol from sugarcane tops.

Scientists with strong academic background and practical experiences have shared their knowledge in this book. We trust that the book will enrich the foresight of current researchers who are dedicatedly working on biofuels from lignocellulosic biomass.

Abu Yousuf
Sylhet, Bangladesh

Domenico Pirozzi
Naples, Italy

Filomena Sannino
Naples, Italy

Acknowledgments

We are expressing our gratitude to all the distinguished authors for their thoughtful contribution for making the book project successful. Their patience and diligence in revising the first draft of the chapters after assimilating the suggestions and comments are highly appreciated.

We would like to acknowledge the solicitous contributions of all the reviewers who spent their valuable time in constructive and professional manner to improve the quality of the book.

We are grateful to staff at Elsevier particularly Ms. Raquel Zanol (Acquisitions Editor), who supported us tremendously throughout the project and for her great and encouraging mind. We also acknowledge Thomas van der Ploeg and Mariana Pizzolatto Henriques (Editorial Project Manager) and Swapna Praveen (Copyrights Coordinator) for their cordial handling of the book.

CHAPTER 1

Fundamentals of lignocellulosic biomass

Abu Yousuf[1], Domenico Pirozzi[2] and Filomena Sannino[3]
[1]Department of Chemical Engineering & Polymer Science, Shahjalal University of Science and Technology, Sylhet, Bangladesh
[2]Department of Chemical, Materials and Production Engineering, University of Naples "Federico II", Naples, Italy
[3]Department of Agriculture, University of Naples "Federico II", Naples, Italy

Contents

1.1 Introduction

Lignocellulosic biomass (LCB), also known as lignocellulose, is the most abundant biorenewable material on the earth [1], produced from atmospheric CO_2 and water using the sunlight energy through the photosynthesis process. It is a complex matrix, mainly made of polysaccharides, phenolic polymers, and proteins that constitute the essential part of woody cell walls of plants. LCB has a complex spatial structure, in which cellulose (a carbohydrate polymer) is wrapped by the dense structure formed by hemicellulose (another carbohydrate polymer) and lignin (aromatic polymer).

Lignocellulosic Biomass to Liquid Biofuels
DOI: https://doi.org/10.1016/B978-0-12-815936-1.00001-0

LCB is usually categorized into three types of waste: biomass, virgin biomass, and energy crops. Trees, bushes, and sand grasses are placed into virgin biomass class, whereas agricultural residue, stover, and bagasse are placed in waste biomass class. Energy crops are raw materials used for the production of second-generation biofuels as they offer high biomass productivity.

LCB has a long history as an energy source: for many centuries, wood has been the most widely used raw material to burn fire. During the Industrial Revolution, due to the increase in energy needs, wood was progressively replaced by fossil fuels. However, from the middle of the 20th century, problems rose from pollution and the exhaustion of fossil fuels has increased the demand of biomass for the production of energy [2].

The first biofuels to be developed have been bioethanol, initially obtained from starch and sugars, and biodiesel, obtained from fats and oils. However, the diffusion of these products, so-called first-generation biofuels, has been limited as they cause direct competition between biofuel and food production.

More recently, second-generation biofuels were developed, based on the conversion of LCB components to liquid fuels. Second-generation biofuels allow the utilization of the entire plants, such as woody crops, agricultural residues, or waste, as well as dedicated nonfood energy crops grown on marginal land, thus allowing a dramatic increase of the productivity.

The production of biofuels and energy from LCB is based on two main routes. Biochemical processes are typically carried out with LCB having C/N ratio lower than 30 and humidity at collection higher than 30%. These processes are based on chemical reactions carried out thanks to the exploitation of enzymes, mushrooms, and microorganisms. An alternative is offered by thermochemical processes, used when the LCB available has C/N ratio higher than 30 and humidity content below 30%. In the last years, novel biofuels have been produced from LCB, such as bio-H_2, butanol, dimethylfuran, and gamma-valerolactone [3,4].

Some technological barriers still arise in the production of biofuels from LCB, but robust research is going on to overcome those obstacles. One of such problems is that LCB has evolved to resist deprivation and to deliberate hydrolytic stability and structural toughness to the cell walls of the plants. This robustness is attributable to the cross-linking between the polysaccharides and the lignin via ether and ester linkages [5].

Nevertheless, LCB is considered as the most promising raw material for the renewable biofuel production as it is readily available, low cost, and environment friendly.

Before in-depth discussion of processes for the production of biofuels, this chapter is intended to give some fundamentals of LCB. It also incorporates the concept of LCB biorefinery to produce fuels and chemicals.

1.2 Components of lignocellulosic biomass

The main components of LCB are cellulose, hemicellulose, and lignin (Table 1.1). Other minor parts are ash and extractive components.

Cellulose molecules are arranged in regular bundles, forming crystalline regions, or in random geometry forming amorphous regions. Microfibrils of cellulose polymers are linked by hydrogen and van der Waals bonds and are protected by hemicellulose and lignin.

Carbohydrates, which are the main component of cellulose and hemicellulose, make up for about 70% of the dry weight of lignocellulose biomass and represent the feedstock for nearly all of the most promising bio-based building blocks and chemical intermediates, whatever the conversion technology (biological or thermochemical) adopted. Lignin is another important constituent of lignocellulose biomass making up for about 25% of its weight and is by far the most important natural resource of aromatics, beside a good solid biofuel.

Lignocellulose can play a vital role in the energy sector as different forms of energy products can be obtained from LCB, such as solid (briquettes), liquid (bioethanol and biodiesel), and gas (biogas and bio-H_2).

1.2.1 Cellulose

Cellulose is an organic compound composed of polysaccharides, which consists of a straight chain of D-glucose molecules linked through β-(1→4) glycosidic bonds with the formula $(C_6H_{10}O_5)_n$. It is

Table 1.1 Major components of various lignocellulosic materials [3,6,7].

Raw material	Cellulose (%)	Hemicellulose (%)	Lignin (%)
Energy crops	21–54	5–30	5–10
Grasses	25–40	25–50	10–30
Softwoods	45–50	25–35	25–35
Hardwoods	45–55	24–40	18–25

the main structural constituent of the primary cell wall of green plants, algae, and oomycetes.

Cellulose is the most common and available organic polymer material in the world. Cotton fiber is containing 90% of cellulose content, wood is 40%–50%, and dried hemp is containing approximately 57%. Higher amounts of cellulose are contained in wood pulp and cotton for the industrial use. Cellulose is frequently used to yield paperboard and paper-type materials. A prospective characteristic of cellulose is crystallinity. Cellulose is transformed into amorphous solid when the reactor environment is controlled as 25 MPa pressure and the temperature of water as 320°C. Several environment-friendly biofuels can be derived from conversion of cellulosic materials, such as agricultural residues and energy crops.

1.2.2 Hemicellulose

Hemicellulose is a branched heteropolymer containing approximately 500–3000 sugar units [8]. It consists in different sugar units, with a prevalence of pentose components (xylose and arabinose) together with hexoses (mannose, glucose, galactose, and rhamnose) and acetylated sugars. Hemicellulose cross-links with either cellulose or lignin, strengthening the cell wall (Fig. 1.1). Although hemicelluloses are widely available, their utilization is more difficult in comparison to cellulose, due to its structural diversity, and also because the enzymatic hydrolysis of pentose sugar units is less simple. However, hemicelluloses offer more possibilities for regioselective chemical and enzymatic modifications in comparison to cellulose, due to the variability in sugar constituents, glycosidic linkages, and structure of glycosyl side chains as well as two reactive hydroxyl groups at the xylose repeating unit. In this view a bog effort is being made in research activity.

1.2.3 Lignin

Lignin is the third major component of LCB having polymeric complex structure, which is responsible for some of structural materials in the particular types of tissues of vascular plants and some of algae [10]. It is an inevitable part of plant cell wall, especially in bark and wood. Because of cross-linked phenolic polymers in its structure, it shows rigidity and hard quality. It is mainly amorphous (noncrystalline). Lignin is branched long-chain polymer made up of three types of monomers, such as primarily

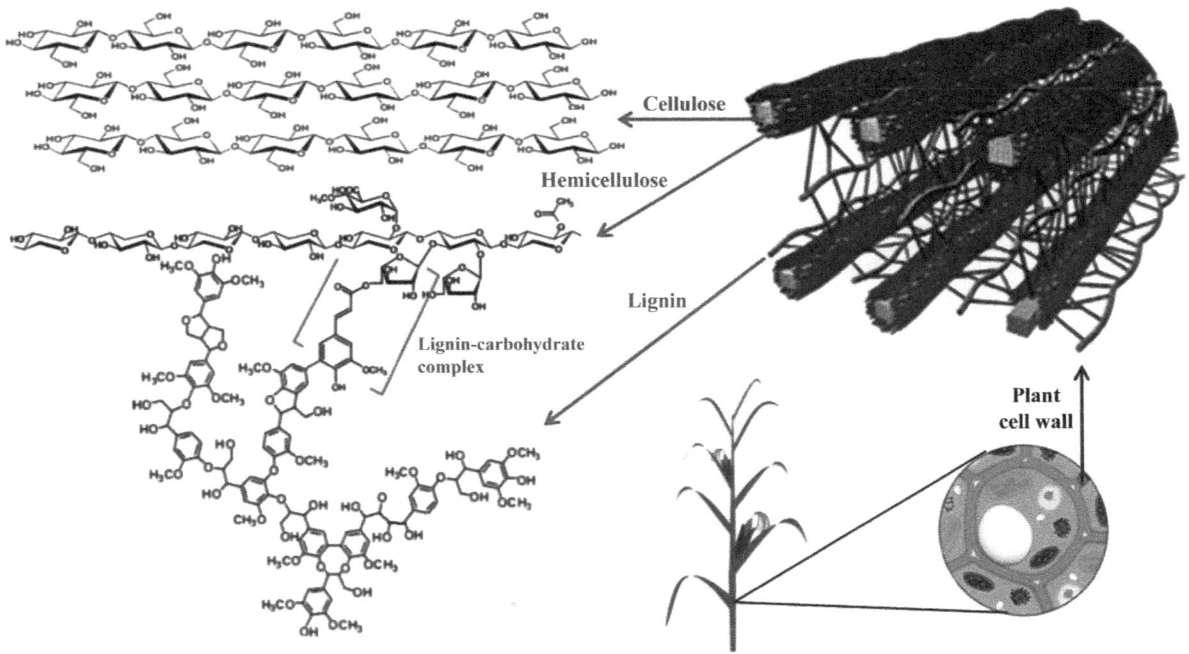

Figure 1.1 Structure and components of lignocellulose [9].

three-dimensional polymer of 4-propenyl phenol, 4-propenyl-2-methoxy phenol, and 4-propenyl-2.5-dimethoxyl phenol [7].

1.3 Sources of lignocellulosic biomass

1.3.1 Annual and perennial energy grasses

Urbane grasses are well-thought-out as the most prominent sources of cellulosic biomass. The major herbaceous energy crops that play a vital role as the source of LCB are as follows: switchgrass, *Miscanthus*, canary grass, giant reed, alfalfa (*Medicago sativa*), and Napier grass (*Pennisetum purpureum*). Besides circumventing "food versus fuel" conflict, they are considered to have energetic, economic, and environmental advantages over food crops for ethanol manufacturing [11,12].

1.3.2 Woody biomass

The forest trees, which are fast growing and have short rotation periods, are major source of cellulosic biomass. Though forests are unequally distributed, they play a vital role on reducing the landslide, redundancy of CO_2 into our atmosphere, and maintain an equilibrium condition between human being and wild life. Meanwhile, every year approximately 370 million tons of LCB are producing from forest in the United States [13]; other forest-rich countries are Canada, Russian, China, and Brazil. The collected LCB from the abovementioned countries is the half of the amount of whole world. Sources of woody biomass are as follows:

- Natural forest residues
- Forestry wastes: sawdust from sawmills, wood chips, and branches from dead trees
- Tree bark
- Wood shavings
- Sawdust
- Low-grade lumber
- Rejected part from sawmills
- Rejected log from plywood mills
- Rejected trunk/log/branch from pulp mills

1.3.3 Nonwoody biomass

The lower lignin content biomasses are known as nonwoody biomass. They contain comparatively low energy and bulky in size. However, this type of biomass is abundant and cheap. They can be collected from a wide range of sources, such as

- agriculture field wastes (paddy husks, straw, grasses, crop stubble, and trash) and
- agricultural processing wastes (palm oil waste, sugarcane bagasse and animal paunch waste, cotton gin trash, etc.).

1.4 Chemistry of lignocellulosic biomass

Before studying the biorefinery of LCB, it is essential to understand the chemistry of individual biomass because the structure or composition of biomass affects the chemical pretreatment, enzymatic digestibility, and the generation of compounds inhibiting fermentative microorganisms used to produce the final fuel or chemical. Chemical or spectroscopic analysis can determine the percentages of individual sugars, protein, uronic acids, and lignin. For example, during the conversion of corn stover, hardwoods, or rice straw, we are in fact working primarily with the plant's structural parts, most of which are cell walls. Therefore more knowledge is required about the natural composition and structure of polymers and chemicals in plant tissue.

Cellulose is a six-carbon compound formed by a plant or microbial cell. Glucose is consisted by carbon, hydrogen, and oxygen. Although glucose ($C_6H_{12}O_6$) is the smallest unit (monomer) that can be isolated from cellulose degradation, cellobiose (Scheme 1.1) is normally the fundamental building block of cellulose that is nothing but a dimer of anhydride. Each monomer unit of glucose is named as glucan $(C_6H_{10}O_5)_n$ formed by losing one molecule of water, which makes polymer long chain [14].

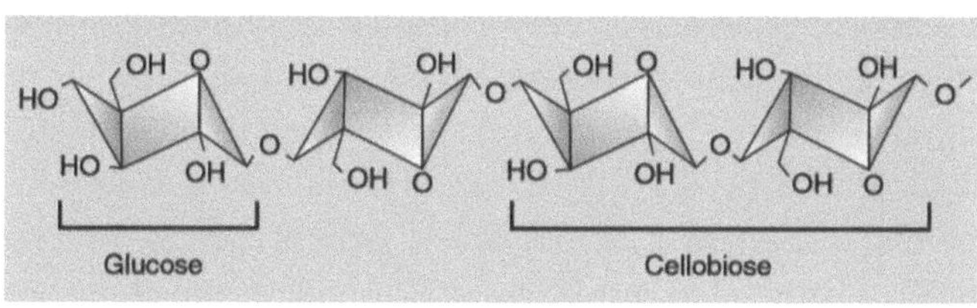

Scheme 1.1 Molecular chain structure of cellulose.

The cellulose polymer chain is a rectilinear polymer, which has no branch. The number of glucan present in a polymer chain is the measuring unit for the determination of extent of a particular polymeric cellulose usually expressed as $(C_6H_{10}O_5)_n$, where n is the degree of polymerization (DP). DP of cellulose hangs on the type of plant or microorganism from which it is isolated and also the method of isolation. For some of the native polymers the DP is estimated to be from 2000 to 14,000 glucan units, DP of the wood pulp in the range of 650–1500 units per glucan chain.

Due to the impending of monomer unit of cellulose to formulate three hydrogen bonds with a monomer of neighboring chain, the chains are fit tightly together for the formation of larger units known as microfibrils. The result is a very stable configuration—essentially free of interstitial spaces, making it anhydrous and quite recalcitrant to hydrolysis by acid, base, or enzyme action. In a polymer chain of cellulose, low chain disturbance is occurred in the crystalline region named micelles. Regions without extensive interchain hydrogen bonding are consequently less structured (amorphous). Amorphous parts are less susceptible to hydrolysis comparative to crystalline part. Natural cellulose consists of cellulose fibrils bound together by an amorphous matrix comprising pectin, hemicellulose, and lignin [14].

The composition of hemicellulose depends on the source biomass or species type. Hardwoods, annual plants, and cereal are predominated by xylan hemicellulose, whereas softwoods are predominated by glucomannan hemicellulose [15]. Hemicelluloses are linked with cellulose by inter- and intramolecular hydrogen bonds and with lignin by ester and ether bonds [16]. Hemicellulose also forms covalent associations with lignin, a complex aromatic polymer, whose structure and organization within the cell wall are not completely understood yet. However, chemical properties of several hemicelluloses of LCB s have been discovered as shown in Table 1.2.

Lignin molecule is made by phenylpropane units allied in a three-dimensional structure, which is very complex. Lignin is considered to be the glue holding the cellulose fibrils together and difficult to completely remove. Lignin provides both stiffness to the cell and protection against the microbial attack because lignin molecules contain several types of chemical bonds and linear polymer chain which increase the mechanical properties [14]. Plant lignin can be broadly divided into three classes: softwood (gymnosperm), hardwood (angiosperm), and grass or annual plant (graminaceous) lignin. Three different phenylpropane units, or monolignols, are responsible for lignin biosynthesis. Guaiacyl lignin is composed principally of coniferyl alcohol units, while guaiacyl–syringyl lignin

Table 1.2 Chemical characteristics of hemicellulose for several lignocellulose plants [17].

Polysaccharide type	Biological origin	Percentage (%) dry biomass	Backbone	Side chains	Linkage	DP
Arabinogalactan	Softwoods	1–3; 35[a]	β-D-Galp	β-D-Galp α-L-Araf β-L-Arap	β-(1→6) α-(1→3) β-(1→3)	100–600
Xyloglucan	Hardwoods, grasses	2–25	β-D-Glcp β-D-Xylp	β-D-Xylp β-D-Galp α-L-Araf α-L-Fucp Acetyl	β-(1→4) α-(1→3) β-(1→2) α-(1→2) α-(1→2)	
Galactoglucomannan	Softwoods	10–25	β-D-Manp β-D-Glcp	β-D-Galp Acetyl	α-(1→6)	40–100
Glucomannan	Softwoods and hardwoods	2–5	β-D-Manp β-D-Glcp	40–70		
Glucuroxylan	Hardwoods	15–30	β-D-Xylp	4-O-Me-α-D-GlcpA Acetyl	α-(1→2)	100–200
Arabino Glucuronoxylan	Grasses, cereals, and softwoods	5–10	β-D-Xylp	4-O-Me-α-D-GlcpAβ-L-Araf	α-(1→2) α-(1→3)	50–185
Arabinoxylans	Cereals	0.15–30	β-D-Xylp	α-L-ArafFeruloy	α-(1→2) α-(1→3)	
Glucuronoarabinoxylans	Grasses and cereals	15–30	β-D-Xylp	α-L-Araf 4-O-Me-α-D-GlcpA Acetyl	α-(1→2) α-(1→3)	
Homoxylans	Algae	β-D-Xylp				

Araf, Arabinofuranose; *Arap*, arabinopyranose; *DP*, degree of polymerization; *Galp*, galactopyranose; *Glcp*, glucopyranose; *Manp*, mannopyranose; *Me-α-D-GlcpA*, glucuronic acid; *Xylp*, xylopyranose.
[a]Larchwood.

contains monomeric units from coniferyl and sinapyl alcohol (Scheme 1.2). In general, guaiacyl lignin is found in softwoods, while guaiacyl–syringyl lignin is present in hardwoods. Graminaceous lignin is composed mainly of *p*-coumaryl alcohol units [18].

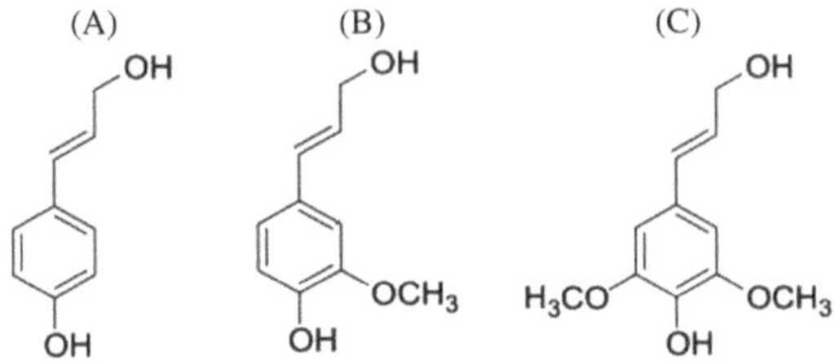

Scheme 1.2 Monolignols: (A) *p*-coumaryl alcohol (hydroxyphenyl unit), (B) coniferyl alcohol (guaiacyl units), and (C) sinapyl alcohol (syringyl unit).

Lignin is bonded to the cellulose and hemicellulose by polysaccharides and serves as a strengthening and hydrophobic agent, complicating biomass breakdown. The schematic diagram (Scheme 1.3) shows possible covalent cross-links between polysaccharides and lignin in cell walls. Since lignin is the hardest part of the plant cell wall and not easy to convert to biofuels, it is essential to find out alternative uses to maximize the energy yield from biomass.

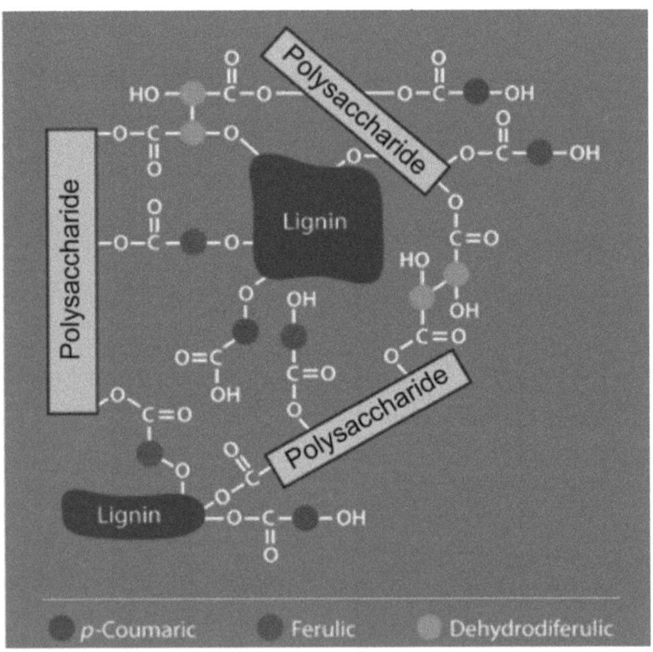

Scheme 1.3 Linkage of lignin with polysaccharides.

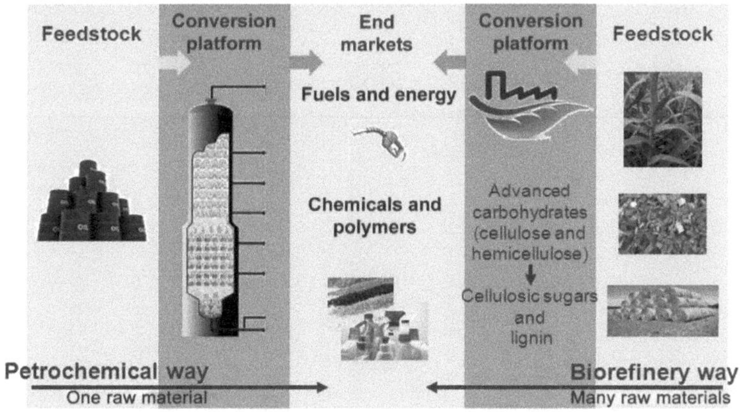

Figure 1.2 Conceptual difference between petroleum refinery and the biorefinery.

1.5 Biorefinery of lignocellulosic biomass

The biorefinery is a facility that integrates biomass conversion processes and equipment to produce fuels, power, and chemicals from biomass, using a variety of different technologies. Consequently, the concept of biorefinery, now widely accepted, is analogous to that of petroleum refinery (Fig. 1.2), except that it makes use of renewable plant-derived materials (derived from photosynthesizing plants), whereas an oil refinery uses nonrenewable fossil-derived petroleum.

The technologies based upon the concept of biorefinery can provide a wide range of bio-based products that include biofuels (bioethanol, biodiesel, and biomethane), biomaterials (fibers, pulp for paper manufacture, composite compounds, lignin-based hydrogels, etc.), and a host of biochemicals through downstream fermentation and refining processes. Therefore LCB has a wide range of application as the alternative to petroleum for the generation of biofuels and chemicals, which makes it a recognized and valuable commodity all over the world.

Effectively breaking lignocellulose recalcitrance, releasing monomer from the long-chain polymeric sugars, and coutilization of lignocellulose components are the largest technical and economical challenges for the biorefinery of LCB. Fractionation of LCB is really a tough task because of its complex structure and composition constituent [19]. Moreover, it is the first step of biorefinery to get fuels, chemicals, or materials from LCB. The fractionation routes and their process conditions are summarized in Table 1.3.

Organosolv fractionation (OF) is a unique and one-step process for fractionating the LCB to produce cellulosic fibers, hemicellulose sugar,

Table 1.3 Recently developed fractionation process of lignocellulosic biomass.

Fractionation	Process conditions	Products	References
Pyrolysis	A long vapor residence time (>10 min) and a low operating temperature (290°C–400°C) for slow pyrolysis	30 wt.% of liquids, 35 wt.% of char, and 35 wt.% of gas	[20]
	A hot-vapor residence time of 10–30 s at moderate temperature (around 500°C) for intermediate pyrolysis	50 wt.% of liquids, 25 wt.% of char, and 25 wt.% gas	[20]
	A short hot-vapor residence time (w1 s) also at a moderate temperature of 500°C for fast pyrolysis	Biooil (75 wt.% of liquids, 12 wt.% of char, and 13 wt.% of gas)	[20]
Microwave-assisted DES	45 s with microwave irradiation at 800 W using a DES composed of choline chloride and lactic acid (ChCl:LA)	Removing lignin and xylan while retaining most of cellulose in the pretreated solids	[21]
Aldehydes	Requires 6–7 h	Stabilized lignin	[22]
OrganoCat	A biphasic system comprising water and bio-based 2-methyltetrahydrofuran (2-MeTHF) as solvents and oxalic acid as catalyst at mild temperatures (up to 140°C)	• Selective hemicellulose depolymerize to form an aqueous stream of the corresponding carbohydrates • Solid cellulose pulp remains suspended and the disentangled lignin is extracted in situ with the organic phase	[23]
Hydrothermal and delignification	At 180°C under nonisothermal conditions	Cellulose and hemicelluloses	[24]
Organosolv	Phosphoric acid (80%, v/v), 50°C and 100 rpm for 24 h	• Cellulose as a solid residue • Lignin • Hemicellulose	[24]

DES, Deep eutectic solvents.

and low molecular weight lignin. In this process the raw material (LCB) is mixed with particular organic solvent, with or without acid, catalyst, and other necessary constituents for the separating of lignin, hemicellulose, and cellulose in solid phase. Here, catalysts are generally used for the increasing solubilization of hemicellulose and digestibility of the other substrate, whereas organic solvents (acetone, ethyl alcohol, ethylene glycol, aqueous phenol, *n*-butanol, etc.) are used to enhance the process readily [25].

This process is most feasible for the fractionation of LCB because of the utilization of all the biomass components of LCB. However, some drawbacks are associated with OF, such as high cost of organic solvent, need extra caring to handle because of the extreme volatility of organic solvent. No leakage is allowed here due to the possibility of heavy explosion [26]. Its successful commercialization will depend on the development of high-value coproducts from lignin and hemicelluloses.

The LCB biorefineries are used to separate or extract a wide range of finished product by using combination of different technologies. Fig. 1.3 shows the probable finished products (fuel and chemicals) that

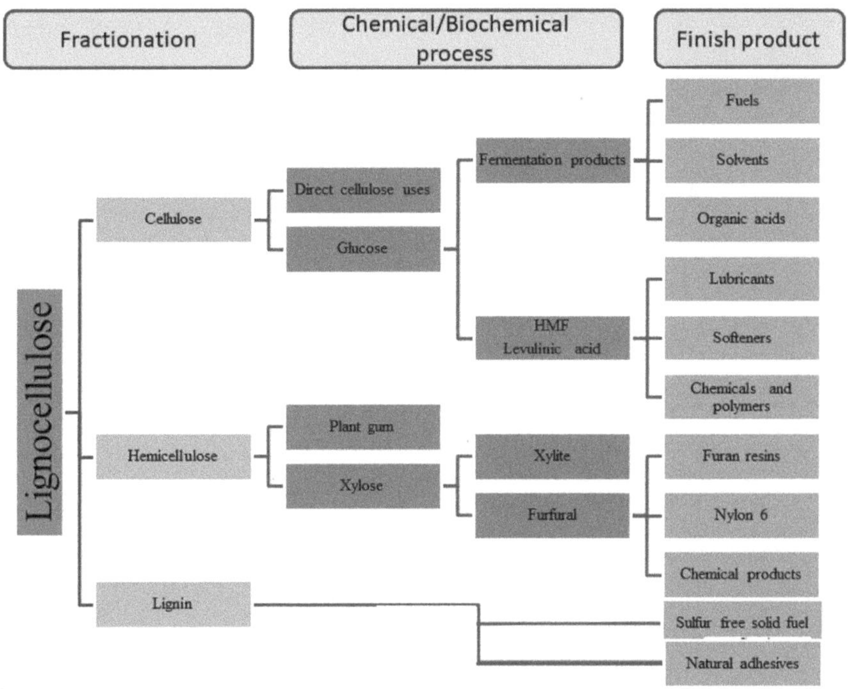

Figure 1.3 Biorefinery of LCB for obtaining different finished products. *LCB*, Lignocellulosic biomass.

can be obtained from the fractioned major components of hemicellulose (5-carbons polymer), cellulose (6-carbons polymer), and lignin (polymer of phenol) through the biorefinery process.

1.6 Conclusion

The concept of LCB to liquid fuels is really a broad issue for the modern age. There is very narrow field in our surroundings, where cellulosic materials are used as feedstock of liquid biofuels. The crystallinity of cellulose and its association with hemicellulose and lignin are two key challenges that prevent the efficient breakdown of cellulose into glucose molecules that can be converted to liquid biofuels. This chapter is just trying to give an updated research finding about the source, chemistry, and biorefinery of this valuable material. Biorefineries may play a vital role in tackling climate change by reducing the demand of fossil fuel and offering sustainable way to generate renewable biofuels, chemicals, and materials. At the same time, LCB-based biorefinery is being popular because of its great contribution on bioeconomy.

References

[1] C.-H. Zhou, X. Xia, C.-X. Lin, D.-S. Tong, J. Beltramini, Catalytic conversion of lignocellulosic biomass to fine chemicals and fuels, Chem. Soc. Rev. 40 (2011) 5588–5617.

[2] A. Carroll, C. Somerville, Cellulosic biofuels, Annu. Rev. Plant. Biol. 60 (2009) 165–182.

[3] A. Yousuf, Biodiesel from lignocellulosic biomass—prospects and challenges, Waste Manage. 32 (2012) 2061–2067.

[4] A. Barbara, Tokay "Biomass chemicals", Ullmann's Encyclopedia of Industrial Chemistry, Wiley-VCH, Weinheim, 2002.

[5] J. Houghton, S. Weatherwax, J. Ferrell, Breaking the Biological Barriers to Cellulosic Ethanol: A Joint Research Agenda, EERE Publication and Product Library, 2006.

[6] W. Betts, R. Dart, A. Ball, S. Pedlar, Biosynthesis and Structure of Lignocellulose, Biodegradation, Springer, 1991, pp. 139–155.

[7] T. Kikas, M. Tutt, M. Raud, M. Alaru, R. Lauk, J. Olt, Basis of energy crop selection for biofuel production: cellulose vs. lignin, Int. J. Green Energy 13 (2016) 49–54.

[8] L.J. Gibson, The hierarchical structure and mechanics of plant materials, J. R. Soc. Interface 9 (2012) 2749–2766.

[9] B. Volynets, F. Ein-Mozaffari, Y. Dahman, Biomass processing into ethanol: pretreatment, enzymatic hydrolysis, fermentation, rheology, and mixing, Green Process. Synth. 6 (2017) 1–22.

[10] P.T. Martone, J.M. Estevez, F. Lu, K. Ruel, M.W. Denny, C. Somerville, et al., Discovery of lignin in seaweed reveals convergent evolution of cell-wall architecture, Curr. Biol. 19 (2009) 169–175.
[11] A.K. Chandel, O.V. Singh, Weedy lignocellulosic feedstock and microbial metabolic engineering: advancing the generation of 'Biofuel', Appl. Microbiol. Biotechnol. 89 (2011) 1289–1303.
[12] J. Hill, E. Nelson, D. Tilman, S. Polasky, D. Tiffany, Environmental, economic, and energetic costs and benefits of biodiesel and ethanol biofuels, Proc. Natl. Acad. Sci. U.S.A. 103 (2006) 11206–11210.
[13] R.D. Perlack, Biomass as Feedstock for a Bioenergy and Bioproducts Industry: The Technical Feasibility of a Billion-Ton Annual Supply, Oak Ridge National Laboratory, 2005.
[14] P. Mäki-Arvela, E. Salminen, T. Riittonen, P. Virtanen, N. Kumar, J.-P. Mikkola, The challenge of efficient synthesis of biofuels from lignocellulose for future renewable transportation fuels, Int. J. Chem. Eng. 2012 (2012).
[15] R. Sun, Cereal Straw as a Resource for Sustainable Biomaterials and Biofuels: Chemistry, Extractives, Lignins, Hemicelluloses and Cellulose, Elsevier, 2010.
[16] L. Viikari, A. Suurnäkki, S. Grönqvist, L. Raaska, A. Ragauskas, Forest Products: Biotechnology in Pulp and Paper Processing, Encyclopedia of Microbiology, Academic Press, 2009, pp. 80–94.
[17] F.M. Gírio, C. Fonseca, F. Carvalheiro, L.C. Duarte, S. Marques, R. Bogel-èukasik, Hemicelluloses for fuel ethanol: a review, Bioresour. Technol. 101 (2010) 4775–4800.
[18] P. Fardim, N. Durán, Modification of fibre surfaces during pulping and refining as analysed by SEM, XPS and ToF-SIMS, Colloids Surf., A: Physicochem. Eng. Aspects 223 (2003) 263–276.
[19] Y. Sun, J. Cheng, Hydrolysis of lignocellulosic materials for ethanol production: a review, Bioresour. Technol. 83 (2002) 1–11.
[20] F.-X. Collard, M. Carrier, J. Görgens, Fractionation of Lignocellulosic Material With Pyrolysis Processing, Biomass Fractionation Technologies for a Lignocellulosic Feedstock Based Biorefinery, Elsevier, 2016, pp. 81–101.
[21] Z. Chen, C. Wan, Ultrafast fractionation of lignocellulosic biomass by microwave-assisted deep eutectic solvent pretreatment, Bioresour. Technol. 250 (2018) 532–537.
[22] M.T. Amiri, G.R. Dick, Y.M. Questell-Santiago, J.S. Luterbacher, Fractionation of lignocellulosic biomass to produce uncondensed aldehyde-stabilized lignin, Nat. Protoc. 14 (2019) 921–954.
[23] P.M. Grande, J. Viell, N. Theyssen, W. Marquardt, P.D. de María, W. Leitner, Fractionation of lignocellulosic biomass using the OrganoCat process, Green Chem. 17 (2015) 3533–3539.
[24] N. Smichi, Y. Messaoudi, M. Gargouri, Lignocellulosic biomass fractionation: production of ethanol, lignin and carbon source for fungal culture, Waste Biomass Valor. 9 (2018) 947–956.
[25] H. Cheng, L. Wang, Lignocelluloses Feedstock Biorefinery as Petrorefinery Substitutes, Biomass Now-Sustainable Growth and Use, IntechOpen, 2013.
[26] S. Aziz, K. Sarkanen, Organosolv pulping—a review, Tappi J. 72 (1989) 169–175.

CHAPTER 2

Pretreatment of lignocellulosic biomass for efficient enzymatic saccharification of cellulose

Jingzhi Zhang[1,*], Haifeng Zhou[2,*], Dehua Liu[1] and Xuebing Zhao[1]
[1]Key Laboratory for Industrial Biocatalysis, Ministry of Education of China, Institute of Applied Chemistry, Department of Chemical Engineering, Tsinghua University, Beijing, China;
[2]Key Laboratory of Low Carbon Energy and Chemical Engineering, College of Chemical and Environmental Engineering, Shandong University of Science and Technology, Qingdao, China

Contents

2.1 Introduction

Lignocellulosic biomass has been considered as one of the most promising feedstocks for a biorefinery to produce various biofuels, chemicals, and materials. However, owing to the strong recalcitrant structure of biomass

* Both authors contributed equally to this work.

Lignocellulosic Biomass to Liquid Biofuels
DOI: https://doi.org/10.1016/B978-0-12-815936-1.00002-2

to biodegradation of cell wall structure by enzymatic and microbial attack, a suitable pretreatment process is required to facilitate the enzymes' access to the plant polysaccharides [1]. Pretreatment is a prerequisite step for bioconversion of lignocellulosic biomass and has a significant impact on the cost of downstream processing including detoxification, enzymatic hydrolysis, waste treatment, and other variables [2]. It has been reported that pretreatment constitutes more than 40% of the total conversion process cost in some cases [2]. Many pretreatment methods have been developed to remove the hemicelluloses and/or lignin, break down cell wall structure, increase specific surface area, create more pores, decrease cellulose crystallinity, and so on, in order to promote the enzymatic hydrolysis or any other microbial decomposition of cellulose [3]. Therefore pretreatment has been considered as a key technical barrier and economic consumption contributor to biomass bioconversion. A number of research works have been published to improve the techno-economic efficiency of pretreatment for cost-competitive production of biofuels and chemicals from biorefining of biomass [4].

The different nature of biomass feedstocks makes the working mechanisms of pretreatment techniques quite different. There is no gold universal pretreatment for all biomass. Therefore a variety of pretreatment methods have been tested during the last decades but each method still has its own drawbacks and advantages. In general, pretreatment of biomass can be classified into four categories according to their function mode and the reagent used, namely, physical, chemical, biological, and combined (hybrid) pretreatment, such as physicochemical pretreatment or subsequent use of physical, chemical, and/or biological pretreatments.

The ideal pretreatment should be effective and of low cost. There are several important aspects that should be taken into consideration for pretreatment process as follows [5,6]:

(1) Wide adaptation for different raw materials, including crops species, sites ages, and harvesting times.

(2) Highly digestible pretreated solid. Cellulose from pretreated substrates should be easily, effectively, and highly hydrolyzed with yield of more than 90%.

(3) No significant sugar degradation with the minimum formation of toxic compounds. The yields of fermentable cellulosic and hemicellulosic sugars achieved during the pretreatment steps should be close to 100%. The liquid from the pretreatment process should be fermentable with a low-cost and high yield.

(4) No requirement of biomass size processing to avoid extra energy input for size reduction before the pretreatment.

(5) Operation in reasonable size and moderate cost reactors. Pretreatment reactors should be of low cost by minimizing their volumes, using appropriate materials, and keeping operating pressure reasonable.

(6) Nonproduction of solid residues. Solid-waste formed during the preparation of hydrolyzate should not present processing or disposal challenges.

(7) Effectiveness at low moisture content. The utilization of low moisture content raw material would reduce the energy consumption during pretreatment.

(8) Obtaining high sugar concentration and fermentation compatibility. The concentration of sugars from pretreatment process and enzyme hydrolysis should be above 10% to ensure an adequate concentration of fermentation product, for example, ethanol, and keep recovery and other downstream costs manageable.

(9) Lignin recovery. Lignin and other compositions should be recovered and make sure the downstream processing simply and for conversion into valuable coproducts.

(10) Minimum heat and power requirements. Heat and power are the main energy cost for pretreatment, and low energy consumption is very important for pretreatment.

However, it should be noted that to meet all of the abovementioned "strict criteria" is impossible, and compromise usually has to be made because each pretreatment has its own disadvantages and advantages. To combine the merits of different pretreatments, hybrid pretreatment can be employed; however, such combination may increase the complicity of the process and the cost. In the following sections, the operation modes, mechanisms, and enzymatic digestibility of the pretreated substrates will be discussed for different pretreatment technologies.

2.2 Physical pretreatment and mechanisms

Physical pretreatment aims to increase the surface accessibility of lignocellulosic biomass to enzymes by reducing the size of the materials or destructing the cell wall structure. There are many types of physical pretreatments, but they are generally classified into mechanical comminution (chipping, cutting, grinding, or milling) and irradiation [gamma rays, electron beam (EB), or microwave] pretreatments [1,7].

2.2.1 Operation model of physical pretreatment

Mechanical comminution can reduce the size of lignocellulosic biomass from centimeters to smaller sizes usually ranging from micrometers to millimeters [3]. Usually, the energy consumption for efficient physical pretreatment is high and is closely related to the final particle size [1,8]. It has been reported that for the same size reduction, 70% more energy input was needed for mechanical comminution than steam explosion (SE) pretreatment [9]. Biomass characteristics, type of the pretreating machine, and final particle size are the dominant factors for energy requirement [4]. For example, the comminution of hardwood needs more energy consumption than agricultural residues [10]. For this reason, mechanical pretreatment is usually combined with chemical or other pretreatment processes, before or after, to reduce the particle size [4].

2.2.1.1 Mechanical comminution

Comminution is used to reduce the particle size of biomass and increase accessible surface area, reducing crystallinity and degree of polymerization (DP), thus increasing the biomass biodegradability [4]. Typical mechanical comminution includes chipping and milling as shown in Table 2.1. Chipping is usually necessary to reduce the size of raw lignocellulosic biomass for further processing, for example, to make log to wood chips. Milling process that can be used for biomass pretreatment involves various types, such as ball milling, hammer milling, knife milling, vibro milling, tow-roll milling, colloid milling, wet-disk, and attrition millings [4]. As indicated by Agbor et al., harvesting and preconditioning can reduce lignocellulosic biomass from logs to coarse sizes (about 10–50 mm), and chipping can reduce size to 10–30 mm, while grinding and milling can reduce the size to 0.2–2 mm [11]. Grinding and milling are also effective to alter the inherent ultrastructure of biomass, such as crystallinity and DP of cellulose [7]. As shown in Table 2.1, vibratory ball milling has been found to be more efficient in reducing cellulose crystallinity of spruce and aspen chips than ordinary ball milling [12]. Disk milling has been reported to achieve higher enzymatic hydrolysis yield than hammer milling [13]. The moisture content of biomass should be also considered in the selection of milling methods [4]. For dry biomass, hammer and knife millings

Table 2.1 Primary types and functions of mechanical comminution for physical pretreatment of biomass.

Items	Types	Size reduction	Description	References
Chipping	For harvesting and preconditioning	From logs to coarse sizes (about 10–50 mm)	Usually used to reduce the size of raw lignocellulosic biomass for further processing	[11]
	Chipping for size reduction	Can reduce size to 10–30 mm		
Grinding and milling	Ball milling	0.2–2 mm	Can well reduce cellulose crystallinity of lignocellulose (vibratory ball milling)	[8,12]
	Hammer milling		Being suitable for dry biomass	[7,14]
	Knife milling			
	Vibro milling		Being suitable for both dry and wet biomass (ball and vibro milling)	[7,14]
	Colloid milling		–	–
	Tow-roll milling		–	–
	Wet-disk milling		Can achieve higher enzymatic hydrolysis yield than hammer milling (disk milling)	[13]
	Attrition milling		–	–

are more suitable, while ball and vibro milling are more widely used and suitable for both dry and wet biomass [7,14].

Size reduction by milling also has been employed independently to improve the biodegradability of biomass conversion to biogas, bioethanol, and biohydrogen. When rice straw was subjected to disk and ball milling, the yields of glucose and xylose by enzymatic hydrolysis reached up to 89% (from 78%) and 54% (from 41%), respectively [15]. Glucose and xylose yield increased to 40% and 32% by wet disk milling for pretreated sugarcane bagasse and rice straw, respectively [16]. An obvious advantage of mechanical comminution for size reduction is that no inhibitors are generated during the process. However, various inhibitors are generated in chemical pretreatments, such as hydroxymethylfurfural

(HMF), furfural and acetic acid, and so on, which are toxic to yeast or other microorganisms.

2.2.1.2 Irradiation pretreatment

As shown in Fig. 2.1, irradiations including gamma ray, ultrasound, microwave, and EB have been used for biomass pretreatment to improve the cellulose digestibility [4]. Gamma irradiation can deconstruct the cell wall structure with depolymerization of cellulose and lignin thus improving the biodegradability of the pretreated materials [17]. Yin et al. found that gamma irradiation combined with alkaline pretreatment of wheat straw could well enhance the accessible surface area for enzymes [18]. Xiang et al. applied cobalt-60 gamma-ray irradiation for pretreatment of hybrid poplar sawdust (with 300 kGy irradiation pretreatment) and the enzymatic saccharification (reducing sugar yield) at the first 12 h increased by 16.7% (from 71.2 to 83.1 mg/g) with 20 FPU/g enzyme loading at 45°C [19]. Gamma irradiation also can be used to assist the fungal degradation of the polypropylene/biomass composites [20]. Wang et al. compared gamma irradiation and SE pretreatment for ethanol production from agricultural residues. For the irradiation-pretreated rice straw, the cellulose, hemicellulose, and lignin were much more easily degraded, and the glucose yield was enhanced from 6.58% to 47.44%, being better than SE pretreated results [21].

Ultrasound can disrupt the structure of cell wall, increase the specific surface area, and reduce the DP of cell wall components [22]. Yu et al. employed ultrasound-assisted ionic liquid (IL) pretreatment to enhance the enzymatic and acid hydrolysis of sugarcane bagasse and wheat straw [23]. The results showed that the enzymatic hydrolysis of bagasse and wheat straw pretreated with IL resulted in the maximal glucose yield at ultrasound of 20 kHz (40.32% and 53.17%), and the maximal glucose yields of acid hydrolysis were 33.32% and 48.07% when pretreated at ultrasound of 40 kHz [24]. Ultrasound pretreatment also has an influence

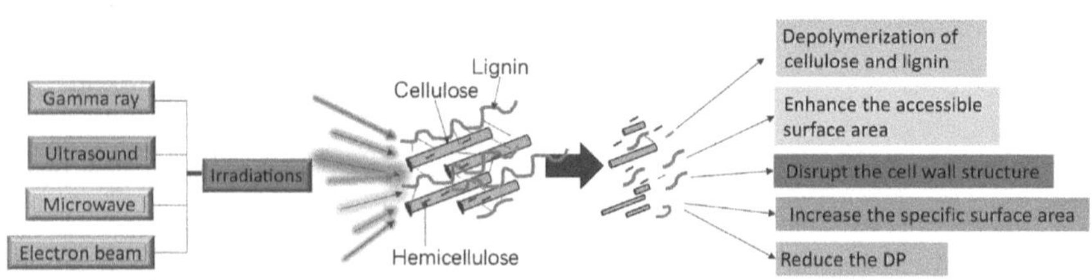

Figure 2.1 Actions of irradiation-based physical treatment for biomass.

on wood physiochemical structure changes. He et al. studied the effect of ultrasound in pretreating eucalyptus wood with 300 W and frequency of 28 kHz. The results showed that ultrasound was effective in modifying the physiochemical structure of eucalyptus wood [24]. Besides, ultrasound has been found to be applied in some other bioprocessing of biomass, such as anaerobic digestion of microalgal biomass for improving methane yield [25], ultrasound-assisted alkali pretreatment for bioethanol and xylanases production from chili residue [26], and so on.

Microwave pretreatment relies on the creation of heat by direct interaction between a heated object and an applied electromagnetic field [27]. Therefore the heating efficiency of microwave is higher with easier operation than the conventional conduction/convection heating (such as air bath heating, water bath heating, oil bath heating, and oven heating) [28,29]. However, microwave pretreatment is usually used as a heating method for other pretreatment, especially for chemical pretreatments. Microwave irradiation with water to pretreat switchgrass obtained 53% higher total sugar than that treated with conventional heating pretreatment [29]. As found by Zhao et al. [30], alkali pretreatment combined with microwave irradiation of rice hulls could significantly increase the production of cellulase and reducing sugars by *Trichoderma* sp. 3.2942. The maximum filter paper activity, carboxymethylcellulose, and reducing sugar content (RSC) were increased by 35.2%, 21.4%, and 13%, respectively, compared with those of untreated rice hulls, which was primarily due to the rupture of the rigid structure of rice hulls by the pretreatment. Jin et al. studied microwave-$Ca(OH)_2$ pretreatment for catalpa sawdust and found that the optimal condition was with 40 mesh particle size, 2.25% (w/v) $Ca(OH)_2$ dosage, 400 W microwave power, and 6 min pretreatment time. When the pretreated substrates were enzymatically hydrolyzed with 175 FPU/g enzyme loading for 96 h, the reducing sugar yield reached 402.73 mg/g, being increased by 682.15% compared with that of raw catalpa sawdust [27].

EB pretreatment relies on high energy electrons to create reactive radical species within the biomass and the secondary reactions of these radicals typically lead to a series of bond reactions within cell wall polymers [31]. EB has been found to have the function of reducing the DP of cellulose and causing alteration of lignin–hemicellulose matrix [26]. EB pretreatment also has been verified to show beneficial effects on the subsequent treatment including acetic acid pretreatment, SE pretreatment, hot water extraction, etc. When switchgrass was pretreated by EB irradiation, it was found that the molecular weight and crystallinity of cellulose were reduced, resulting in a well-improved cellulose accessibility. When the biomass was pretreated with hot water extraction and 1000 kGy EB, the hemicellulose content was reduced by 13.3% and the glucose yield by enzymatic hydrolysis was enhanced by more than 4 times [26].

2.2.1.3 Extrusion pretreatment

Extrusion is a continuous and highly versatile pretreatment method and has been considered as a promising thermo-mechanical process. Biomass structure is altered due to the shearing, mixing, and heating actions, which are mainly promoted in the mixing zone of the screw formed by successive kneading elements connected with a slight deviation angle between them. Therefore extrusion pretreatment can result in the enhancement of cellulose exposure ratio and increasing porosity and surface area [32,33]. There are two types of extrusion equipment, referring to single-screw and twin-screw extruders [32]. Single-screw extruders only have one single solid piece, while twin-screw extruders are composed of two small screw elements and assembled in the shafts. The screw elements of twin-screw extruders have three main types including kneading blocks, reverse screws, and conveying screws. In addition, there are two classifications of twin-screw extruders, such as counter-rotating and corotating [32]. The different elements in the extruder have different functions including transport, mixing, and shearing to disintegrate biomass.

The screw elements' arrangements are also named as the screw profile, which defines the position of each element, with the characteristics in terms of pitch, stagger angle, length, and how the different elements are spaced [29]. Screw profile is very important in extrusion processing, which affects the transport of the material in twin-screw extruders significantly [32]. Different types of extrusion machines have different mixing ways and flow patterns depending on the inside of the barrel structures. Single-screw extruders produce mainly a distributive mixing in the blending agent with spreading through the lignocellulose matrix and obtain a good spatial distribution. The force of single-screw extruder makes the material exclusively frictional or viscous and the screw is one-piece element [32]. For twin-screw extruder it produces both distributive and dispersive mixing, reducing particle size, and thus altering the physical properties of biomass [4]. In corotating twin-screw extruders the shearing and plasticizing effect is axial (the maximum velocity being achieved at the intermeshing zone), while in counter-rotating equipment, the effect is radial (highest velocity achieved at the screw tips) [11].

Screw configuration characteristic is another important factor for improving the pretreatment efficiency. The addition of one reverse element at the third mixing zone end can greatly improve the pretreatment efficiency. Some researchers investigated the extrusion pretreatment combined with chemical treatment (sodium hydroxide) on lignin removal. For corncobs the glucose conversion of the lignin-removed substrates could reach up to 56%, 64%, and 68% by conveying, kneading, and reverse screw elements, respectively [11]. For pretreatment of sugarcane bagasse, extrusion pretreatment did not change the CrI of the substrate, being 54.1% and 57.3% for raw and pretreated bagasse, respectively [33].

Extrusion can be operated at high solid loading, moderate temperature, and pH, and the pretreated substrates usually show improved enzyme hydrolysis [34]. However, extrusion is usually combined with chemical pretreatment methods, such as SE or acid pretreatment [33]. Compared to batch pretreatment system, extrusion can be operated continuously with higher processing capacity [35]. Besides, like other physical pretreatment processes, there are no or only a few inhibitors formed by extrusion pretreatment itself [34]. However, some additives for changing the flow properties of lignocellulose are needed to add into the system in order to avoid clogging of the equipment, but these additives may also cause the degradation of the materials [36]. It has been reported that the substances with cellulose affinity are good options as extrusion additives for biomass pretreatment, such as glycerol, ethylene glycol, and dimethyl sulfoxide. They can effectively fibrillate biomass cell walls, opening structure, and lowering the equipment torque [36]. Moro et al. investigated the extrusion pretreatment of sugarcane bagasse and straw using a corotating twin-screw extruder with water, glycerol, Tween 80, and ethylene glycol as additives. The results showed that when the ratio between biomass and glycerol was 1:0.75 for bagasse and 1:0.53 for straw, the enzymatic hydrolysis of pretreated solid increased to 68.2% after a multiple extrusion with the insertion of a reverse element in the screw configuration [33].

2.2.1.4 Pulsed electric field pretreatment

Pulsed electrical field (PEF) is relatively new for biomass pretreatment. In PEF pretreatment process a high voltage strikes the samples placed between two electrodes with a very short burst at about 100 μm [4].

High-strength external electric field leads to a rapid electric potential across the cell membrane with a rapid electrical breakdown of plant cell wall [37]. The loss of membrane barrier functions is the primary impact of PEF [38]. The advantages of PEF are of low energy and normal operation temperatures and pressure [4]. PEF has been used in pretreating rapeseed green biomass (stems) to enhance pressing and extractives recovery. For the effects of pressure, when electric field strength and pulse number were employed for juice expression, the juice yield increased from 34% to 81% [38]. Besides, PEF can be also applied in nonthermal inactivation of microbial cells, electroporation of cell membranes, softening tissues and peeling, and extraction of thermal labile biocompounds [37]. However, more investigation on the pretreatment of lignocellulosic biomass for cellulose hydrolysis has to be further performed.

2.2.2 Energy and cost consideration on physical pretreatment

Relatively high energy consumption is the main limitation of physical pretreatment by grinding or milling. Compared with herbaceous biomass, pretreatment of woody biomass consumes more energy [39]. It has been reported that the energy consumption for milling wood chips into fibers is about 500−800 W h/kg [39,40]. As found by Zhu et al., for ethanol production from wood with a yield of 300 L/t dry wood, the conversion efficiency from thermal energy (stored in ethanol) to electric-mechanical energy is only about 30%, and the thermal energy in ethanol produced is just sufficient for wood-size reduction [39]. The milling process is also known as a pretreatment process with low efficiency [41]. In most cases, a combination of chemical or other physical pretreatments in order to improve the milling efficiency and overcome the energy barrier is needed. Using a low liquid-to-solid ratio (L/W) is an advantageous way to reduce thermal energy consumption in physical pretreatment process [42]. Postchemical pretreatment after size-reduction is another method, and there are several advantages for this combination. For example, it can avoid energy-intensive operation of mixing high-consistency pulp with chemicals and improve the permeation of chemicals in the biomass particle thus increasing pretreatment efficiency [42].

For extrusion pretreatment the energy cost mainly comes from electricity consumption for providing thermal energy to reach and keep the operation temperature, and mechanical energy of the motor to rotate the screws for extrusion machines [32]. Therefore choosing appropriate temperature and screw speed is important to reduce the energy consumption. Some researchers have explored the applications of biomass extrusion at low temperatures (less than 100°C) to get the promising results [43]. For mechanical energy consumed in extrusion pretreatment, torque is one important factor to be considered. The torque is related to screw speed, loading, and viscosity of the material inside the extruder. Lower torque is required for thermo-softened biomass and operation at higher temperature which reduces the viscosity. The relationship between temperature and torque is negative. To compare the energy consumption under different condition a parameter termed specific mechanical energy factor has been defined by estimating the mechanical energy required to produce a unit weight of pretreated biomass in a particular extrusion machine [44]. The screw speed appears to be more important than temperature. When the extrusion pretreatment was carried out at low temperature and high screw speed, the lowest energy consumption of biomass was found to be 20.0 kJ/g [33]. Energy consumption can be reduced when the equipment is running at low screw speed, but if the screw speed is too low, corresponding feeding rate decreases with smaller processing capacity and leading to higher energy consumption per gram of biomass. For example, the energy consumption for 100 rpm (20.0 kJ/g at 30°C and 27.8 kJ/g at 150°C) was lower than for 20 rpm (81.1 kJ/g at 30°C and 123 kJ/g at 150°C) [33]. Therefore to reduce energy consumption, extrusion pretreatment is recommended to be working at low temperatures and low values of torque to minimize the mechanical energy spent.

2.3 Chemical pretreatment

2.3.1 Liquid hot water pretreatment

Liquid hot water (LHW) pretreatment, also known as hydrothermal pretreatment, is a process using water as a heating medium at high temperature (usually from 130°C to 240°C) and high pressure to maintain water in liquid phase without any chemicals added [4,45]. At high temperature, water shows acid properties and can work as an acidic catalyst [46]. LHW pretreatment as an effective process can be used as an alternative to dilute acid pretreatment of herbaceous biomass, such as corn fiber [46]. Compared to dilute acid, the primary advantage of LHW pretreatment refers to the avoided use of mineral acid with lower sugar degradation [46]. LHW pretreatment has been found to partially hydrolyze hemicellulose and disrupt the lignin and cellulose structures [47]. During the pretreatment, the formation of monomeric sugars could be changed by controlling the pH of the aqueous phase [4]. Thus the toxic inhibitors, such as furfural and HMF formatted by sugar degradation reactions, can be limited by maintaining pH from 4 to 7 [47]. Another advantage of LHW pretreatment is the low cost of reaction medium (only water), and thus the need for washing or neutralization of pretreated biomass and the inhibitor concentration in hydrolyzates are low. The biomass feedstock usually does not need preprocessing for size reduction, and the material for reactor construction is not expensive. After LHW pretreatment, the hemicellulose recovery is relatively high. However, LHW also shows some disadvantages including relatively high energy consumption, low concentration of hemicellulosic sugar in the pretreatment hydrolyzate, and a large amount of waste water in downstream processing [4].

Researches on LHW pretreatment of various biomass feedstocks can be found in literatures. When LHW was used to pretreat microalgae for bioethanol production with solid-to-liquid ratio of 1:13 (w/v) at 147°C for 40 min, the glucose concentration and recovery were 14.2 g/L and 89.32%, respectively, which were up to fivefold higher than that without LHW pretreatment [48]. LHW was also used for bamboo pretreatment to enhance the enzymatic hydrolyzability [49]. Weinwurm et al. combined LHW with ethanol organosolv (EO) pretreatment, and the maximum delignification and carbohydrate removal were achieved when EO treatment at 200°C with 20% ethanol was used after LHW treatment [50]. Efficient improvement of cellulose digestibility by LHW pretreatment was also found for other biomass feedstocks, including wheat straw [51], giant reed [52], sugarcane bagasse [53], and so on.

2.3.2 Acid-catalyzed chemical pretreatment

2.3.2.1 Dilute acid prehydrolysis

Dilute acid prehydrolysis was reported as early as the 19th century, with commercial applications from the beginning of the 20th century. It has been used for pretreatment of a wide range of biomass including herbaceous crops, agricultural residues, hardwood, and softwoods. The effective concentration of acid for pretreatment is usually below 4%. It can be performed at high temperature (e.g., 180°C) for a short period of time or at low temperature (e.g., 120°C) for a longer retention time (30–90 min) [54]. The main objective of the dilute acid pretreatment is to solubilize the hemicellulose fraction thereby improving the enzymatic accessibility of cellulose [6]. Sulfuric acid, hydrochloric acid, phosphoric acid, and nitric acid have been used for dilute acid pretreatment, while sulfuric acid is studied the most because it is highly effective and inexpensive [55]. Phosphoric acid has also been widely used since it is relatively cheap and can hydrolyze biomass efficiently. Moreover, using NaOH or ammonia to neutralize the phosphoric acid, hydrolyzate produces sodium or ammonia phosphate, which could be used as a nutrient for microorganisms in fermentation step [56].

For dilute acid pretreatment of rice straw, a maximal sugar yield of 83% could be achieved by 1% (w/w) sulfuric acid for a reaction time of 1–5 min at 160°C or 180°C, followed by enzymatic hydrolysis [57]. Pretreatment of olive tree with 1.4% sulfuric acid at 210°C resulted in a maximum enzymatic hydrolysis yield (76.5%), while the maximum sugar yield (36.3 g sugar/100 g raw material) was achieved with 1% sulfuric acid at 180°C, representing 75% of all sugars in the raw material [58]. Compared to dilute sulfuric acid pretreatment of rye straw, dilute nitric acid pretreatment was found to give higher glucose concentration [59]. Nevertheless, it is difficult to remove the by-products from nitric acid pretreatment by washing the pretreated substrates [60]. The applications of diluted phosphoric acid for corn stover achieved 85% glucose yield with 0.5% (v/v) H_3PO_4 and 10% solid loading at 180°C for 15 min [61].

Although dilute acid pretreatment presents the advantage of solubilizing hemicelluloses, the acid must be neutralized for the downstream enzymatic hydrolysis or fermentation processes. In addition, the formation of some sugar degradation compounds, for example, furfural and HMF, inhibits the microorganism metabolism [62]. Corrosion caused by dilute acid pretreatment also mandates expensive construction material.

2.3.2.2 Acid catalyzed organosolv pretreatment

Organosolv pretreatment refers to the biomass pretreatment process with organic solvents with or without the addition of catalyst. When no external catalyst is used, organosolv pretreatment usually requires high organic solvent concentration (≥60%) and high pretreatment temperature (160°C–220°C) [63], which is also termed auto-catalyzed organosolv pretreatment. In this process the formed organic acid, such as acetic acid, can play as a catalyst. However, to promote the pretreatment efficiency, external catalysts, such as mineral or strong organic acids, are usually used. Acid-catalyzed organosolv pretreatment is a composite process involving lignin and hemicellulose degradation, solvation, and solubilization of lignin fragments. The supplementary of acid catalyst could increase the solubilization degree of lignin and hemicelluloses under less severe conditions [64]. Cleavage of β-O-4 linkages is primarily responsible for the depolymerization of lignin macromolecule during acid-catalyzed organosolv pretreatment, which generally proceeds by formation of ω-guaiacylacetone followed by formation of Hibbert's ketones (Fig. 2.2A) or elimination of formaldehyde (Fig. 2.2B) [65].

Various organic solvents have been used for organosolv pretreatment under the catalysis of different acids in temperature range from room temperature to 250°C [66], for which alcohols and organic acids are the most frequently used. Lower molecular weight aliphatic alcohols, such as ethanol and methanol, with low boiling points, are the most favored due to

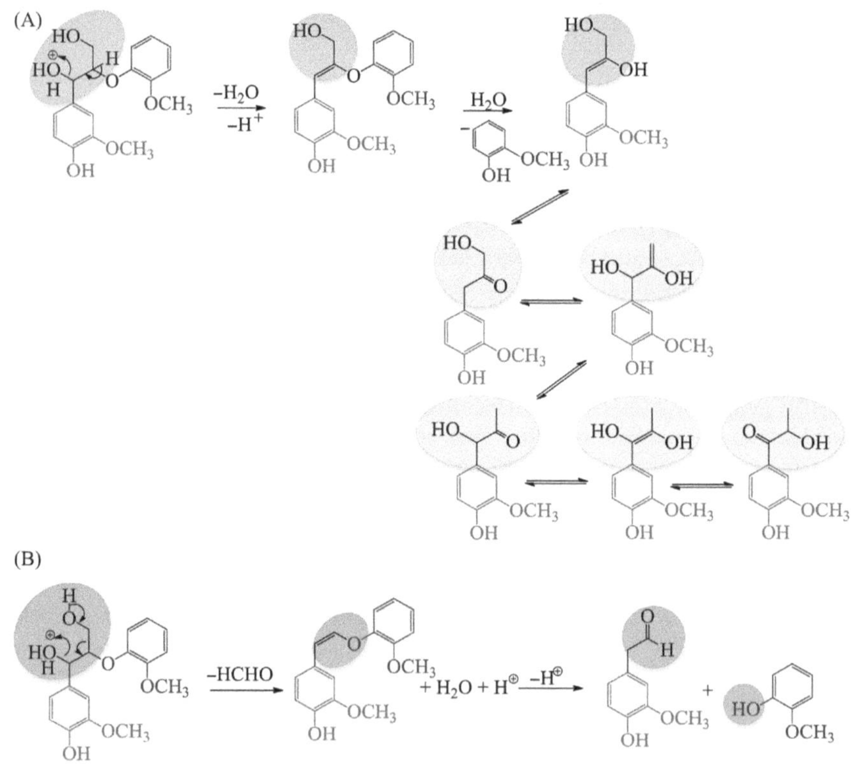

Figure 2.2 The main reactions of acid catalyzed organosolv pretreatment. (A) β-O-4 linkages cleavage by formation of ω-guaiacylacetone followed by formation of Hibbert's ketones and (B) elimination of formaldehyde. *Adapted from Z. Zhou, F. Lei, P. Li, J. Jiang, Lignocellulosic biomass to biofuels and biochemicals: a comprehensive review with a focus on ethanol organosolv pretreatment technology, Biotechnol. Bioeng. 115 (2018) 2683–2702 [65].*

the low cost and ease of recovery. High boiling point alcohols, usually poly-hydroxy alcohols, such as ethylene glycol and glycerol, present an advantage of performing under atmospheric pressure [67]. However, the relatively high price of solvents and high energy consumption for solvent recovery significantly decrease the attractiveness of using organic solvent for biomass pretreatment [68]. The acid catalysts used include organic (e.g., formic acid, acetic acid, and oxalic acid) or inorganic acids (e.g., sulfuric acid and hydrochloric acid) [69]. Zhao and Liu [70] have developed an organosolv pretreatment based on formic acid delignification followed by alkaline deformylation, namely, the Formiline process. The final ethanol concentrations obtained by simultaneous saccharification and fermentation (SSF) of the Formiline-pretreated sugarcane bagasse reached 55.4 and 80.1 g/L, respectively, at initial solid consistencies of 15% and 20%. Goh et al. [71] optimized the H_2SO_4-catalyzed ethanol pretreatment of empty palm fruit bunch, and the optimal condition was found to be ethanol/water ratio of 65:35 (v/v), sulfuric acid concentration of 2.0% (w/w), at 160°C for 78 min.

By acid-catalyzed organosolv pretreatment, fractionation of cellulose, hemicellulose, and lignin can be achieved. Cellulose is recovered as a solid while most lignin and hemicellulose are dissolved in organic solvent [72]. High-quality lignin, sulfur-free with high purity and low molecular weight, can be isolated from the organic solvents recovery process as a solid material, while the sugars as a syrup [73]. The removal of lignin and hemicellulose can substantially deconstruct the biomass recalcitrance, therefore improving the enzymatic digestibility and fermentability of cellulose [74]. Nevertheless, there are inherent drawbacks for acid-catalyzed organosolv pretreatment. It is necessary to wash the pretreated solids with organic solvent previous water washing in order to avoid the reprecipitation of dissolved lignin. Although the organic solvents can be easily recovered by distillation or evaporation for reuse, the cost of chemicals and energy consumption for solvent recovery is still higher than those of other leading pretreatment processes. Moreover, organosolv pretreatment must be carried out under extremely tight and efficient control because of the volatility of organic solvents, and thus no digester leaks can be tolerated due to inherent fire, explosion hazards, and environmental and health and safety concerns [68]. Organic acid pretreatment also causes acylation of cellulose which may significantly reduce the cellulose digestibility by decreasing the molecular recognition of cellulose by cellulases.

2.3.3 Alkali-catalyzed pretreatment

2.3.3.1 Alkaline delignification

Alkaline delignification is one of the most viable pretreatment approaches because of its strong pretreatment efficiency and relatively simple process scheme. It utilizes various alkalis, such as sodium hydroxide, sodium carbonate, calcium hydroxide (lime), and ammonia (aqueous, liquid, and gaseous), as active chemicals for delignification. The alkali digests the lignin structure and breaks the linkages between lignin and carbohydrate, which makes the carbohydrates in the biomass more accessible for enzymatic degradation [75]. Alkali pretreatment can also make the biomass swelling, hence increasing the internal surface area and porosity of the biomass and decreasing both the DP and cellulose crystallinity [76]. Generally speaking, alkali delignification is more effective for grass species with low lignin content than woody species, especially softwood [77]. In addition, alkaline delignification pretreatment is not as energy-intensive as some of the other pretreatments since it can be carried out at lower temperatures and pressures [78]. Alkaline delignification can be performed at ambient conditions; however, longer pretreatment times, usually hours or days rather than minutes or seconds, may be needed to achieve the same level of digestibility [79]. Compared with acid pretreatment, alkaline delignification has less influence on sugar degradation, but a disadvantage occurs since some of the alkalis are converted to irrecoverable salts or incorporated as salts into the biomass during the pretreatment [76]. In alkaline systems, β-aryl ether bonds in both phenolic arylpropane units and nonphenolic arylpropane units can be cleaved. Alkaline cleavage of β-aryl ether bonds is shown in Fig. 2.3 [65,80].

Pretreatment of biomass with NaOH has been studied for a long time and received much attention. NaOH pretreatment is usually performed with concentration ranging from 0.5% to 10% for 1–24 h at room temperature to 180°C [81]. Compared with 20% saccharification (g sugar/g stover) of untreated corn stover, 52% saccharification of the pretreated substrates during enzymatic hydrolysis was obtained by 2% NaOH pretreatment at 150°C for 5 min. Soto et al. found that more delignification was achieved when using higher NaOH concentration; nevertheless, saccharification was negligibly affected by alkali concentration [82]. Using lime instead of NaOH may reduce the cost of the alkali agent, and lime seems to be especially suitable for agricultural residues, such as corn stover or hardwood materials, for example, poplar [83]. Lime is an inexpensive and safe alkali. It could be recovered as calcium carbonate through an

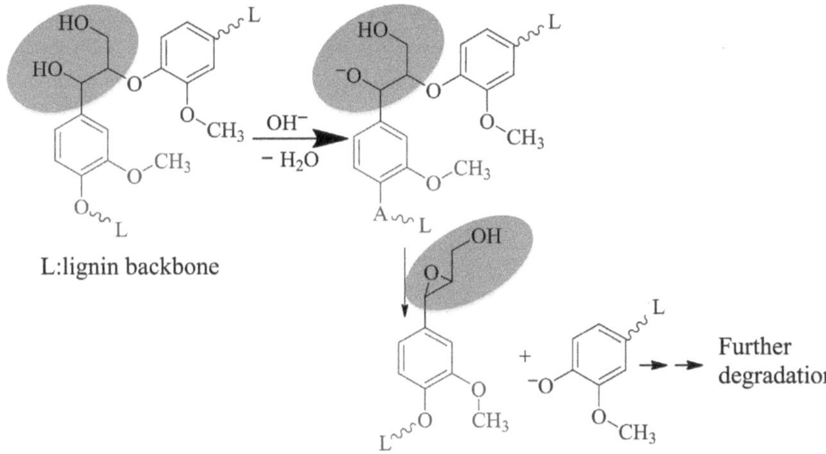

Figure 2.3 Cleavage of β-aryl ether bonds in alkaline conditions. *Adapted from R. Rinaldi, R. Jastrzebski, MT. Clough, J. Ralph, M. Kennema, PC. Bruijnincx, et al., Paving the way for lignin valorisation: recent advances in bioengineering, biorefining and catalysis, Angew. Chem. Int. Ed. 55 (2016) 8164–8215 [80].*

aqueous reaction system with inexpensive carbon dioxide. Subsequently, calcium hydroxide could be regenerated by existing lime kiln technology [84]. Compared to untreated corn stover, pretreatment with lime enhanced the enzymatic saccharification by a factor of nine [85]. For industrial purposes, the recommended pretreatment conditions were pretreatment at 120°C for 4 h with a water loading of 5 g/g and a lime loading of 0.075 g/g (based on dry corn stover). Comparison of different kinds of alkali chemical pretreatments of rice straw was reported, and among all the alkali chemicals, including NaOH, $Ca(OH)_2$, and KOH at 25°C for 24 h, NaOH (6 wt.% loading of dry rice straw) pretreatment could enhance the enzymatic hydrolysis by 85% [86].

AFEX (Ammonia Fiber Explosion) is an alkaline thermal pretreatment using liquid ammonia and SE concept [79]. AFEX process is usually performed at moderate temperature (60°C–120°C) and high pressure (1–5.2 MPa) followed by a rapid pressure release, and the residence time may be low (5–10 min) to moderate (30 min) depending on the degree of saturation of the biomass [87]. The liquid ammonia loading is approximately 1–2 kg ammonia/kg dry biomass [88]. AFEX is basically a "dry to dry" process and processes only solid material. Therefore the pretreated biomass is suitable for long-term storage and can be fed at a desired solid loading in subsequent enzymatic saccharification and fermentation, including very high solids loadings [89]. During AFEX pretreatment, cellulose and hemicellulose are well preserved since only a small amount of biomass is solubilized. AFEX results in decrystallization of the cellulose. Furthermore, hemicellulose is partially hydrolyzed and deacetylated [90]. Some lignin–carbohydrate bonds are disrupted, and the material structure is rigorously altered, which results in the increased water-holding capacity and higher digestibility [91]. It is not necessary for neutralization prior to the enzymatic saccharification of AFEX pretreated biomass. In addition, there is almost no inhibitor formed for microbial fermentation [92]. As reported, compared with dilute acid pretreatment, the formation of furans was 36-fold lower for AFEX, while the yield of carboxylic acids (e.g., lactic and succinic acids) was 100–1000-fold lower than sodium hydroxide pretreatment [93]. The hydrolyzate can be directly used for fermentation without any detoxification. Furthermore, the residual nitrogenous salt is an important nitrogen source for microbial fermentation [94]. AFEX has been reported to achieve high conversion rates (80%–97%) of different kinds of biomass, such as corn stover [89], switchgrass [95], wheat straw [96], rice straw [97], birch, and willow [98]. However, AFEX is less effective for the biomass with relatively high lignin content, such as aspen and newspaper [99]. Another advantage of AFEX is that the ammonia used in the pretreatment can be recovered and recycled. Nevertheless, the associated complexity and costs of ammonia recovery are high. Therefore the process still needs integration and optimization to achieve the industrialization of AFEX [100].

2.3.3.2 Alkali-catalyzed organosolv pretreatment

Alkalis are also used to facilitate the organosolv pretreatment process. Using alkaline catalysts, such as alkalis and neutral alkali earth metal, can assist organosolv pretreatment under mild conditions. With regards to the issues derived from acid catalyzed organosolv pretreatment, such as being corrosive, hazardous, and inhibitory characteristics, alkali-catalyzed organosolv pretreatment has recently attracted considerable attention [101]. In order to increase the enzymatic hydrolysis, the removal of lignin seems to be more helpful than the removal of hemicellulose under a mild condition (<100°C) [102]. Alkaline catalysts, such as sodium hydroxide, potassium hydroxide, ammonia, and lime, have been used for pretreatment due to their lignin removal effectiveness [101].

Ethanosolv pretreatment catalyzed by NaOH was applied to improve the enzymatic hydrolysis of moso bamboo [103]. The cellulose to glucose conversion yield was significantly improved from 2.4% to 45.1%. In addition, the concentrations of fermentation inhibitors, for instance HMF and furfural, were negligible in the spent liquor, which were much lower than those from sulfuric acid catalyzed water or ethanosolv pretreatments. The organosolv pretreatment of *Liriodendron tulipifera* with sodium hydroxide could enhance the enzymatic hydrolysis and ethanol yield at lower temperature [74]. Wang et al. compared the effects of various catalysts (formic acid, triethylamine, and sodium hydroxide) and solvents (methanol, ethanol, *n*-propanol, and *n*-butanol) on the substrate's physicochemical characteristics under mild organosolv pretreatment [104]. They found that the pretreatment with NaOH was more efficient on hemicelluloses dissolution, and therefore higher glucose yield was obtained than formic acid and triethylamine catalyzed pretreatments.

2.3.4 Oxidative pretreatment

Oxidative pretreatment refers to using oxidant to pretreat the biomass, in which the biomass components, especially lignin, undergo oxidative depolymerization. Various oxidants have been used for oxidative delignification; however, the oxidants to be used for pretreatment must be cheap but effective. Promising oxidative pretreatments include wet oxidation, alkaline H_2O_2, and peracid pretreatments.

2.3.4.1 Wet oxidation

Wet oxidation process was proposed to remove organic compounds in the liquid phase by Zimmerman in the year of 1950, in which oxygen or air was used as an oxidant to oxidize the organic matters completely to carbon dioxide and water [105]. The wet oxidation pretreatment involves treating biomass with water and oxygen or air at elevated temperatures (120°C–238°C) and pressure (45–480 psi) [106]. The first reaction during wet oxidation refers to the formation of acids. Due to the deesterification of the acetate groups in hemicellulose and oxidation of some hemicellulose fragments, the hemicelluloses are solubilized with formation of organic acids. With the increase in acid concentration, the hydrolytic reactions are facilitated. More and more hemicelluloses are broken down into lower molecular weight fragments and dissolved in the water [107]. The wet oxidation not only shows effects on the hemicelluloses but also on the cellulose and lignin fractions. When wet oxidation is performed under a relatively mild condition (120°C–172°C), hemicellulose and partial lignin are solubilized, resulting in a solid fraction enriched in cellulose

[106,107]. At a higher temperature, such as 172°C–227°C, the oxidation is easier to occur, resulting in the formation of a large amount of fragmentations, which are further decomposed to carbon dioxide, water, and carboxylic acids [77]. During wet oxidation process, phenolic compounds are further oxidized to carboxylic acids [6]. In addition, furfural and HMF are not produced [86], which are potential inhibitors in the subsequent fermentation process.

Wet oxidation has been successfully used for the pretreatment of wheat straw, newspaper waste, maize silage, sugarcane bagasse, rice husk, spruce, etc. [108]. Pedersen and Meyer [109] studied the effects of substrate particle size and wet oxidation on physical surface structures and enzymatic hydrolysis of wheat straw. The result showed that the released glucose from the smallest particles achieved 90% of the theoretical maximum after 24 h enzymatic hydrolysis. Wet oxidation pretreatment could tear up the surface structures of the particles, increasing the enzymatic xylose and glucose yields by 5.4 and 1.8 times, respectively. After a wet oxidation pretreatment at 200°C for 10 min with 12 bars of oxygen, the sugar yield in a 72 h enzymatic hydrolysis could reach 79% of theoretical for a bark-free Norway spruce, a softwood [110]. However, compared with many other delignification pretreatments, the lignin obtained in wet oxidation pretreatment cannot be used as a solid fuel, decreasing the potential profit from by-products in large-scale production [91]. Moreover, the capital and maintenance cost for this process are high primarily due to the high temperature and pressure used in the process.

2.3.4.2 Alkaline hydrogen peroxide

Alkaline peroxide pretreatment majorly refers to pretreating lignocellulosic biomass with peroxide, typically hydrogen peroxide, under alkaline condition. This pretreatment results in delignification of lignocellulosic materials with a cellulosic residue highly susceptible to enzymatic hydrolysis. For alkaline hydrogen peroxide pretreatment the delignification reaction is strongly pH-dependent with an optimum pH of 11.5–11.6, which is the p*Ka* for the dissociation of H_2O_2 [111]. As a result of its decomposition, the highly reactive oxygen species superoxide ($O_2^- \cdot$) and hydroxyl radical ($HO \cdot$) are produced, which are the primary species to oxidize lignin [112]. The H_2O_2-derived radicals promote the depolymerization of lignin by attacking the lignin side chains and fragmenting the lignin macrostructure into low molecular weight compounds [113]. A number of chemical changes occur in the alkaline peroxide pretreatment of biomass including a partial reduction of esterified acetic acid, ferulic acid, and *p*-coumaric acids as well as a significant cleavage of etherified linkages between lignin and hemicelluloses [114]. In fact, almost no significant changes in the chemical structure of hemicelluloses might be observed, because the oxidation reaction only occurs in the aliphatic part of the macromolecule [115].

Alkaline peroxide pretreatment has been applied to various biomass feedstock including wheat straw [116], softwood [117], sugarcane bagasse [118], bamboo [119], corn stover [120], Miscanthus [121], and other herbaceous and woody biomass [122]. After alkaline peroxide pretreatment, the pretreated wheat straw presents 96.7% enzymatic hydrolysis yield. During the pretreatment, no measurable quantities of furfural and HMF are produced. The ethanol yield could be 0.46 g/g of available sugars (0.29 g/g straw) [123]. Recently, the alkaline peroxide pretreatment at low peroxide loadings has been used to enhance the enzymatic hydrolysis of softwood [117]. It was found that the presence of H_2O_2 is essential for this pretreatment. In order to achieve a better substrate hydrolyzability, it is more effective to maintain a higher alkaline loading than increasing the H_2O_2 loadings. When Douglas-fir was pretreated at 180°C for 30 min with a low peroxide loading of 0.10 (mass fraction of H_2O_2 to biomass), a cellulose to glucose yield of 95% could be achieved [117]. The key factor is the balance between alkaline and peroxide loadings during the pretreatment.

2.3.4.3 Peracid oxidation

As powerful oxidants, peracids, such as peroxymonosulfuric acid, peroxyformic acid, and peracetic acid (PAA), also have been used for biomass delignification and pretreatment [68]. Among the peracids used, PAA is the most promising because it is more stable than performic acid [124] and easier to recover and prepare than peroxymonosulfuric acid [125]. PAA has been found to be very selective to oxidize lignin. It oxidizes the hydroxyl groups in lignin side chains to carbonyl groups [126] and cleaves β-aryl ether bonds, resulting in the decrease of lignin molecular weight and introduction of hydrophilic groups [127]. In addition, PAA oxidation increases the water solubility of lignin via several reactions: formation of hydroquinones by phenolic rings hydroxylation [128], oxidation of hydroquinones to quinones which undergo ring opening to produce water-soluble muconic, maleic, and fumaric acid derivatives [129]. The main reaction types of PAA with lignin-related structures using lignin models are shown in Fig. 2.4 [68]. The removal of lignin can well increase the accessibility of cellulose. It has been found that the improvement of enzymatic hydrolysis of sugarcane bagasse by PAA is mainly due to the delignification and an increase in the surface area with exposure of cellulose fibers [102]. However, because of the partial removal of lignin and hemicelluloses, the crystallinity increased after PAA pretreatment, but it seemed not to significantly impact cellulose digestibility [68].

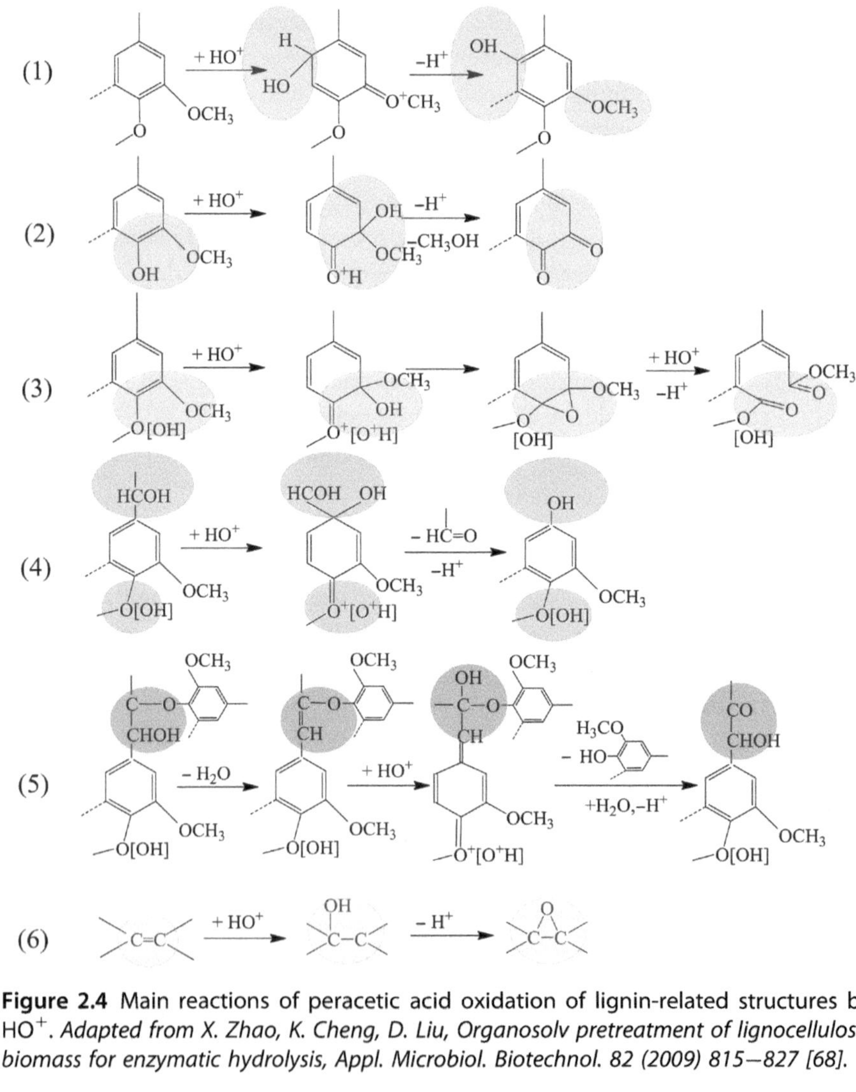

Figure 2.4 Main reactions of peracetic acid oxidation of lignin-related structures by HO^+. *Adapted from X. Zhao, K. Cheng, D. Liu, Organosolv pretreatment of lignocellulosic biomass for enzymatic hydrolysis, Appl. Microbiol. Biotechnol. 82 (2009) 815–827 [68].*

Some reported works have shown that well-improved enzymatic digestibility can be achieved by PAA pretreatment [130]. When bagasse was pretreated with 50% PAA (based on raw biomass) and liquid-to-solid ratio of 6:1 at 80°C for 2 h, more than 80% of cellulose to glucose conversion was obtained [131]. Compared with acid and alkaline pretreatment performed at the same conditions, PAA pretreatment was more efficient for achieving higher sugar yield for Crofton weed stem [132]. In order to reduce the amount of expensive PAA used in a single-stage pretreatment, combined pretreatment with PAA and other chemicals, especially alkalis, was reported. By a first-step pre-pretreatment with NaOH to swell the biomass and partially removing lignin, PAA loading used for the second-step treatment can be significantly decreased to achieve similar or even better cellulose digestibility [130].

PAA pretreatment can be carried out at a relatively low temperature and atmospheric pressure, thereby avoiding the degradation of monosaccharide to form inhibitors such as furfural and HMF [133]. In addition, the PAA itself degrades into acetic acid, water, and oxygen. Therefore the washed biomass after PAA pretreatment can be efficiently fermented to make ethanol [134]. However, PAA is too expensive for commercial use [135]. It is also corrosive to equipment and is explosive in a concentrated form. This safety hazard results in the increased cost of storage and transportation of PAA.

2.3.5 Cellulose solvent–based pretreatment

Cellulose solvent–based pretreatment refers to the process using cellulose-dissolving solvents to pretreat biomass, thus increasing cellulose accessibility. A schematic process using cellulose solvents to produce bioethanol from lignocellulosic materials is shown in Fig. 2.5. The most commonly used cellulose solvents are *N*-methylmorpholine *N*-oxide (NMMO), concentrated acids, such as concentrated phosphoric acids (CPAs) and ILs.

2.3.5.1 N-Methylmorpholine N-oxide

NMMO contains N–O bonds with high polarity, which makes it capable of being used as a cellulose solvent. NMMO is industrially used in the Lyocell process. NMMO can disrupt hydrogen bonds in cellulose and form a new hydrogen bond with cellulose [136]. After dissolution in NMMO, cellulose can be regenerated by rapid precipitation with water. The regenerated cellulose is more accessible to cellulases because the crystalline structure of cellulose changes from cellulose I into II [137].

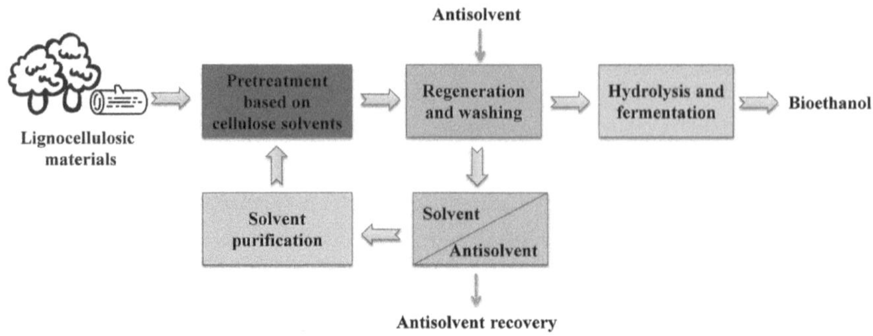

Figure 2.5 A schematic process using cellulose solvents to produce bioethanol from lignocellulosic materials.

Nevertheless, the dissolution mechanism is dependent on the water content of the mixture. There are mainly three states for cellulose in NMMO solution depending on water content, including dissolution without noticeable swelling (water content between 13% and 17%), dissolution by swelling and ballooning (water content between 19% and 24%), and swelling by ballooning without dissolution (water content above 25%) [138]. In the case of ethanol production the NMMO concentration of 83%–87%, usually 85%, is found to be more efficient than lower concentration to increase the biomass digestibility [139].

NMMO has been reported for efficient pretreatment of different lignocellulosic biomass [140]. NMMO pretreatment is usually performed at relatively low temperatures (90°C–120°C) under atmospheric pressures for a relatively short time (0.5–3 h) [139]. During the NMMO pretreatment the production of inhibitory compounds, such as HMF and furfural, is negligible. The toxicity of NMMO is less than that of ethanol, which is an environment-friendly solvent [141]. In addition, NMMO can be recovered almost completely [142]. With 85 wt.% NMMO pretreatment at 130°C for 3 h the ethanol yield of hardwood and softwood could be increased from 18.6% and 6.8% to 85.4% and 89%, respectively [143]. The comparison between NMMO and IL (BMIM-OAc) pretreatments of rice straw at 120°C for 5 h was reported by Poornejad et al. [144]. It was found that the glucan conversion was complete for IL pretreatment, while reaching 96% for NMMO pretreatment. However, in terms of SSF, the ethanol yield of NMMO pretreated samples was 93.3%, higher than that of IL pretreatment (79.7%).

NMMO pretreatment still has some drawbacks, such as inhibitory effects of NMMO on microorganisms when its concentration was higher than 2% [145]. The toxicity of NMMO is observed by increasing the glycerol yield during fermentation, resulting in the decrease of ethanol fermentation rate and final ethanol yield [146]. In order to reduce the cost, efficient recycling of NMMO is necessary. However, some side reactions, which are present in the Lyocell process, might also affect the pretreatment of lignocellulose and result in the acceleration of NMMO degradation [147]. It was reported that the recycled NMMO was not as effective as fresh NMMO for pretreating forest residues with a high lignin content and bark after five cycles [138].

2.3.5.2 Concentrated acid for biomass hydrolysis and cellulose decrystallization

Concentrated acid, usually >30% of acid concentration, can be used for biomass hydrolysis at moderate temperatures (<100°C) under atmospheric pressures to achieve a high sugar yield without subsequent enzymatic hydrolysis [54]. Concentrated acid pretreatment inevitably leads to hydrolysis of polysaccharides by three sequential stages: (1) prehydrolysis of hemicellulose fraction, (2) the main hydrolysis stage to dissolve cellulose and convert it into soluble cello-oligosaccharides, and (3) posthydrolysis to convert cello-oligosaccharides into monosaccharides [148]. In the initial two stages, "solubilization" is responsible for the decrystallization of the lignocellulose structure, while in the last stage, monosaccharides are liberated from the fragments of cellulose and hemicelluloses [149]. The sugar yields of this pretreatment are affected by operational variables, primarily including acid concentration, solid loading, process temperature, and pretreatment time [59]. Compared with dilute acid saccharification, the sugar yields of concentrated acid hydrolysis are much higher due to the decreased degradation of sugars [150]. Furthermore, because of the low temperature and atmospheric pressure used, inhibitors from sugar degradation are less than those from dilute acid pretreatment [148]. Sulfuric acid is so far the most commonly used for concentrated acid hydrolysis of biomass since it is inexpensive and highly effective, but other mineral acids, such as hydrochloric, nitric, phosphoric, and trifluoroacetic acid, are also studied. Concentrated sulfuric acid saccharification of wood chip, corn stover, and bamboo has been reported [150], and the obtained sugars show excellent fermentability. The formation of furfural and HMF is low, and the main by-products in the hydrolyzate are organic acids [150].

Concentrated acid, such as CPA (> 80 wt.%), is also an efficient cellulose solvent for biomass pretreatment to increase cellulose hydrolyzability [151]. It can be combined with an organic solvent (e.g., acetone or ethanol) to fractionate lignocellulose under mild reaction conditions [152]. CPA has been used to [153] (1) completely dissolve cellulose fibers by disrupting orderly hydrogen bonds network of crystalline cellulose, providing better cellulose accessibility; (2) break up the lignin–carbohydrate complex bonds, slightly hydrolyze cellulose and hemicellulose, and remove acetyl groups from hemicellulose to eliminate the major obstacles to hydrolysis. The mild pretreatment conditions, typically at 50°C and atmospheric pressure for 20–60 min, can reduce the utility consumption [154]. Moreover, the sugar degradation is less during this pretreatment, thereby the inhibitors are less as well [155]. CPA pretreatment has been demonstrated to be efficient for various biomass, for instance poplar [156], bamboo [157], elephant grass [158], switchgrass [153], common reed [159], and corn stover [160]. Moxley et al. [155] investigated the effect of pretreatment conditions (phosphoric acid concentration, reaction temperature, and duration time) on sugar release of industrial hemp hurds. The glucan digestibility was 96% at 24 h with a cellulose loading of 15 FPU/g glucan after 84.0% CPA pretreatment at 50°C for 60 min. It was found that phosphoric acid can efficiently disrupt the recalcitrant lignocellulose structures only when the acid concentration was above 83% for a sufficient duration time. CPA posttreatment after Formiline pretreatment of wheat straw was reported [161]. Using this combined pretreatment, the ethanol concentration achieved 41.6 g/L with a yield of 91.2% at a relatively low cellulose loading (5 FPU/g solid) within 24 h SSF. In order to achieve a fractionation of biomass, subsequent washing steps are needed after CPA pretreatment with a first organic solvent washing to remove lignin and a second water washing to remove partially hydrolyzed hemicellulose fragments [160]. Therefore although this method is very efficient to overcome biomass recalcitrance, a large amount of phosphoric acid, organic solvents, and water are usually needed [152].

However, the major drawback of concentrated acid hydrolysis and pretreatment lies on the corrosiveness, toxicity, and hazard of the acid [162]. Therefore reactors made from expensive materials resistant to corrosion are necessary, such as acid-resistant alloys or ceramics. In addition, acid recycle is necessary for the economic feasibility of this process. However, the recovery of mineral acid from the hydrolyzate is usually complicated and energy intensive. Extraction, distillation, ionic chromatography, and electrodialysis have been employed for acid recovery. Fortunately, in the last 50 years, the acid recovery yield has significantly increased from 80% to 97% [163]. However, the cost still should be reduced for a commercial application.

2.3.5.3 Ionic liquid

ILs are molten salts, which consist of paired ions and have tunable properties. ILs are considered "green solvents" due to their negligible vapor pressure, nonflammability, and good thermal and chemical stability [164]. ILs containing imidazolium or pyridinium cations paired with chloride, carboxylates, phosphates, and phosphonates are able to dissolve cellulose through high hydrogen bond basicity [165]. However, low-basicity anions, such as [DCA] and [Tos], are not effective for cellulose dissolution [166]. The dissolution of lignocellulose in ILs disrupts the three-dimensional network, therefore making biomass more digestible by enzymes [167]. A number of researches have been reported for the pretreatment of kinds of lignocellulose with various ILs [168].

Liu and Chen [169] obtained a significant improvement of enzymatic hydrolysis yield using [BMIM]Cl to treat raw and steam-exploded wheat straw. The degree of hydrolysis of pretreated wheat straw reached 70.4% while the pretreated steam-exploded wheat straw could be completely hydrolyzed. It has been found that the hydrolyzability improvement by IL pretreatment is attributed to the decrease in the DP of cellulose and hemicellulose and decrystallization of cellulose resulting in the increased accessibility. Compared to dilute acid pretreatment, switchgrass by [EMIM]Ac pretreatment showed greatly increased surface area, reduced cellulose crystallinity, and lignin content [170]. In addition, after 12 h enzymatic hydrolysis, the glucose yield of [EMIM]Ac pretreated samples reached 90%, while the glucose yield of dilute acid pretreated samples was only 80% after 72 h saccharification. Cox et al. [171] used 1-*H*-3-methylimidazolium chloride for pretreatment of yellow pine wood chips at 110°C–150°C for up to 5 h. Dissolution of hemicelluloses and lignin from cell walls of pine wood was observed. Although the dissolution was faster at higher temperatures, significant cellulose degradation was also found at the highest temperatures tested. The cellulose-rich fraction was easily saccharified by enzymatic hydrolysis with cellulases from *Trichoderma viride*, and higher glucose yields were obtained at longer pretreatment duration at 130°C. Another approach was developed in which simultaneous pretreatment and saccharification of biomass in ILs were conducted [172]. However, in this integrated process, enzymatic hydrolysis is performed with the presence of ILs, and therefore, the cellulase may be easily deactivated. Therefore improving the enzyme stability and activity is crucial to this integrated process.

Application of ILs for biomass pretreatment has provided a new path for effective utilization of biomass. However, there are still many challenges for industrialization of IL-based pretreatment. The most important challenge is the high cost of ILs, and thus to find economical solutions for the reuse and recycle of ILs is of great importance. How to eliminate the toxicity of the residual ILs to cellulases and microorganisms is also important.

2.4 Biological pretreatment

Biological pretreatment refers to using microorganisms or enzymes to pretreat various lignocellulosic materials prior to enzymatic hydrolysis of polysaccharide [1]. The primary microorganisms used mainly include white- (e.g., *Phanerochaete chrysosporium* and *Ceriporia lacerata*), brown- (e.g., *Serpula lacrymans* and *Coniophora puteana*), and soft-rot (e.g., *Paecilomyces* sp. and *Cadophora* spp.) fungi as well as bacteria (e.g., *Bacillus circulans* and *Sphingomonas paucimobilis*) to degrade lignin, partially hydrolyze hemicelluloses, and reduce *DP* of cellulose [4]. It was reported that white-rot fungi are the most widely used and effective for biological pretreatment [7]. The advantages of biological pretreatment are low energy requirement, no or little chemicals requirement, mild pretreatment conditions, and little environmental pollution. However, the primary disadvantage of biological pretreatment is the low efficiency [7]. Different fungi or bacteria secrete different enzyme systems and have different effects on pretreatment. White- and soft-rot fungi secrete enzyme systems comprising xylanase, lignin peroxidases, polyphenol oxidases, manganese-dependent peroxidases, and laccases as the main compositions and can catalyze the degradation of lignin and hemicellulose [4].

Sindhu et al. investigated biological pretreatment of paddy straw with white-fungus (*Trametes hirsuta*) and compared with SE pretreatment at 121°C. The results indicated that biological pretreatment had higher lignin removal than SE. The highest saccharification efficiency observed after 24 h for biological pretreatment (76.5%) was also somewhat higher than that of steam pretreatment (74.1%), with a maximum production of sugar (52.91 g/L) observed for biologically pretreated biomass at 10% glucan loading after 24 h enzymatic hydrolysis [2]. Gao et al. compared *Trametes versicolor* 52J (TV52J), *T. versicolor* m4D (TVm4D), and *P. chrysosporium* to pretreat wheat straw pellets using solid-state fermentation. The results demonstrated that the longest time (35 days) was needed for *P. chrysosporium*, while shortest for TV53J (21 days) to achieve a good pretreatment efficiency. After biological pretreatment the combination and connection between single fibers became loose, which were advantageous to improve the porosity of the substrates [173]. Saha et al. compared 26 white-rot fungal strains to evaluate the fermentable sugar production from corn stover under solid-state cultivation at 74% moisture and 28°C for 30 days. Pretreatment with *Cyathus stercoreus* NRRL-6573 obtained the highest sugar yield of 394 ± 13 mg/g of pretreated stover, followed by treatment with *Pycnoporus sanguineus* FP-10356-Sp (393 ± 17 mg/g) and *Phlebia brevispora* NRRL-13108 (383 ± 13 mg/g) [173]. Biological pretreatment is also combined with other pretreatment processes, such as chemical pretreatment and physicochemical pretreatment. Yan et al. studied biological pretreatment combined with dilute acid pretreatment and found that bacteria could enhance dilute acid pretreatment. This was because the used fungi *Cupriavidus basilensis* B-8 could work on the lignin droplets formed in dilute acid pretreatment thus recovering cracks and holes on rice straw surface and leaving an open and porous structure for the easy access of enzymes to the inner cellulose. The enzymatic digestibility of rice straw was increased by 35%–70% and 173%–244% in combined pretreatment process compared to dilute acid pretreated and untreated rice straw, respectively [174]. Biological pretreatment is also applied in improving biogas production and other bioconversion processes [173]. However, process optimization and intensification are still needed to make the biological pretreatment more efficient [4].

2.5 Combined pretreatments

As aforementioned, each pretreatment has its own merits and drawbacks. In addition, the recalcitrance of different raw biomass is different; and thus, no single pretreatment process is universally effective for all kinds of biomass. Therefore to achieve a high pretreatment efficiency, different pretreatments are usually combinedly used. Generally, single mechanical milling pretreatment is energy intensive to achieve a high cellulose digestibility; biological pretreatment is limited by its low efficiency and long pretreatment time; dilute acid pretreatment is effective for herbaceous and hardwood, but poor for softwood with remaining lignin; alkaline pretreatment process works well with hardwood, agricultural residues, and also

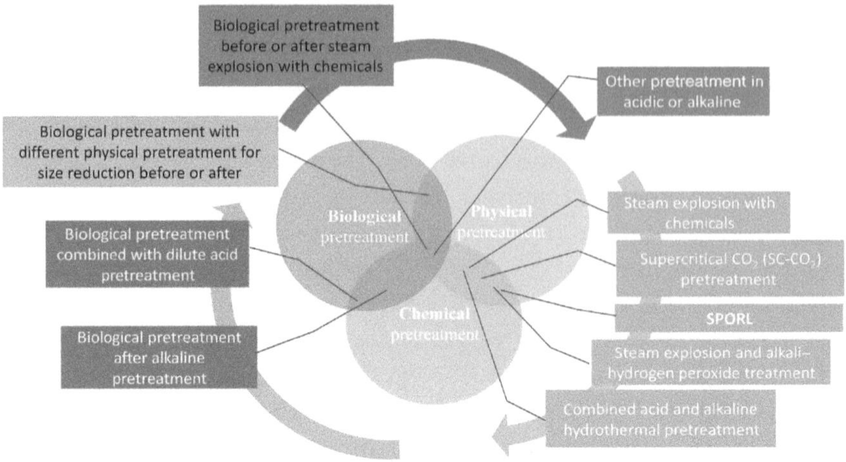

Figure 2.6 Schematic diagram of different combined pretreatment of lignocellulosic biomass.

for softwood but alkalis are usually expensive; and organosolv and cellulose solvent–based pretreatments are limited by the relatively high cost of chemicals and energy consumption for solvent recovery [4]. Therefore, combinations of different pretreatments, for example, physical–physical, physical–chemical, chemical–chemical, and biological–chemical (shown in Fig. 2.6), have been found to be efficiently performed with lower pretreatment severity, shorter grinding time, decrease in the formation of inhibitors, enhanced enzymatic hydrolysis efficiency, and reduction of energy consumption [4].

2.5.1 Physicochemical pretreatment

2.5.1.1 Steam explosion

SE has a long history for biomass pretreatment and pulping. During steam explosion, biomass substrates are heated with high-pressure saturated steam for a short time, from few seconds to several minutes, and then the pressure is swiftly reduced. By this sudden release of pressure, there is an explosive decompression happening to biomass. Although chemicals can be used to promote SE pretreatment, the SE process without the addition of any other external chemicals is termed autohydrolysis [22]. Residence time, temperature (pretreatment pressure), particle size, and moisture content are the key parameters affecting SE pretreatment efficiency [4]. Pretreatment temperature and pressure for typical SE range from 160°C to 260°C and 0.69 to 4.83 MPa, respectively [175]. By SE pretreatment, the size of biomass fibers is reduced, hemicelluloses are degraded, and lignin undergoes transformation, removal, and redistribution at high temperature, and thus the cellulose hydrolyzability is increased [22]. Water can work as an acid at high temperature, and acetic acid produced from the acetyl groups of hemicelluloses can play as catalysts to facilitate the degradation of biomass component [4]. SE is usually effective for grass biomass and hardwood, but not so satisfying to softwood biomass unless strong catalysts, such as H_2SO_4, are used under higher pretreatment severity [175].

The advantages of SE include relatively low cost, low energy consumption, being suitable for big particle size, no external chemical addition, low environmental impact, and also high sugar recovery [22]. However, the inhibitors, such as furfural/HMF and phenolic compounds from sugar and lignin degradation at high temperature, are the primary drawback of SE pretreatment. To remove the inhibitors, a large amount of water is usually needed to wash the pretreated biomass, but washing with water may lead to loss of sugar and reprecipitation of lignin [4]. SE has been proved to be an effective pretreatment process for the production of bioethanol, biogas, or other bioproducts. SE pretreated floodplain meadow hay at 200°C resulted in the highest glucose yield up to 85% with ethanol yield of 97% [176]. Zhao et al. [177] employed SE to pretreat corn stover and the pretreated solid was then digested by cow rumen fluid for anaerobic gas production in vitro. The best pretreatment condition was found to be 1.5 MPa, 10% moisture content for 180 s of pretreatment time [177]. Lizasoain et al. (2016) studied biogas production from reed biomass by using SE pretreatment. When pretreatment condition was 200°C for 15 min, the methane yield could reach up to 89% [178]. *Cynara cardunculus* pretreated by SE at 235°C for 1 min in a 0.5 L reactor with a receiving chamber of 30 L, followed by fermentation with SSF processes, the highest ethanol concentration could reach up to 18.7 g/L with fermentation efficiency of 66.6% and ethanol yield of 10.1 g ethanol/100 g untreated cardoon [178]. More applications of SE pretreatment can be found for spruce wood to improve enzyme hydrolysis [179] and for wheat straw to increase methane yield by 20%–30% [180]. SE is recognized as one of the most effective and promising pretreatment processes to pretreat hardwoods and agricultural residues with the industrial application [175].

2.5.1.2 Steam explosion promoted by addition of chemical additives

To improve the pretreatment efficiency or reduce the required pretreatment severity for SE, different chemicals have been added in the pretreatment process depending on the feedstocks. Since a large number of enzymatic hydrolysis and fermentation inhibitors can be formed under severe SE conditions, particularly at high temperature, some additional acids, such as SO_2, H_3PO_4, and H_2SO_4, can decrease temperature needed [1]. Tooyserkani et al. studied the effects of SO_2-catalyzed SE pretreatment on softwood pellets. They found that the addition of SO_2 substantially reduced the particle size of the woodchips and eliminated further grinding [181]. The addition of acid catalysts in SE process can increase the recovery of hemicellulose sugars, decrease the formation of inhibitors, and further improve the enzymatic hydrolysis of the pretreated solid residues [175]. External H_2SO_4 addition can promote the solubilization of hemicelluloses, lower pretreatment temperature, but also partially hydrolyze cellulose especially the amorphous region [182].

2.5.2 Supercritical CO_2 pretreatment

Supercritical CO_2 (SC-CO_2) can diffuse like a gas into the interspaces of biomass and dissolve like a liquid as a part of biomass materials [1]. SC-CO_2 pretreatment is also usually termed CO_2 explosion pretreatment which has a similar mechanism to that of SP. In the operation of the process, SC-CO_2 is fed into a high-pressure vessel containing biomass followed by heating to the required temperature and keeping for several minutes at high temperatures. In this process, CO_2 diffuses into the biomass and forms carbonic acid to hydrolyze hemicelluloses. The pressure is then suddenly released and the biomass structure is disrupted thus increasing cellulose accessibility.

SC-CO_2 pretreatment has a couple of advantages, including low cost, nontoxicity, nonflammability, easy recovery, and low environmental impact [7]. Compared with thermal treatments, SC-CO_2 can be operated at lower temperatures (but higher pressure), which reduced the formation of inhibitors from hemicelluloses and lignin degradation. Moreover, SC-CO_2 has higher diffusivity and low viscosity so that it may work more efficiently for wet biomass [183]. It was found that the enzymatic digestibility of SC-CO_2 pretreated aspen and southern yellow pine increased as the moisture content of the raw biomass increased; but for absolutely dried aspen and southern yellow pine, SC-CO_2 showed no significant improvement [183]. SC-CO_2 also can be combined with other pretreatment to further improve cellulose digestibility. For SC-CO_2 pretreated sugarcane bagasse the glucose yield of enzymatic hydrolysis could reach to

72%, while the SC-CO_2 combined with alkali pretreatment could achieve 20% higher glucose yield than alkali pretreatment only [183].

2.5.2.1 Sulfite pretreatment to overcome recalcitrance of lignocellulose

Sulfite pretreatment to overcome recalcitrance of lignocellulose (SPORL) pretreatment developed by Zhu et al. [184] is a combined pretreatment of using sulfite to soften wood chips followed by size reduction with a disk milling. SPORL pretreatment showed efficient improvement of cellulose digestibility for woody biomass. In this process, sulfuric acid and aqueous sulfite solution are used as the active chemical agents to cook woody chips at relatively high temperatures (usually between 130°C and 170°C) for different duration time. The cooked chips are then mechanically milled for size reduction [39]. SPORL pretreatment partially removes hemicelluloses and lignin with great modification of lignin structure that dramatically deconstructs cell wall structure and reduces the nonproductive adsorption of cellulase on lignin. However, it has been found that hemicelluloses removal seems to be more crucial than delignification for the improvement of cellulose digestibility. The amounts of inhibitor formed in this process are low, with HMF and furfural of only 5 and 1 mg/g untreated wood, respectively [184]. When spruce chips were pretreated by SPORL process with 8%–10% bisulfite and 1.8%–3.7% sulfuric acid on spruce wood (oven dried) at 180°C for 30 min, more than 90% of cellulose conversion could be obtained. SPORL pretreatment also worked well on Douglas-fir, poplar, and red pine [185,186]. To integrate the influence of various key factors on SPORL pretreatment, a combined hydrolysis factor has been defined and applied to predict xylan hydrolysis, formation of inhibitors, and enzymatic digestibility of pretreated substrates for woody and herbaceous biomass [187]. However, the disadvantage of this pretreatment is the high cost of chemical recovery and effluent of sulfur-containing wastewater if the formed lignosulfonate is not well recovered [64].

2.5.2.2 Subsequent acid and alkaline pretreatments

Acid pretreatment may hydrolyze a considerable part of hemicelluloses, while alkaline pretreatment is efficient to remove lignin. Therefore combination of acid and alkaline pretreatments may achieve a fractionation of lignocellulosic biomass with great exposure of cellulose. Maekawa [188] pretreated a wide range of biomass materials including rice straw, hardwoods, and softwoods using SE and alkaline hydrogen peroxide treatment in succession and found that the enzymatic digestibility was improved by 2–2.5 times compared with the single SE pretreatment. Pan et al. [189] performed a two-stage strategy to pretreat Douglas-fir by steam pretreatment and extraction with cold 1% NaOH. They found that NaOH extraction reduced the lignin content by only approximately 7%, but cellulose to glucose conversion was enhanced by about 30%, reaching 85% at a cellulase loading of 40 FPU/g substrate. Guo et al. developed a combined acid and alkaline hydrothermal pretreatment using dilute sulfuric acid and lime sequentially for pretreating *Miscanthus* in order to obtain high recovery of both hexose and pentose. Under the optimal condition, this two-stage process achieved >80% glucose yield and >70% xylose recovery, and ethanol yield of 0.145 g ethanol/g biomass in a subsequent fermentation. These works have demonstrated that two-stage methods combining acid and alkaline treatments can achieve higher sugar recovery than single stage processes and require less enzyme loading for efficient hydrolysis of cellulose. However, the process complexity is also increased so that the economic feasibility of this combined process still needs evaluation.

2.6 Concluding remarks and prospective

Due to the biomass recalcitrance of lignocellulose, pretreatment to disrupt the cell wall structure to increase cellulose accessibility is a necessary step for biological conversion of the structural polysaccharides to biofuels and biochemicals. An ideal pretreatment process should be of high lignin removal, high hemicellulose removal, minimal carbohydrate loss, suitable for various lignocellulose raw materials with little formation of inhibitors, without using expensive chemicals and a good potential for large-scale application. Various pretreatment methods have been developed in the past decades. Most of them are efficient in laboratory scale to enhance cellulose digestibility; however, only a few of the pretreatment processes seem to be economically feasible in commercial scale. The mode of action, mechanisms, advantages, and disadvantages of different pretreatments are summarized and compared in Table 2.2. Each pretreatment has its own merits and drawbacks. Currently, versatile pretreatment which can well improve the digestibility of both woody and grass biomass with low energy consumption and cost has not been available yet. Combined use of different pretreatments may provide new insights to

Table 2.2 General modes of action, mechanisms, advantages, and disadvantages of different pretreatment methods for increasing biomass enzymatic digestibility.

Pretreatment methods	Mode of action and mechanisms to improve cellulose digestibility	Advantages	Disadvantages
Physical pretreatment			
Mechanical comminution	By using mechanical force to reduce the particle size thus increasing specific surface area. Milling can also disrupt cell wall structure and cause depolymerization and decrystallization of cellulose	Mechanical size reduction is generally a prerequisite step for most pretreatment process. No or few of inhibitors are formed with no discharge of wastes	Energy consumption is relatively high, and the cellulose digestibility is generally not high enough
Irradiation pretreatment	By using gamma ray, ultrasound, microwave, or electron beam to disrupt cell wall structure, degrade hemicelluloses and lignin and decrease cellulose polymerization	Simple and clean. No or only a few of the inhibitors are formed with little discharge of wastes	Pretreatment efficiency still needs to be improved for independent use. Usually combinedly used with other pretreatment
Extrusion pretreatment	To deconstruct biomass structure by shearing, mixing, and heating action	Can achieve continuous operation with little formation of inhibitors	Cannot be independently used. Usually combined with hydrothermal, acid, or alkaline pretreatments
Pulsed electric field pretreatment	To breakdown cell wall by using high-strength external electric field to form electric potential across the cell wall	Simple and clean. No inhibitors are formed with little discharge of wastes	Usually used for fresh biomass to breakdown plant cell wall. Applicability and efficiency for dry lignocellulose biomass should be improved

(*Continued*)

Table 2.2 (Continued)

Pretreatment methods	Mode of action and mechanisms to improve cellulose digestibility	Advantages	Disadvantages
Chemical pretreatment			
Liquid hot water (hydrothermal) pretreatment	Using liquid hot water to pretreat biomass at high temperature and high pressure without any external chemicals added to pretreat biomass. Water plays as an acidic acid at high temperature to hydrolyze a considerable part of hemicelluloses and disrupt the lignin and cellulose structures	Low cost of reaction medium (water). Only a little formation of inhibitors. Hemicellulosic sugar recovery is high. Cellulose digestibility is well improved	Relatively high energy consumption, low concentration of hemicellulosic sugar in the pretreatment hydrolyzate as well as a large amount of waste water in downstream processing
Dilute acid prehydrolysis	Using mineral acid to hydrolyze hemicelluloses and modify lignin structure as well as reduce biomass particle size and increase porosity	Efficiently hydrolyze hemicellulose and recover it as monosaccharides. The used chemical is relatively cheap. Cellulose digestibility is well improved	Formation of inhibitors. Acid must be neutralized for the downstream enzymatic hydrolysis or fermentation processes with formation of solid waste. Corrosion by the acid mandates expensive construction material. The residual lignin shows strong nonproductive adsorption of cellulase enzymes
Alkaline pretreatment	Using alkalis to digest the lignin structure and break the linkages between lignin and carbohydrate resulting in delignification and partial removal of hemicelluloses. Cell wall structure thus is deconstructed with great increase in porosity and surface area.	Lignin can be efficiently removed and the pretreatment can be operated at low temperature. Cellulose digestibility is well improved	Alkalis are usually relatively expensive. Waste water is formed when alkalis and the dissolved lignin are not recovered

Catalyzed organosolv pretreatment	Using organic solvent to pretreat biomass with addition of acid or alkaline catalysts. Biomass is delignified with significant removal of hemicelluloses. Cell wall structure is deconstructed with reduced particle size and liberation of cellulose fiber	Can achieve a fractionation of biomass to obtain cellulose, hemicellulosic sugars, and lignin. Cellulose digestibility is greatly improved. Organic solvents can be recycled by distillation or extraction	Organic solvents are usually expensive. Severe corrosion occurs when acid catalyst is used. Energy consumption is high for solvent recovery. Formation of inhibitors is inevitable at high temperature
Oxidative pretreatment	Using oxidants to oxidize lignin and partially degrade hemicelluloses, thus reducing particle size and increasing the porosity. Cellulose is liberated at a high degree of delignification	Can achieve a high removal of lignin and significantly expose of cellulose. Cellulose digestibility is greatly enhanced	Oxidants are usually consumed and the chemical cost is high. Washing is necessary to remove the residual oxidants and biomass degradation products, which causes wastewater effluent
Cellulose solvent–based pretreatment	Using cellulose solvents to dissolve biomass followed by precipitation of the cellulose by addition of antisolvents. Biomass undergoes delignification, removal of hemicelluloses, and cellulose is decrystallized and depolymerized	Can achieve a significant removal of lignin and hemicelluloses and deconstruction of cellulose hydrogen-bonding networks, thus converting crystalline cellulose to amorphous cellulose. Can be operated under mild pretreatment conditions. Cellulose hydrolyzability is greatly increased. No or little formation of inhibitors	Recovery of the solvents is usually complicated with high energy consumption. The residual solvent must be washed out to eliminate its inhibition to enzymatic hydrolysis or fermentation
Biological pretreatment			
Microbial pretreatment	Using microorganisms, such as white-, brown-, and soft-rot fungi to pretreat biomass with degradation of lignin and hemicelluloses, thus increasing the substrates porosity	Low energy requirement, no or little chemicals requirement, mild pretreatment conditions, and no environmental pollution	Long pretreatment period with low efficiency. Cellulose also undergoes significant degradation

(Continued)

Table 2.2 (Continued)

Pretreatment methods	Mode of action and mechanisms to improve cellulose digestibility	Advantages	Disadvantages
Combined pretreatment			
Physicochemical pretreatment (steam explosion, CO_2 explosion, etc.)	Biomass substrates are heated with high-pressure saturated steam or supercritical CO_2 for a short duration of time and then the pressure is swiftly reduced causing explosive decompression to deconstruct cell wall structure with significant removal of hemicelluloses and modification of lignin structure (for stem explosion)	Efficient to improve cellulose digestibility with relatively low cost, low energy consumption, being suitable for big particle size and low environmental impact	Significant decomposition of sugars to form inhibitors. High pressure is needed. Steam explosion is usually effective for grass biomass and hardwood, but not so satisfying to softwood biomass. CO_2 explosion is not effective to dry biomass
SPROL	Using sulfite to soften wood chips followed by size reduction with a disk milling. A considerable part of hemicelluloses is removed with partial delignification. Lignin structure is greatly modified to make it more hydrophilic. Cell wall structure is greatly deconstructed with increase in cellulose accessibility	Particularly effective to woody biomass. Less formation of inhibitors compared with other thermochemical pretreatments. High cellulose digestibility. Low nonproductive adsorption of cellulase enzyme on lignin	High cost of chemical recovery and effluent of sulfur-containing wastewater if the formed lignosulfonate is not well recovered
Subsequent acid and alkaline pretreatment	Combination of acid and alkaline pretreatment to remove hemicelluloses and lignin respectively in a multistage process	High sugar recovery for both hexose and pentose compared with single acid or alkaline pretreatment. Cellulose digestibility is well enhanced with less enzymes required	Process complexity is increased. The issues encountered in the single-stage pretreatment also exit in the subsequent acid and alkaline pretreatment. Relatively high operation costs

improving the efficiency and reducing energy consumption. However, the process complexity and cost should be considered for such combinations. To develop new pretreatment technology or improve the currently existing technologies, it is recommended that the molecular mechanisms of biomass recalcitrance should be investigated in depth, particularly to elucidate how hemicelluloses and lignin protect cellulose at microscale. Such elucidation may guide the improvement of pretreatment process. Moreover, process integration also should be considered to reduce water and energy consumption, and generation of waste solid and water. One example of such process integration is the consolidated bioprocessing (CBP), which integrates the conversion of lignocellulose into desired products in one step without added enzymes. CBP has been widely recognized as the ultimate configuration for low-cost hydrolysis and fermentation of cellulosic biomass [190]. Another promising way is to integratedly utilize biomass components for production of different products. To achieve this objective, fractionation of biomass is the most crucial step. Organosolv and combination of acid and alkaline pretreatment may obtain good fractionation of biomass. However, more evaluation and optimization should be done to increase the economic feasibility.

Acknowledgments

The authors are grateful for the supports of this work by National Natural Science Foundation of China (Nos. 21808123, 21506117, and 21878176) and the National Energy Administration Project (No. NY20130402).

References

[1] S. Sun, S. Sun, X. Cao, R. Sun, The role of pretreatment in improving the enzymatic hydrolysis of lignocellulosic materials, Bioresour. Technol. 199 (2016) 49–58.
[2] R. Sindhu, P. Binod, A. Pandey, Biological pretreatment of lignocellulosic biomass—an overview, Bioresour. Technol. 199 (2016) 76–82.
[3] H. Ma, W.-W. Liu, X. Chen, Y.-J. Wu, Z.-L. Yu, Enhanced enzymatic saccharification of rice straw by microwave pretreatment, Bioresour. Technol. 100 (2009) 1279–1284.
[4] Y. Zheng, J. Shi, M. Tu, Y.-S. Cheng, Principles and development of lignocellulosic biomass pretreatment for biofuels, Adv. Bioenergy 2 (2017) 1–68.
[5] B. Yang, C.E. Wyman, Pretreatment: the key to unlocking low-cost cellulosic ethanol, Biofuels Bioprod. Biorefin. 2 (2008) 26–40.
[6] P. Alvira, E. Tomas-Pejo, M. Ballesteros, M.J. Negro, Pretreatment technologies for an efficient bioethanol production process based on enzymatic hydrolysis: a review, Bioresour. Technol. 101 (2010) 4851–4861.

[7] M.J. Taherzadeh, K. Karimi, Pretreatment of lignocellulosic wastes to improve ethanol and biogas production: a review, Int. J. Mol. Sci. 9 (2008) 1621–1651.
[8] R. Martin-Sampedro, I. Filpponen, I.C. Hoeger, J.Y. Zhu, J. Laine, O.J. Rojas, Rapid and complete enzyme hydrolysis of lignocellulosic nanofibrils, ACS Macro Lett. 1 (2012) 1321–1325.
[9] M.T. Holtzapple, A.E. Humphrey, J.D. Taylor, Energy requirements for the size-reduction of poplar and aspen wood, Biotechnol. Bioeng. 33 (1989) 207–210.
[10] L. Cadoche, G.D. Lopez, Assessment of size-reduction as a preliminary step in production of ethanol from lignocellulosic wasters, Biol. Wastes 30 (1989) 153–157.
[11] V.B. Agbor, N. Cicek, R. Sparling, A. Berlin, D.B. Levin, Biomass pretreatment: fundamentals toward application, Biotechnol. Adv. 29 (2011) 675–685.
[12] M.A. Millett, A.J. Baker, L.D. Satter, Physical and chemical pretreatments for enhancing cellulose saccharification, Biotechnol. Bioeng. 6 (1976) 125–153.
[13] J.Y. Zhu, G.S. Wang, X.J. Pan, R. Gleisner, Specific surface to evaluate the efficiencies of milling and pretreatment of wood for enzymatic saccharification, Chem. Eng. Sci. 64 (2009) 474–485.
[14] L. Kratky, T. Jirout, Biomass size reduction machines for enhancing biogas production, Chem. Eng. Technol. 34 (2011) 391–399.
[15] A. Hideno, H. Inoue, K. Tsukahara, S. Fujimoto, T. Minowa, S. Inoue, et al., Wet disk milling pretreatment without sulfuric acid for enzymatic hydrolysis of rice straw, Bioresour. Technol. 100 (2009) 2706–2711.
[16] A.S.A. da Silva, H. Inoue, T. Endo, S. Yano, E.P.S. Bon, Milling pretreatment of sugarcane bagasse and straw for enzymatic hydrolysis and ethanol fermentation, Bioresour. Technol. 101 (2010) 7402–7409.
[17] A. Charlesby, The degradation of cellulose by ionizing radiation, J Polymer Sci. 15 (1955) 263–270.
[18] Y. Yin, J. Wang, Enhancement of enzymatic hydrolysis of wheat straw by gamma irradiation-alkaline pretreatment, Radiat. Phys. Chem. 123 (2016) 63–67.
[19] Y. Xiang, Y. Xiang, L. Wang, Cobalt-60 gamma-ray irradiation pretreatment and sludge protein for enhancing enzymatic saccharification of hybrid poplar sawdust, Bioresour. Technol. 221 (2016) 9–14.
[20] E. Butnaru, R.N. Darie-Niţă, T. Zaharescu, T. Balaeş, C. Tănase, G. Hitruc, et al., Gamma irradiation assisted fungal degradation of the polypropylene/biomass composites, Radiat. Phys. Chem. 125 (2016) 134–144.
[21] K.Q. Wang, Xy Xiong, Jp Chen, L. Chen, Xj Su, Y. Liu, Comparison of gamma irradiation and steam explosion pretreatment for ethanol production from agricultural residues, Biomass Bioenergy 46 (2012) 301–308.
[22] Y. Zheng, J. Zhao, F. Xu, Y. Li, Pretreatment of lignocellulosic biomass for enhanced biogas production, Progr. Energy Combust. Sci. 42 (2014) 35–53.
[23] X. Yu, X. Bao, C. Zhou, L. Zhang, A.E.A. Yagoub, H. Yang, et al., Ultrasound-ionic liquid enhanced enzymatic and acid hydrolysis of biomass cellulose, Ultrason. Sonochem. 41 (2018) 410–418.
[24] P.B. Subhedar, P.R. Gogate, Chapter 6—Use of ultrasound for pretreatment of biomass and subsequent hydrolysis and fermentation, in: S.I. Mussatto (Ed.), Biomass Fractionation Technologies for a Lignocellulosic Feedstock Based Biorefinery, Elsevier, Amsterdam, 2016, pp. 127–149.
[25] F. Passos, S. Astals, I. Ferrer, Anaerobic digestion of microalgal biomass after ultrasound pretreatment, Waste Manage. 34 (2014) 2098–2103.
[26] R. Sindhu, P. Binod, A.K. Mathew, Development of a novel ultrasound-assisted alkali pretreatment strategy for the production of bioethanol and xylanases from chili post harvest residue, Bioresour. Technol. 242 (2017) 146–151.

[27] S. Jin, G.M. Zhang, P.Y. Zhang, F. Li, S.Q. Wang, S.Y. Fan, et al., Microwave assisted alkaline pretreatment to enhance enzymatic saccharification of catalpa sawdust, Bioresour. Technol. 221 (2016) 26–30.
[28] P. Binod, K. Satyanagalakshmi, R. Sindhu, K.U. Janu, R.K. Sukumaran, A. Pandey, Short duration microwave assisted pretreatment enhances the enzymatic saccharification and fermentable sugar yield from sugarcane bagasse, Renew. Energy 37 (2012) 109–116.
[29] Z. Hu, Z. Wen, Enhancing enzymatic digestibility of switchgrass by microwave-assisted alkali pretreatment, Biochem. Eng. J. 38 (2008) 369–378.
[30] X.B. Zhao, Y.J. Zhou, G.J. Zheng, D.H. Liu, Microwave pretreatment of substrates for cellulase production by solid-state fermentation, Appl. Biochem. Biotechnol. 160 (2010) 1557–1571.
[31] T. Leskinen, S.S. Kelley, D.S. Argyropoulos, E-beam irradiation & steam explosion as biomass pretreatment, and the complex role of lignin in substrate recalcitrance, Biomass Bioenergy 103 (2017) 21–28.
[32] A. Duque, P. Manzanares, M. Ballesteros, Extrusion as a pretreatment for lignocellulosic biomass: fundamentals and applications, Renew. Energy 114 (2017) 1427–1441.
[33] M.K. Moro, M.K. Moro, R.S.S. Teixeira, A.A. da Silva, M.D. Fujimoto, P.A. Melo, et al., Continuous pretreatment of sugarcane biomass using a twin-screw extruder, Ind. Crop Prod. 97 (2017) 509–517.
[34] C. Karunanithy, K. Muthukumarappan, Optimization of extruder and prairie cord grass parameters for maximum sugar recovery through enzymatic hydrolysis, J Biobased Mater. Bioenergy 5 (2011) 520–531.
[35] C. Karunanithy, K. Muthukumarappan, Influence of extruder and feedstock variables on torque requirement during pretreatment of different types of biomass—a response surface analysis, Biosyst. Eng. 109 (2011) 37–51.
[36] S.H. Mood, A.H. Golfeshan, M. Tabatabaei, G.S. Jouzani, G.H. Najafi, M. Gholami, et al., Lignocellulosic biomass to bioethanol, a comprehensive review with a focus on pretreatment, Renew. Sustain. Energy Rev. 27 (2013) 77–93.
[37] C.M.R. Rocha, Z. Genisheva, P. Ferreira-Santos, R. Rodrigues, A.A. Vicente, J.A. Teixeira, et al., Electric field-based technologies for valorization of bioresources, Bioresour. Technol. 254 (2018) 325–339.
[38] X. Yu, T. Gouyo, N. Grimi, O. Bals, E. Vorobiev, Pulsed electric field pretreatment of rapeseed green biomass (stems) to enhance pressing and extractives recovery, Bioresour. Technol. 199 (2016) 194–201.
[39] J.Y. Zhu, X.J. Pan, Woody biomass pretreatment for cellulosic ethanol production: technology and energy consumption evaluation, Bioresour. Technol. 101 (2010) 4992–5002.
[40] D.J. Schell, C. Harwood, Milling of lignocellulosic biomass-results of pilot-scale testing, Appl. Biochem. Biotechnol. 45–46 (1994) 159–168.
[41] R. Singh, B.B. Krishna, J. Kumar, T. Bhaskar, Opportunities for utilization of non-conventional energy sources for biomass pretreatment, Bioresour. Technol. 199 (2016) 398–407.
[42] J.Y. Zhu, X. Pan, R.S. Zalesny, Pretreatment of woody biomass for biofuel production: energy efficiency, technologies, and recalcitrance, Appl. Microbiol. Biotechnol. 87 (2010) 47–857.
[43] A. Duque, P. Manzanares, I. Ballesteros, M.J. Negro, J.M. Oliva, F. Saez, et al., Optimization of integrated alkaline-extrusion pretreatment of barley straw for sugar production by enzymatic hydrolysis, Process Biochem. 48 (2013) 775–781.
[44] B. Lamsal, J. Yoo, K. Brijwani, S. Alavi, Extrusion as a thermo-mechanical pretreatment for lignocellulosic ethanol, Biomass Bioenergy 34 (2010) 1703–1710.

[45] N. Akhtar, K. Gupta, D. Goyal, A. Goyal, Recent advances in pretreatment technologies for efficient hydrolysis of lignocellulosic biomass, Environ. Prog. Sustain. Energy 35 (2016) 489–511.
[46] B.S. Dien, Enzymatic saccharification of hot-water pretreated corn fiber for production of monosaccharides, Enzyme Microb. Technol. 39 (2006) 1137–1144.
[47] N.S. Mosier, R. Hendrickson, M. Brewer, N. Ho, M. Sedlak, R. Dreshel, et al., Industrial scale-up of pH-controlled liquid hot water pretreatment of corn fiber for fuel ethanol production, Appl. Biochem. Biotechnol. 125 (2005) 77–97.
[48] T. Yuan, X.K. Li, S.Y. Xiao, Y. Guo, Wzh Zhou, J.L. Xu, et al., Microalgae pretreatment with liquid hot water to enhance enzymatic hydrolysis efficiency, Bioresour. Technol. 220 (2016) 530–536.
[49] J. Liu, R. Li, L. Shuai, J. You, Y. Zhao, L. Chen, Comparison of liquid hot water (LHW) and high boiling alcohol/water (HBAW) pretreatments for improving enzymatic saccharification of cellulose in bamboo, Ind. Crop Prod. 107 (2017) 139–148.
[50] F. Weinwurm, T. Turk, J. Denner, K. Whitmore, A. Friedl, Combined liquid hot water and ethanol organosolv treatment of wheat straw for extraction and reaction modeling, J. Cleaner Prod. 165 (2017) 1473–1484.
[51] C. Huang, X. Wu, Y. Huang, Ch Lai, X. Li, Q. Yong, Prewashing enhances the liquid hot water pretreatment efficiency of waste wheat straw with high free ash content, Bioresour. Technol. 219 (2016) 583–588.
[52] D. Jiang, X. Ge, Q. Zhang, Y. Li, Comparison of liquid hot water and alkaline pretreatments of giant reed for improved enzymatic digestibility and biogas energy production, Bioresour. Technol. 216 (2016) 60–68.
[53] Q. Yu, X. Zhuang, Z. Yuan, W. Qi, W. Wang, Q. Wang, et al., Pretreatment of sugarcane bagasse with liquid hot water and aqueous ammonia, Bioresour. Technol. 144 (2013) 210–215.
[54] H. Zabed, J.N. Sahu, A. Suely, A.N. Boyce, G. Faruq, Bioethanol production from renewable sources: current perspectives and technological progress, Renew. Sustain. Energy Rev. 71 (2017) 475–501.
[55] N. Mosier, C. Wyman, B. Dale, R. Elander, Y.Y. Lee, M. Holtzapple, et al., Features of promising technologies for pretreatment of lignocellulosic biomass, Bioresour. Technol. 96 (2005) 673–686.
[56] S. Gámez, J.J. González-Cabriales, J.A. Ramírez, G. Garrote, M. Vázquez, Study of the hydrolysis of sugar cane bagasse using phosphoric acid, J. Food Eng. 74 (2006) 78–88.
[57] T.-C. Hsu, G.-L. Guo, W.-H. Chen, W.-S. Hwang, Effect of dilute acid pretreatment of rice straw on structural properties and enzymatic hydrolysis, Bioresour. Technol. 101 (2010) 4907–4913.
[58] C. Cara, E. Ruiz, J.M. Oliva, F. Sáez, E. Castro, Conversion of olive tree biomass into fermentable sugars by dilute acid pretreatment and enzymatic saccharification, Bioresour. Technol. 99 (2008) 1869–1876.
[59] E.C. Bensah, M. Mensah, Chemical pretreatment methods for the production of cellulosic ethanol: technologies and innovations, Int. J. Chem. Eng. 2013 (2013) 21. Volume 2013, Article ID 719607, 21 pages.
[60] M. Tutt, T. Kikas, J. Olt, Influence of different pretreatment methods on bioethanol production from wheat straw, Agron. Res. 10 (2012) 269–276.
[61] A. Avci, B.C. Saha, B.S. Dien, G.J. Kennedy, M.A. Cotta, Response surface optimization of corn stover pretreatment using dilute phosphoric acid for enzymatic hydrolysis and ethanol production, Bioresour. Technol. 130 (2013) 603–612.
[62] B.C. Saha, L.B. Iten, M.A. Cotta, Y.V. Wu, Dilute acid pretreatment, enzymatic saccharification and fermentation of wheat straw to ethanol, Process Biochem. 40 (2005) 3693–3700.

[63] H. Chen, J. Zhao, T. Hu, X. Zhao, D. Liu, A comparison of several organosolv pretreatments for improving the enzymatic hydrolysis of wheat straw: substrate digestibility, fermentability and structural features, Appl. Energy 150 (2015) 224–232.
[64] P.R. Seidl, A.K. Goulart, Pretreatment processes for lignocellulosic biomass conversion to biofuels and bioproducts, Curr. Opin. Green Sustain. Chem. 2 (2016) 48–53.
[65] Z. Zhou, F. Lei, P. Li, J. Jiang, Lignocellulosic biomass to biofuels and biochemicals: a comprehensive review with a focus on ethanol organosolv pretreatment technology, Biotechnol. Bioeng. 115 (2018) 2683–2702.
[66] I. Salapa, C. Katsimpouras, E. Topakas, D. Sidiras, Organosolv pretreatment of wheat straw for efficient ethanol production using various solvents, Biomass Bioenergy 100 (2017) 10–16.
[67] H. Teramura, K. Sasaki, T. Oshima, F. Matsuda, M. Okamoto, T. Shirai, et al., Organosolv pretreatment of sorghum bagasse using a low concentration of hydrophobic solvents such as 1-butanol or 1-pentanol, Biotechnol. Biofuels 9 (2016) 27.
[68] X. Zhao, K. Cheng, D. Liu, Organosolv pretreatment of lignocellulosic biomass for enzymatic hydrolysis, Appl. Microbiol. Biotechnol. 82 (2009) 815–827.
[69] L. Qin, Z.-H. Liu, B.-Z. Li, B.E. Dale, Y.-J. Yuan, Mass balance and transformation of corn stover by pretreatment with different dilute organic acids, Bioresour. Technol. 112 (2012) 319–326.
[70] X. Zhao, D. Liu, Fractionating pretreatment of sugarcane bagasse by aqueous formic acid with direct recycle of spent liquor to increase cellulose digestibility—the Formiline process, Bioresour. Technol. 117 (2012) 25–32.
[71] C.S. Goh, H.T. Tan, K.T. Lee, N. Brosse, Evaluation and optimization of organosolv pretreatment using combined severity factors and response surface methodology, Biomass Bioenergy 35 (2011) 4025–4033.
[72] K. Zhang, Z. Pei, D. Wang, Organic solvent pretreatment of lignocellulosic biomass for biofuels and biochemicals: a review, Bioresour. Technol. 199 (2016) 21–33.
[73] C. Arato, E.K. Pye, G. Gjennestad, The lignol approach to biorefining of woody biomass to produce ethanol and chemicals, Appl. Biochem. Biotechnol. 123 (2005) 871–882.
[74] B.-W. Koo, H. Kim, N. Park, S. Lee, H. Yeo, I. Choi, Organosolv pretreatment of *Liriodendron tulipifera* and simultaneous saccharification and fermentation for bioethanol production, Biomass Bioenergy 35 (2011) 1833–1840.
[75] A. Pandey, C.R. Soccol, P. Nigam, V.T. Soccol, Biotechnological potential of agro-industrial residues. I: sugarcane bagasse, Bioresour. Technol. 74 (2000) 69–80.
[76] P. Bajpai, Pretreatment of lignocellulosic biomass, Pretreatment of Lignocellulosic Biomass for Biofuel Production, Springer, Singapore, Singapore, 2016, pp. 17–70.
[77] A.B. Bjerre, A.B. Olesen, T. Fernqvist, A. Plöger, A.S. Schmidt, Pretreatment of wheat straw using combined wet oxidation and alkaline hydrolysis resulting in convertible cellulose and hemicellulose, Biotechnol. Bioeng. 49 (1996) 568–577.
[78] H. Jørgensen, J.B. Kristensen, C. Felby, Enzymatic conversion of lignocellulose into fermentable sugars: challenges and opportunities, Biofuels Bioprod. Biorefin. 1 (2007) 119–134.
[79] M. Balat, H. Balat, C. Öz, Progress in bioethanol processing, Prog. Energy Combust. 34 (2008) 551–573.
[80] R. Rinaldi, R. Jastrzebski, M.T. Clough, J. Ralph, M. Kennema, P.C. Bruijnincx, et al., Paving the way for lignin valorisation: recent advances in bioengineering, biorefining and catalysis, Angew. Chem. Int. Ed. 55 (2016) 8164–8215.
[81] J.K. Xu, R.C. Sun, Chapter 19—Recent advances in alkaline pretreatment of lignocellulosic biomass, in: S.I. Mussatto (Ed.), Biomass Fractionation Technologies for a

Lignocellulosic Feedstock Based Biorefinery, Elsevier, Amsterdam, 2016, pp. 431–459.
[82] M.L. Soto, H. Domínguez, M.J. Núñez, J.M. Lema, Enzymatic saccharification of alkali-treated sunflower hulls, Bioresour. Technol. 49 (1994) 53–59.
[83] S. Kim, M.T. Holtzapple, Delignification kinetics of corn stover in lime pretreatment, Bioresour. Technol. 97 (2006) 778–785.
[84] V.S. Chang, B. Burr, M.T. Holtzapple, Lime pretreatment of switchgrass, Appl. Biochem. Biotechnol. 63 (1997) 1–19.
[85] W.E. Kaar, M.T. Holtzapple, Using lime pretreatment to facilitate the enzymic hydrolysis of corn stover, Biomass Bioenergy 18 (2000) 189–199.
[86] R. Singh, A. Shukla, S. Tiwari, M. Srivastava, A review on delignification of lignocellulosic biomass for enhancement of ethanol production potential, Renew. Sustain. Energy Rev. 32 (2014) 713–728.
[87] N. Sarkar, S.K. Ghosh, S. Bannerjee, K. Aikat, Bioethanol production from agricultural wastes: an overview, Renew. Energy 37 (2012) 19–27.
[88] H. Alizadeh, F. Teymouri, T.I. Gilbert, B.E. Dale, Pretreatment of switchgrass by ammonia fiber explosion (AFEX), Appl. Biochem. Biotechnol. 124 (2005) 1133–1141.
[89] F. Teymouri, L. Laureano-Perez, H. Alizadeh, B.E. Dale, Optimization of the ammonia fiber explosion (AFEX) treatment parameters for enzymatic hydrolysis of corn stover, Bioresour. Technol. 96 (2005) 2014–2018.
[90] S.P.S. Chundawat, B. Venkatesh, B.E. Dale, Effect of particle size based separation of milled corn stover on AFEX pretreatment and enzymatic digestibility, Biotechnol. Bioeng. 96 (2007) 219–231.
[91] M. Galbe, G. Zacchi, Pretreatment of lignocellulosic materials for efficient bioethanol production, in: L. Olsson (Ed.), Biofuels, Springer Berlin Heidelberg, Berlin, Heidelberg, 2007, pp. 41–65.
[92] D. Kim, Physico-chemical conversion of lignocellulose: inhibitor effects and detoxification strategies: a mini review, Molecules 23 (2018). 309, 1–21.
[93] S.P.S. Chundawat, R. Vismeh, L.N. Sharma, J.F. Humpula, L.C. Sousa, C.K. Chambliss, et al., Multifaceted characterization of cell wall decomposition products formed during ammonia fiber expansion (AFEX) and dilute acid based pretreatments, Bioresour. Technol. 101 (2010) 8429–8438.
[94] M.W. Lau, B.E. Dale, Cellulosic ethanol production from AFEX-treated corn stover using *Saccharomyces cerevisiae* 424A (LNH-ST), Proc. Natl. Acad. Sci. U.S.A. 106 (2009) 1368–1373.
[95] B. Bals, C. Rogers, M. Jin, V. Balan, B. Dale, Evaluation of ammonia fibre expansion (AFEX) pretreatment for enzymatic hydrolysis of switchgrass harvested in different seasons and locations, Biotechnol. Biofuels 3 (2010) 1–11.
[96] F. Talebnia, D. Karakashev, I. Angelidaki, Production of bioethanol from wheat straw: an overview on pretreatment, hydrolysis and fermentation, Bioresour. Technol. 101 (2010) 4744–4753.
[97] C. Zhong, M.W. Lau, V. Balan, B.E. Dale, Y.J. Yuan, Optimization of enzymatic hydrolysis and ethanol fermentation from AFEX-treated rice straw, Appl. Microbiol. Biotechnol. 84 (2009) 667–676.
[98] K. Sakuragi, K. Igarashi, M. Samejima, Application of ammonia pretreatment to enable enzymatic hydrolysis of hardwood biomass, Polym. Degrad. Stabil. 148 (2018) 19–25.
[99] P. Kumar, D.M. Barrett, M.J. Delwiche, P. Stroeve, Methods for pretreatment of lignocellulosic biomass for efficient hydrolysis and biofuel production, Ind. Eng. Chem. Res. 48 (2009) 3713–3729.
[100] J.S. Kim, Y.Y. Lee, T.H. Kim, A review on alkaline pretreatment technology for bioconversion of lignocellulosic biomass, Bioresour. Technol. 199 (2015) 42–48.

[101] C. Tang, J. Shan, Y. Chen, L. Zhong, T. Shen, C. Zhu, et al., Organic amine catalytic organosolv pretreatment of corn stover for enzymatic saccharification and high-quality lignin, Bioresour. Technol. 232 (2017) 222–228.
[102] X.B. Zhao, L. Wang, D.H. Liu, Peracetic acid pretreatment of sugarcane bagasse for enzymatic hydrolysis: a continued work, J. Chem. Technol. Biotechnol. 83 (2008) 950–956.
[103] Z. Li, Z. Jiang, B. Fei, Ethanosolv with NaOH pretreatment of moso bamboo for efficient enzymatic saccharification, Bioresources 8 (2013) 4711–4721.
[104] K. Wang, H. Yang, S. Guo, Y. Tang, J. Jiang, F. Xu, et al., Organosolv fractionation process with various catalysts for improving bioconversion of triploid poplar, Process Biochem. 47 (2012) 1503–1509.
[105] F.J. Zimmermann, Waste Disposal, US Patent 2665249, 1950.
[106] G.D. McGinnis, W.W. Wilson, S.E. Prince, C.C. Chen, Conversion of biomass into chemicals with high-temperature wet oxidation, Ind. Eng. Chem. Prod. Res. Dev. 22 (1983) 633–636.
[107] G.D. McGinnis, W.W. Wilson, C.E. Mullen, Biomass pretreatment with water and high-pressure oxygen. The wet-oxidation process, Ind. Eng. Chem. Prod. Res. Dev. 22 (1983) 352–357.
[108] E. Arvaniti, A.B. Bjerre, J.E. Schmidt, Wet oxidation pretreatment of rape straw for ethanol production, Biomass Bioenergy 39 (2012) 94–105.
[109] M. Pedersen, A.S. Meyer, Influence of substrate particle size and wet oxidation on physical surface structures and enzymatic hydrolysis of wheat straw, Biotechnol. Progr. 25 (2009) 399–408.
[110] H. Palonen, A.B. Thomsen, M. Tenkanen, A.S. Schmidt, L. Viikari, Evaluation of wet oxidation pretreatment for enzymatic hydrolysis of softwood, Appl. Biochem. Biotechnol. 117 (2004) 1–17.
[111] J.M. Gould, Studies on the mechanism of alkaline peroxide delignification of agricultural residues, Biotechnol. Bioeng. 27 (1985) 225–231.
[112] S.M. Lewis, D.P. Holzgraefe, L.L. Berger, G.C. Fahey Jr., J.M. Gould, G.F. Fanta, Alkaline hydrogen peroxide treatments of crop residues to increase ruminal dry matter disappearance in sacco, Anim. Feed Sci. Technol. 17 (1987) 179–199.
[113] E. Cabrera, M.J. Muñoz, R. Martín, I. Caro, C. Curbelo, A.B. Díaz, Alkaline and alkaline peroxide pretreatments at mild temperature to enhance enzymatic hydrolysis of rice hulls and straw, Bioresour. Technol. 167 (2014) 1–7.
[114] M.J. Selig, T.B. Vinzant, M.E. Himmel, S.R. Decker, The effect of lignin removal by alkaline peroxide pretreatment on the susceptibility of corn stover to purified cellulolytic and xylanolytic enzymes, Appl. Biochem. Biotechnol. 155 (2009) 94–103.
[115] J.X. Sun, F.C. Mao, X.F. Sun, R. Sun, Comparative study of hemicelluloses isolated with alkaline peroxide from lignocellulosic materials, J. Wood Chem. Technol. 24 (2005) 239–262.
[116] B. Qi, X. Chen, F. Shen, Y. Su, Y. Wan, Optimization of enzymatic hydrolysis of wheat straw pretreated by alkaline peroxide using response surface methodology, Ind. Eng. Chem. Res. 48 (2009) 7346–7353.
[117] C. Alvarez-Vasco, X. Zhang, Alkaline hydrogen peroxide (AHP) pretreatment of softwood: enhanced enzymatic hydrolysability at low peroxide loadings, Biomass Bioenergy 96 (2017) 96–102.
[118] S.C. Rabelo, N.A. Amezquita Fonseca, R.R. Andrade, R. Maciel Filho, A.C. Costa, Ethanol production from enzymatic hydrolysis of sugarcane bagasse pretreated with lime and alkaline hydrogen peroxide, Biomass Bioenergy 35 (2011) 2600–2607.

[119] Y. Yamashita, M. Shono, C. Sasaki, Y. Nakamura, Alkaline peroxide pretreatment for efficient enzymatic saccharification of bamboo, Carbohydr. Polym. 79 (2010) 914–920.

[120] G. Banerjee, S. Car, J.S. Scott-Craig, D.B. Hodge, J.D. Walton, Alkaline peroxide pretreatment of corn stover: effects of biomass, peroxide, and enzyme loading and composition on yields of glucose and xylose, Biotechnol. Biofuels 4 (2011) 16.

[121] B. Wang, X. Wang, H. Feng, Deconstructing recalcitrant *Miscanthus* with alkaline peroxide and electrolyzed water, Bioresour. Technol. 101 (2010) 752–760.

[122] A. Hideno, Short-time alkaline peroxide pretreatment for rapid pulping and efficient enzymatic hydrolysis of rice straw, Bioresour. Technol. 230 (2017) 140–142.

[123] B.C. Saha, M.A. Cotta, Ethanol production from alkaline peroxide pretreated enzymatically saccharified wheat straw, Biotechnol Progr 22 (2006) 449–453.

[124] X.Y. Sun, X.B. Zhao, W. Du, D.H. Liu, Kinetics of formic acid-autocatalyzed preparation of performic acid in aqueous phase, Chin. J. Chem. Eng. 19 (2011) 964–971.

[125] X. Zhao, T. Zhang, Y. Zhou, D. Liu, Preparation of peracetic acid from hydrogen peroxide—Part 1: Kinetics for peracetic acid synthesis and hydrolysis, J. Mol. Cat. A—Chem. 271 (2007) 246–252.

[126] G.X. Pan, L. Spencer, G.J. Leary, Reactivity of ferulic acid and its derivatives toward hydrogen peroxide and peracetic acid, J Agric. Food Chem. 47 (1999) 3325–3331.

[127] K. Sakai, K. Kuroda, S. Kishimoto, Delignification in peracid bleaching. VI. Reaction of the beta-aryl ether type model compounds with peracetic acid, Tappi Tech. Assoc. Pulp Pap. Ind. 5 (1972) 1702–1706.

[128] Y.Z. Lai, K.V. Sarkanen, Delignification by peracetic acid. II. Comparative study on softwood and hardwood lignins, Tappi J. 51 (1968) 449–453.

[129] D. Yin, Q. Jing, W. Dajani, S. Duncan, U. Tschirner, J. Schilling, et al., Improved pretreatment of lignocellulosic biomass using enzymatically-generated peracetic acid, Bioresour. Technol. 102 (2011) 5183–5192.

[130] X. Zhao, Y. Song, D. Liu, Enzymatic hydrolysis and simultaneous saccharification and fermentation of alkali/peracetic acid-pretreated sugarcane bagasse for ethanol and 2,3-butanediol production, Enzyme Microb. Tech. 49 (2011) 413–419.

[131] X.B. Zhao, L. Wang, Dh Liu, Effect of several factors on peracetic acid pretreatment of sugarcane bagasse for enzymatic hydrolysis, J. Chem. Technol. Biotechnol. 82 (2007) 1115–1121.

[132] X.B. Zhao, L.H. Zhang, D.H. Liu, Comparative study on chemical pretreatment methods for improving enzymatic digestibility of crofton weed stem, Bioresour. Technol. 99 (2008) 3729–3736.

[133] S.I. Mussatto, I.C. Roberto, Alternatives for detoxification of diluted-acid lignocellulosic hydrolyzates for use in fermentative processes: a review, Bioresour. Technol. 93 (2004) 1–10.

[134] L.C. Teixeira, J.C. Linden, H.A. Schroeder, Simultaneous saccharification and cofermentation of peracetic acid-pretreated biomass, Appl. Biochem. Biotechnol. 84 (2000) 111–127.

[135] S. Duncan, Q. Jing, A. Katona, R.J. Kazlauskas, J. Schilling, U. Tschirner, et al., Increased saccharification yields from aspen biomass upon treatment with enzymatically generated peracetic acid, Appl. Biochem. Biotechnol. 160 (2010) 1637–1652.

[136] H. Zhao, J.H. Kwak, Y. Wang, J.A. Franz, J.M. White, J.E. Holladay, Interactions between cellulose and *N*-methylmorpholine-*N*-oxide, Carbohydr. Polym. 67 (2007) 97–103.

[137] C. Cuissinat, P. Navard, Swelling and dissolution of cellulose. Part 1: Free floating cotton and wood fibres in *N*-methylmorpholine-*N*-oxide–water mixtures, Macromol. Symp. 244 (2006) 1–18.

[138] R. Wikandari, R. Millati, M.J. Taherzadeh, Chapter 12—Pretreatment of lignocelluloses with solvent *N*-methylmorpholine *N*-oxide, in: Solange I. Mussatto (Ed.), Biomass Fractionation Technologies for a Lignocellulosic Feedstock Based Biorefinery, Elsevier, Amsterdam, 2016, pp. 255−280.
[139] A. Jeihanipour, K. Karimi, M.J. Taherzadeh, Enhancement of ethanol and biogas production from high-crystalline cellulose by different modes of NMO pretreatment, Biotechnol. Bioeng. 105 (2010) 469−476.
[140] A. Goshadrou, K. Karimi, M.J. Taherzadeh, Ethanol and biogas production from birch by NMMO pretreatment, Biomass Bioenergy 49 (2013) 95−101.
[141] M. Gao, S. Chen, J. Han, D. Luo, L. Zhao, Q. Zheng, Effects of a pretreatment with *N*-methylmorpholine-*N*-oxide on the structures and properties of ramie, J. Appl. Polym. Sci. 117 (2010) 2241−2250.
[142] M.E. Hall, A.R. Horrocks, H. Seddon, The flammability of Lyocell, Polym. Degrad. Stab. 64 (1999) 505−510.
[143] M. Shafiei, K. Karimi, M.J. Taherzadeh, Pretreatment of spruce and oak by *N*-methylmorpholine-*N*-oxide (NMMO) for efficient conversion of their cellulose to ethanol, Bioresour. Technol. 101 (2010) 4914−4918.
[144] N. Poornejad, K. Karimi, T. Behzad, Improvement of saccharification and ethanol production from rice straw by NMMO and [BMIM][OAc] pretreatments, Ind. Crop Prod. 41 (2013) 408−413.
[145] M.M. Kabir, M.D.P. Castillo, M.J. Taherzadeh, I.S. Horváth, Effect of the *N*-methylmorpholine-*N*-oxide (NMMO) pretreatment on anaerobic digestion of forest residues, Bioresources 8 (2015) 5409−5423.
[146] P.R. Lennartsson, C. Niklasson, M.J. Taherzadeh, A pilot study on lignocelluloses to ethanol and fish feed using NMMO pretreatment and cultivation with zygomycetes in an air-lift reactor, Bioresour. Technol. 102 (2011) 4425−4432.
[147] T. Rosenau, A. Potthast, I. Adorjan, A. Hofinger, H. Sixta, H. Firgo, et al., Cellulose solutions in *N*-methylmorpholine-*N*-oxide (NMMO) − degradation processes and stabilizers, Cellulose 9 (2002) 283−291.
[148] Y.H. Jung, K.H. Kim, Acidic pretreatment, in: S. Negi, P. Binod, C. Larroche (Eds.), Pretreatment of Biomass, Elsevier, Amsterdam, 2015, pp. 27−50.
[149] D.J. Hayes, An examination of biorefining processes, catalysts and challenges, Catal. Today 145 (2009) 138−151.
[150] Z.-S. Liu, X.-L. Wu, K. Kida, Y.-Q. Tang, Corn stover saccharification with concentrated sulfuric acid: effects of saccharification conditions on sugar recovery and by-product generation, Bioresour. Technol. 119 (2012) 224−233.
[151] Y.H. Zhang, J. Cui, L.R. Lynd, L.R. Kuang, A transition from cellulose swelling to cellulose dissolution by o-phosphoric acid: evidence from enzymatic hydrolysis and supramolecular structure, Biomacromolecules 7 (2006) 644−648.
[152] N. Sathitsuksanoh, A. George, Y.H.P. Zhang, New lignocellulose pretreatments using cellulose solvents: a review, J. Chem. Technol. Biotechnol. 88 (2013) 169−180.
[153] N. Sathitsuksanoh, Z. Zhu, S. Wi, Y.H.P. Zhang, Cellulose solvent-based biomass pretreatment breaks highly ordered hydrogen bonds in cellulose fibers of switchgrass, Biotechnol. Bioeng. 108 (2011) 521−529.
[154] J.D. McMillan, Pretreatment of lignocellulosic biomass, Enzymatic Conversion of Biomass for Fuels Production, vol. 566, American Chemical Society, 1994, pp. 292−324. chap. 15.
[155] G. Moxley, Z. Zhu, Y.H.P. Zhang, Efficient sugar release by the cellulose solvent-based lignocellulose fractionation technology and enzymatic cellulose hydrolysis, J. Agric. Food Chem. 56 (2008) 7885−7890.

[156] N. Sathitsuksanoh, Z. Zhu, Y.H.P. Zhang, Cellulose solvent- and organic solvent-based lignocellulose fractionation enabled efficient sugar release from a variety of lignocellulosic feedstocks, Bioresour. Technol. 117 (2012) 228–233.
[157] N. Sathitsuksanoh, Z. Zhu, T.-J. Ho, M.-D. Bai, Y.-H.P. Zhang, Bamboo saccharification through cellulose solvent-based biomass pretreatment followed by enzymatic hydrolysis at ultra-low cellulase loadings, Bioresour. Technol. 101 (2010) 4926–4929.
[158] X. Ge, V.S. Green, N. Zhang, G. Sivakumar, J. Xu, Eastern gamagrass as an alternative cellulosic feedstock for bioethanol production, Process Biochem. 47 (2012) 335–339.
[159] N. Sathitsuksanoh, Z. Zhu, N. Templeton, J.A. Rollin, S.P. Harvey, Y.H. Zhang, Saccharification of a potential bioenergy crop, *Phragmites australis* (Common Reed), by lignocellulose fractionation followed by enzymatic hydrolysis at decreased cellulase loadings, Ind. Eng. Chem. Res. 48 (2009) 6441–6447.
[160] Z. Zhu, N. Sathitsuksanoh, T. Vinzant, D.J. Schell, J.D. Mcmillan, Y.H.P. Zhang, Comparative study of corn stover pretreated by dilute acid and cellulose solvent-based lignocellulose fractionation: enzymatic hydrolysis, supramolecular structure, and substrate accessibility, Biotechnol. Bioeng. 103 (2009) 715–724.
[161] T. Li, Q. Fang, H. Chen, F. Qi, X. Ou, X. Zhao, et al., Solvent-based delignification and decrystallization of wheat straw for efficient enzymatic hydrolysis of cellulose and ethanol production with low cellulase loadings, RSC Adv. 7 (2017) 10609–10617.
[162] J.J. Vargas Radillo, M.A. RuizLópez, R.R. Macías, L.B. Ramírez, Fermentable sugars from *Lupinus rotundiflorus* biomass by concentrated hydrochloric acid hydrolysis, Bioresources 6 (2011) 344–355.
[163] C.N. Hamelinck, Gv Hooijdonk, A.P.C. Faaij, Ethanol from lignocellulosic biomass: techno-economic performance in short-, middle- and long-term, Biomass Bioenergy 28 (2005) 384–410.
[164] A.W. Bhutto, K. Qureshi, K. Harijan, R. Abro, T. Abbas, A.A. Bazmi, Insight into progress in pre-treatment of lignocellulosic biomass, Energy 122 (2017) 724–745.
[165] A. Pinkert, K.N. Marsh, S. Pang, M.P. Staiger, Ionic liquids and their interaction with cellulose, Chem. Rev. 109 (2009) 6712–6728.
[166] A. Xu, J. Wang, H. Wang, Effects of anionic structure and lithium salts addition on the dissolution of cellulose in 1-butyl-3-methylimidazolium-based ionic liquid solvent systems, Green Chem. 12 (2010) 268–275.
[167] K.M. Torr, K.T. Love, B.A. Simmons, S.J. Hill, Structural features affecting the enzymatic digestibility of pine wood pretreated with ionic liquids, Biotechnol. Bioeng. 113 (2016) 540–549.
[168] Z. Qiu, G.M. Aita, M.S. Walker, Effect of ionic liquid pretreatment on the chemical composition, structure and enzymatic hydrolysis of energy cane bagasse, Bioresour. Technol. 117 (2012) 251–256.
[169] L. Liu, H. Chen, Enzymatic hydrolysis of cellulose materials treated with ionic liquid [BMIM] Cl, Chin. Sci. Bull. 51 (2006) 2432–2436.
[170] C. Li, B. Knierim, C. Manisseri, R. Arora, S. Singh, Comparison of dilute acid and ionic liquid pretreatment of switchgrass: biomass recalcitrance, delignification and enzymatic saccharification, Bioresour. Technol. 101 (2010) 4900–4906.
[171] B.J. Cox, J.G. Ekerdt, Pretreatment of yellow pine in an acidic ionic liquid: extraction of hemicellulose and lignin to facilitate enzymatic digestion, Bioresour. Technol. 134 (2013) 59–65.
[172] A.A. Elgharbawy, M.Z. Alam, M. Moniruzzaman, M. Goto, Ionic liquid pretreatment as emerging approaches for enhanced enzymatic hydrolysis of lignocellulosic biomass, Biochem. Eng. J. 109 (2016) 252–267.

[173] W. Gao, L.G. Tabil, T. Dumonceaux, S. Espinel Rios, R. Zhao, Optimization of biological pretreatment to enhance the quality of wheat straw pellets, Biomass Bioenergy 97 (2017) 77–89.
[174] X. Yan, Z. Wang, K. Zhang, M. Si, M. Liu, L. Chai, Bacteria-enhanced dilute acid pretreatment of lignocellulosic biomass, Bioresour. Technol. 245 (2017) 419–425.
[175] Y. Sun, J.Y. Cheng, Hydrolysis of lignocellulosic materials for ethanol production: a review, Bioresour. Technol. 83 (2002) 1–11.
[176] M. Tutt, Using steam explosion pretreatment method for bioethanol production from floodplain meadow hay, Agron. Res. 12 (2014) 417–424.
[177] S. Zhao, G. Li, N. Zheng, J. Wang, Z. Yu, Steam explosion enhances digestibility and fermentation of corn stover by facilitating ruminal microbial colonization, Bioresour. Technol. 253 (2018) 244–251.
[178] J. Lizasoain, M. Rincón, F. Theuretzbacher, R. Enguídanos, P.J. Nielsen, A. Potthast, et al., Biogas production from reed biomass: effect of pretreatment using different steam explosion conditions, Biomass Bioenergy 95 (2016) 84–91.
[179] T. Pielhop, J. Amgarten, P.R. von Rohr, M.H. Studer, Steam explosion pretreatment of softwood: the effect of the explosive decompression on enzymatic digestibility, Biotechnol. Biofuels 9 (2016) 152.
[180] A. Bauer, C. Leonhartsberger, B. Peter, B. Amon, A. Friedl, T. Amon, Analysis of methane yields from energy crops and agricultural by-products and estimation of energy potential from sustainable crop rotation systems in EU-27, Clean Technol. Environ. Policy 12 (2010) 153–161.
[181] Z. Tooyserkani, L. Kumar, S. Sokhansanj, J. Saddler, X.T. Bi, C.J. Lim, SO_2-catalyzed steam pretreatment enhances the strength and stability of softwood pellets, Bioresour. Technol. 130 (2013) 59–68.
[182] C. Tengborg, K. Stenberg, M. Galbe, G. Zacchi, S. Larsson, E. Palmqvist, Comparison of SO_2 and H_2SO_4 impregnation of softwood prior to steam pretreatment on ethanol production, Appl. Biochem. Biotechnol. 70–72 (1998) 3–15.
[183] A.L. Ferreira Santos, K.Y. Fausta Kawase, G.L. Vieira Coelho, Enzymatic saccharification of lignocellulosic materials after treatment with supercritical carbon dioxide, J. Supercrit. Fluids 56 (2011) 277–282.
[184] J.Y. Zhu, X.J. Pan, G.S. Wang, R. Gleisner, Sulfite pretreatment (SPORL) for robust enzymatic saccharification of spruce and red pine, Bioresour. Technol. 100 (2009) 2411–2418.
[185] Z.J. Wang, J.Y. Zhu, R.S. Zalesny Jr., K.F. Chen, Ethanol production from poplar wood through enzymatic saccharification and fermentation by dilute acid and SPORL pretreatments, Fuel 95 (2012) 606–614.
[186] C. Zhang, C.J. Houtman, J.Y. Zhu, Using low temperature to balance enzymatic saccharification and furan formation during SPORL pretreatment of Douglas-fir, Process Biochem. 49 (2014) 466–473.
[187] C. Zhang, R. Gleisner, C.J. Houtman, X. Pan, J.Y. Zhu, Sulfite pretreatment to overcome the recalcitrance of lignocelluloses for bioconversion of woody biomass, in: Solange I. Mussatto (Ed.), Biomass Fractionation Technologies for a Lignocellulosic Feedstock Based Biorefinery, Elsevier, 2016, pp. 499–541.
[188] E. Maekawa, On an available pretreatment for the enzymatic saccharification of lignocellulosic materials, Wood Sci. Technol. 30 (1996) 133–139.
[189] X.J. Pan, D. Xie, N. Gilkes, D.J. Gregg, J.N. Saddler, Strategies to enhance the enzymatic hydrolysis of pretreated softwood with high residual lignin content, Appl. Biochem. Biotechnol. 121 (2005) 1069–1079.
[190] D.G. Olson, J.E. McBride, A.J. Shaw, L.R. Lynd, Recent progress in consolidated bioprocessing, Curr. Opin. Biotechnol. 23 (2012) 396–405.

CHAPTER 3

Bioconversion of lignocellulosic biomass to bioethanol and biobutanol

Alessandra Verardi[1], Catia Giovanna Lopresto[1], Alessandro Blasi[2], Sudip Chakraborty[1] and Vincenza Calabrò[1]
[1]Department of Computer Engineering, Modeling, Electronics and Systems Engineering (DIMES), University of Calabria, Rende, Italy
[2]ENEA—Italian National Agency for New Technologies, Energy and Sustainable Economic Development, Energy Technologies Department, Research Centre of Trisaia, Rotondella, Italy

Contents

3.1 Introduction

For the last 25 years domestic sources of fuels, need of renewable energy, and chemicals have become highly significant [1]. Bioethanol has the

Lignocellulosic Biomass to Liquid Biofuels
DOI: https://doi.org/10.1016/B978-0-12-815936-1.00003-4

potential to replace petroleum-derived transportation fuels, because bioethanol has broader flammability limits, higher octane number, higher heat of vaporization, and higher flame speeds [2]. The cost involved in the production of ethanol from different biomasses including sugarcane and the food-versus-fuel debate over corn ethanol forced us to search for new feedstocks. Not only that, lignocellulose-rich agriculture waste has also been considered as potential alternative new generation feedstocks. Fast-growing short-rotation forest trees having large amounts of cellulose-rich biomass have emerged as a promising source for bioenergy and at the same time biopolymer production.

The biofuel can be produced from a variety of feedstock, such as plant oils, sugar beets, cereals, organic waste, and the processing of biomass. Out of them, the most promising is lignocellulosic biomass (e.g., wood, straw, and grasses). Bioconversion of straw to bioethanol represents an attractive alternative in comparison with conventional fuel ethanol production from grain [3]. Wheat straw is a promising substrate because it is the largest biomass feedstock in Europe and the second largest in the world after rice straw. Wheat straw has a great potential as feedstock in the future [4]. However, lignocellulose usually carries the structure of the plant biomass and is difficult substrate to degrade. Therefore thermochemical and enzymatic pretreatments are necessary for lignocellulose degradation to make the monomers available for further processing. The hydrolysis step is necessary for the conversion of biomass into monomer sugars for subsequent fermentation into bioethanol [5]. Hydrolysis can be acid or enzymatic. This latter has several advantages over the use of acid, because acid hydrolysis has relatively low yield, no selectivity, and it needs a high process temperature (ranging between 140°C and 160°C) and neutralization after hydrolysis. Enzymatic hydrolysis of cellulose is catalyzed by a class of enzymes called cellulases. There are several factors influencing the efficiency of hydrolysis. The aim of this chapter is to identify the optimum conditions of enzymatic hydrolysis of wheat straw lignocelluloses that remain after furfural production. However, it is very hard to depolymerize lignocellulosic materials because of the presence of complex lignin and hemicelluloses over cellulose. Different chemical pretreatment methods are employed to increase cellulose accessibility [6]. Therefore the production of fermentable sugars from lignocellulosic biomass tends to be complex and capital intensive, with, also, inherent environmental concerns.

Recently, marine macroalgal species have gained considerable global attention as source of third-generation biofuels [7]. The major advantages offered by seaweeds over terrestrial biomass could be synthetized as follows: (1) higher biomass production rate per unit area; (2) do not compete with agricultural plants for land; (3) require no agricultural input such as fertilizer, pesticides, and water; and (4) easier depolymerization as it does not contain lignin in their cell wall [8]. So far, bioethanol production from seaweeds has been mainly confined to a few phycocolloid-yielding species belonging to the genera of *Kappaphycus*, *Gelidium*, *Gracilaria*, *Sargassum*, and *Laminaria*. This strategy would not only affect the existing multibillion hydrocolloid industry but also may lead to another new debate, hydrocolloid versus fuel. Alternatively, the production of bioethanol from cellulosic residue following the extraction of hydrocolloid from seaweed biomass has been demonstrated by Kumar et al. [9]. Nevertheless, the lower cellulose content of residue may prevent it from being a viable feedstock option considering the growing demand for bioethanol. Therefore the selection of seaweed species with higher cellulose content together with higher growth rate is of paramount importance for sustainable bioethanol production. In this context a green seaweed *Ulva fasciata Delile* was selected as a potential feedstock for bioethanol production—following enzymatic hydrolysis. This species in spite of having higher polysaccharide content grows luxuriantly and has a worldwide distribution regardless of geographical barriers. This species is also regarded as an opportunistic species with potentials to form blooms (green tide) when there is a sudden outburst of nutrients in the aquatic streams.

The major compositions of lignocellulosic biomass are lignin, cellulose, hemicelluloses, salt, ash, protein, pectin, etc. But in recent days out of these compositions, cellulose is essential for bioethanol production. Lignin, which is a cross-linked three-dimensional polymer, consists of three monolignols: *p*-coumaryl alcohol, coniferyl alcohol, and sinapyl alcohol [10]. Lignin is covalently bound with polysaccharides in lignocelluloses. It can be described as higher value-added product as it has several applications in different aspects. It is not only used in manufacturing of dispersants, bioplastics, nanoparticles, and composite materials but also for carbon fiber agriculture and biofuel field [11]. So one of the main objectives is to remove lignin from biomass and use it in various applications for the recovery of high value product.

Different physical, biological, and chemical pretreatment methods have been suggested by several researchers to change the structure of lignocellulosic biomass. This includes disruption of the covalent linkage between lignin structure and cellulose, whereas they also alter the structure of lignin. Finally, an efficient method of removing hemicellulose and lignin has been introduced to make the cellulose accessible to hydrolytic enzyme when then can convert it to monosaccharide [12]. There are several limitations of conventional pretreatment methods, such as use of irradiation, chemical cost, and high electricity consumption. In comparison to this the use of different ionic liquids as pretreatment medium can well dissolve cellulose or lignin in it owing to its cationic and anionic structure. Finally, it will remove lignin successfully as well as alter the crystalline structure of cellulose [13]. The property of the solvent can be determined by the way suitable cation and anion is selected; this process transforms it as designer solvent before being implemented in the field of green synthesis chemistry for its commendable performances [14]. Mai et al. [15] summarized different chemical processes, such as distillation (distillation of ionic liquid and distillation of volatile solutes/impurities in ionic liquid), induced phase separation (salting out, introduction of CO_2, and changing the temperature), extraction (solvent extraction and CO_2 extraction), adsorption (adsorption/desorption and chromatography), reverse osmosis, pervaporation, magnetic separation, membrane-based processes [nanofiltration (NF), electrodialysis] and centrifugation, to recover and recycle ionic liquid so that the ionic liquid-based lignin recovery can become an industrially suitable procedure. The main stages involved in lignin recovery route are mixing of biomass powder with ionic liquid in a specific ratio, stirring and heating at high temperature for a certain time period which is then followed by the removal of lignin from ionic liquid-based lignin solution by precipitation. Some organic antisolvents and recycling of ionic liquid are used after evaporation of antisolvent for the recovery of final product. Though lignin can be recovered from poplar wood biomass by only alkaline pretreatment, with certain ionic liquid followed by NaOH, it is possible to gradually increases lignin yield [16]. Lignin is also recovered from softwood (pine) as well as hardwood (*Eucalyptus*) dissolved in imidazolium-based acetate and chloride ionic liquid. Further categorization of recovered lignin showed 31% recovery of lignin [17]. Moghaddam et al. [18] isolated and determined physicochemical characteristics of sugarcane bagasse lignin using IL 1-butyl-3-methylimidazolium chloride, $[b_{mim}]Cl$, and also 1-butyl-3–methylimidazolium methyl sulfonate, $[b_{mim}][CH_3SO_3]$, with HCl as catalyst. The differences in the extraction productivity of organic solvents are due to the changes in their conductivity as well as dielectric constant. NaOH is used to eliminate lignin from carbohydrate-enriched material as alkaline can split the alkali–labile bond between lignin and carbohydrate. 1-Ethyl-3-methylimidazolium acetate, [EMIM]oAc, was also used for softwood (southern yellow pine) treating, and yield of 31% of original lignin as carbohydrate free and 38% as carbohydrate bond was attained. Pinkert et al. [19] reported the process of delignification, isolation of *Pinus radiata* wood lignin, and retention of crystalline structure of cellulose in carbohydrate-rich filtrate by treating the wood flour with imidazolium acesulfamate ionic liquids, that is, [EMIM]Ace and [BMIM]Ace, which are derived from commercial sugar acesulfamate potassium. The impact of different variables, such as wood species, water content, wood particle size, and type of IL cation plus reusing of IL on lignin extraction efficiency (mass of extracted lignin/mass of initial lignin content) with [BMIM]Ace, has been studied. Increasing the extraction temperature from 353 to 416K and extraction time from 1 to 4 h raises the extraction efficiency from 0.32 to 0.81 and 0.05 to 0.07, respectively.

Removal efficacy decreases as the water content in the atmosphere resulting from dry condition (inert gas) to normal atmosphere and growing the moisture content in atmosphere. It increases 13% for [EMIM]Ace treatment than [BMIM]Ace treatment at 373K; 2 h of pretreatment is done due to two reasons: first, the presence of shorter alkyl substituent in [EMIM]Ace as compared to [BMIM]Ace, which declines its toxicity and viscosity resulting increases mobility to enhance IL biomass interface; and second, the existence of more halide impurity in [EMIM]Ace, which has a positive impact on IL lignin interaction. All the studies reveal the efficacy of ionic liquid, as a green and environment favorable solvent to treat the lignocellulosic biomass. Finally, it is being used to remove and recover lignin adequately which is the second most plentiful biological constituent in the world next to cellulose and hemicellulose.

3.2 Suitable strains and their productivity

Bioalcohols, such as bioethanol, biobutanol (or biogasoline), and propanol, are used as liquid transportation fuels obtained through the biomass aerobic or anaerobic fermentation.

The sustainable production of liquid fuels and chemical products through biorefinery approaches is based on feedstocks as lignocellulosic biomass that is widely available from agriculture by-products and industrial residues.

Lignocellulosic biomass mainly comprises cellulose, hemicellulose, and lignin. The high amounts of sugars present in cellulose and hemicellulose fractions can be hydrolyzed and converted into biofuels by a fermentation process. In contrast the recalcitrant structure of lignin is difficult to disrupt, and a pretreatment step is essential to improve its digestibility and increase the release of fermentable sugars.

In most cases, pretreatment technologies based on physicochemical processes involve high-energy demand, high-capital investment, some sugar degradation, and generation of inhibitory compounds that affect the downstream hydrolysis and fermentation steps. In contrast, biological approaches are interesting to improve the efficiency of the bioconversion processes and to overcome barriers in the scale-up and commercialization of renewable biorefineries.

In this context, biodelignification is a promising technology, where the so-called white rot *Basidiomycetes* (*Ceriporiopsis subvermispora*, *Dichomitus squalens*, *Pleurotus ostreatus*, and *Coriolus versicolor*) are able to depolymerize and mineralize lignin efficiently and extensively [20]. Some *Ascomycetes*—such as *Trichoderma reesei* and *Aspergillus terreus*—can also colonize lignocellulosic biomass. Apart from fungi, certain bacterial strains, such as *Bacillus macerans*, *Cellulomonas cartae*, *Cellulomonas uda*, and *Zymomonas mobilis*, have also shown delignification abilities yielding lignin degradation up to 50%. The use of ligninolytic enzymes such as laccases instead of microorganism populations is another feasible alternative for the delignification of lignocellulose [21].

Biodetoxification reduces the amount of inhibitors produced after the physicochemical pretreatment during lignocellulosic bioethanol production. Fungi, such as *T. reesei*, *Coniochaeta ligniaria*, *Amorphotheca resinae* ZN1, and *Aspergillus nidulans* FLZ10, have been studied for microbial detoxification. Bacteria—such as the thermophilic bacterium *Ureibacillus thermophaercus*, *Methylobacterium extorquens*, *Pseudomonas* sp., *Flavobacterium indologenes*, *Acinetobacter* sp., and *Arthrobacter aurescens*—and yeasts—such as *Issatchenkia occidentalis* CCTCC M 206097—have also been used for detoxification purposes to a lesser extent. Besides microbial detoxification, enzymatic detoxification with laccases and peroxidases is one of the main biotechnological methods used to diminish the inhibitory compounds of fermentation broths. Finally, coculture, evolutionary or genetic engineering modifications, cell retention, encapsulation, and flocculation have been developed to increase the intrinsic tolerance or the inherent detoxification capacity of some strains [21].

3.2.1 Bioethanol

An excellent alternative to fossil fuels used in transportation is bioethanol, which is used as a gasoline additive or as a pure fuel with high efficiency and performance.

Cellulosic biomass is available in large amounts as an attractive feedstock, which is composed of a complex mixture of carbohydrate polymers.

Since a significant portion of all process costs is about the feedstock, an economical fermentation process will require rapid and efficient conversion of all sugars present.

3.2.1.1 Saccharification

In the first step, called hydrolysis or saccharification, hydrolytic enzymes—cellulase and hemicellulase—are utilized to degrade cellulose and hemicellulose to monomeric sugars [3]. Some microorganisms (*Clostridium*, *Cellulomonas*, *Trichoderma*, *Penicillium*, *Neurospora*, *Fusarium*, *Aspergillus*, etc.) show a high cellulolytic and hemicellulolytic activity and are able to ferment monosaccharides [22]. Moreover, there are many variations of enzymes that are responsible for polysaccharide hydrolysis, most of which are produced commercially by genetically modified strains of *Saccharomyces cerevisiae* [23], *T. reesei* [24], *Fusarium venenatum*, *Aspergillus oryzae*, and *Aspergillus niger* [25].

Insoluble cellulose can be hydrolyzed by cellulase enzymes into soluble sugar monomers that can be metabolized by microorganisms. Three major groups of enzymes are able to hydrolyze cellulose: (1) endoglucanase (EG) or 1,4-β-D-glucan-4-glucanohydrolase; (2) exoglucanase, including 1,4-β-D-glucan glucanohydrolase (also known as cellodextrinase) and 1,4-β-D-glucan cellobiohydrolase (CBH); and (3) β-glucosidase (BGL) or β-glucoside glucohydrolase. BGL present in cellulose is the key enzyme component and completes the final step during cellulose hydrolysis by converting the cellobiose to glucose. Since it gets inhibited by its product glucose, one of the major challenges in the bioconversion of lignocellulosic biomass into liquid biofuels includes the search for a glucose tolerant BGL [26]. In this context, intensive attention was addressed to fungi belonging to the genus *Trichoderma* and *Aspergillus*, because of their high-level production of secreted cellulases [27,28]. Also, *Clostridiales* (anaerobic) and *Actinomycetales* (aerobic) showed cellulase activity, with different strategies for the degradation of cellulose [29].

Hemicellulases are frequently classified according to their action on distinct substrates and are isolated from a number of bacteria. Commonly occurring hemicelluloses are xylans, arabinoxylan, glucomannan, galacto-glucomann, and so on. In xylan degradation, endo-1,4-β-xylanase, β-xylosidase, α-glucuronidase, α-L-arabinofuranosidase, and acetylxylan esterase act on different heteropolymers, while during glucomannan degradation, β-mannanase and β-mannosidase cleave the polymer backbone [30]. Nevertheless, the knowledge and understanding of hemicellulose systems is far less as compared to the cellulase systems in the thermophilic/thermotolerant bacteria.

Indeed, existing enzymatic hydrolysis technologies—carried out at temperatures lower than 50°C—exhibit slow reaction rates, require high dosages of enzymes, generate low yields of sugars from lignocellulose and are prone to microbial contamination problems. Therefore thermostable enzymes have been investigated to overcome the limitations of existing lignocellulosic biomass conversion processes. Among thermostable enzymes, Bhalla et al. [31] focused on the following groups:

- Cellulases, from thermophilic and mesophilic fungal genera (belonging to the *Aspergillus*, *Chaetomium thermophile*, *C. ligniaria Rhizopus*, *Sclerotium*, *Sporotrichum thermophile*, *Thermoascus thermophile* var. *coprophile*, and *Trichoderma*), thermophilic bacteria (belonging to the genera *Acidothermus*, *Bacillus*, *Caldibacillus*, *Caldocellum*, *Clostridium*, and *Geobacillus*), and hyperthermophilic microorganisms (*Anaerocellum*, *Caldicellulosiruptor obsidiansis* sp. nov., *Rhodothermus*, *Thermotoga*, and archaea *Pyrococcus* and *Sulfolobus*)
- Xylanases, from thermophilic and hyperthermophilic bacteria (*Acidothermus*, *Actinomadura*, *Alicyclobacillus*, *Anoxybacillus*, *Bacillus*, *Cellulomonas*, *Enterobacter*, *Geobacillus*, *Nesterenkonia*, *Paenibacillus*, *Thermoanaerobacterium*, and *Thermotoga*) and fungi (*Laetiporus sulphureus*, *Nonomuraea flexuosa*, *Rhizomucor miehei*, *Talaromyces thermophiles*, *Thermoascus aurantiacus*, *Thermomyces lanuginosus*)

3.2.1.2 Fermentation

Hydrolyzed substrate can be fermented to ethanol by different microorganisms [32], major advantages and drawbacks of which were reported in literature [33]. The search of robust microorganisms is essential to design sustainable processes of second-generation bioethanol. A number of *S. cerevisiae* and *Kluyveromyces marxianus* strains isolated from industrial environments and laboratory background strains were investigated [34].

Ethanologenic thermophilic and thermotolerant microorganisms [35,36] are of current interest due to their ability to ferment various components of cellulosic biomass into chemicals and fuels [37]. Potential thermophilic/thermotolerant ethanologenic bacteria include *Aerobacter* sp., *Aeromonas hydrophila, B. macerans, B. polymyxa, Clostridium acetobutylicum, Clostridium thermohydrosulfurium, Clostridium thermosaccharolyticum, Clostridium thermosulfurogenes, Clostridium tetani, Erwinia* sp., *Escherichia* sp., *Geobacillus* sp., *Klebsiella pneumoniae, K. marxianus, Lactobacillus* sp., *Leuconostoc* sp., *Pichia* sp., *Thermoanaerobacter ethanolicus, Thermoanaerobacterium saccharolyticum,* and *Z. mobilis* [29].

Yeast *S. cerevisiae* is more robust than bacteria and other yeasts, since it presents higher tolerance to ethanol and inhibitors present in hydrolyzates and higher efficiencies of sugar conversion into ethanol. *S. cerevisiae* is widely used in economically feasible biomass-to-ethanol fermentation processes, as it exhibits fast sugar consumption, high yields, and ethanol tolerance. In addition, *S. cerevisiae* displays several advantageous characteristics, including its inherent resistance to low pH, high temperature, and various inhibitors, for which it is used in industrial applications [38]. *S. cerevisiae* has the potential to ferment ethanol from various types of cellulosic biomass and is able to ferment hexoses rapidly and efficiently. In contrast, it cannot naturally ferment pentose sugars such as D-xylose and L-arabinose contained in the hemicellulosic fraction and use these sugars for the growth. The bioconversion of pentoses to ethanol is still one of the major bottlenecks for ethanol commercialization effort.

Xylose-fermenting natural strains of bacteria (*Clostridium* sp., *B. macerans, Clostridium saccharolyticum,* and *T. ethanolicus*) and yeasts (*Pichia stipitis, Pachysolen tannophilus, Candida tropicalis* [39], *Candida shehatae,* and *K. marxianus*) can be used as biocatalysts to convert xylose or cellulose into ethanol. The disadvantages of using bacteria in large-scale fermentation are the low ethanol yields, by-product formation, intolerance to high ethanol concentrations, and growth at narrow and neutral pHs varying from 6.0 to 8.0 [20,40].

Many fungi and yeasts can aerobically assimilate L-arabinose, but most of them are unable to ferment it to ethanol or they exhibit only very low ethanol production rates and yields.

Nevertheless, a lack of microorganisms that will efficiently convert hexoses and pentoses to ethanol is a major constraint to the economical conversion of biomass.

Through metabolic engineering, bacterial and yeast strains have been constructed which feature traits that are advantageous for ethanol production using lignocellulose sugars, in order to minimize by-product formation and to increase the ability to utilize all sugars present in lignocellulose substrate, ethanol yield and productivity, tolerance to ethanol and inhibitors, and tolerance to process hardness [41]. Metabolic engineering is a powerful method to improve, redirect, or generate new metabolic reactions or complete pathways in microorganisms. A number of different strategies have been applied to engineer yeasts capable of efficiently producing ethanol from xylose, including the introduction of initial xylose metabolism and xylose transport, changing the intracellular redox balance, and overexpression of xylulokinase and pentose phosphate (PP) pathways [42,43].

Since many years, *Escherichia coli*, *Klebsiella oxytoca*, and *Z. mobilis* have been genetically engineered to produce ethanol efficiently from all hexose and pentose sugars present in the polymers of hemicellulose [44–49]. Moreover, a recombinant *E. coli* strain from wheat straw was investigated at high-solid loading by both separate hydrolysis and fermentation and fed-batch simultaneous saccharification and fermentation (SSF) [50].

In the last decades, numerous microorganisms used in industry, including *E. coli*, *Bacillus* sp., lactic acid bacteria, *Corynebacterium glutamicum* [51], and *S. cerevisiae*, have been engineered to tolerate toxic compounds and metabolize a range of carbon sources present in hemicellulose [27,29].

The genetic improvement of *S. cerevisiae* strains was widely reviewed [40,52,53]. Genetic and metabolic engineering have been used to insert heterologous genes encoding D-xylose reductase and xylitol dehydrogenase in *S. cerevisiae*, resulting in yeast strains able to utilize the pentose D-xylose and to ferment it to ethanol [54–68].

Moreover, genetic engineering have allowed to select an efficient L-arabinose-fermenting *S. cerevisiae* strain. In particular, *S. cerevisiae* strain was engineered by expression of a bacterial pathway for catabolism of L-arabinose, comprising L-arabinose isomerase, L-ribulokinase, and L-ribulose-5-P 4-epimerase. This has been achieved in order to be able to utilize the pentose sugar L-arabinose for growth and to ferment it to ethanol [69]. In addition, *S. cerevisiae* was engineered to construct an acetate-tolerant strain, because acetate shows a negative effect on the growth of contaminated bacteria and is a safe and inexpensive reagent for inhibiting bacterial growth [70]. *S. cerevisiae* was engineered also to increase its tolerance to high temperatures [71,72], to ethanol [73], to acetic acid and formic acid

(FA) [74], to stress [75] as well as to increase multitolerance to high temperature, acidity and high ethanol production [76].

3.2.1.3 Direct microbial conversion

A method to convert cellulosic biomass to ethanol is the direct microbial conversion (DMC), in which both ethanol and all required enzymes are produced by a single microorganism [23,77,78]. Several strains of *Fusarium oxysporum* can convert D-xylose and cellulose to ethanol in a one-step process, but there is no robust organism available that can produce cellulases or other cell wall degrading enzymes in conjunction with ethanol with a high yield [22].

The development of simultaneous saccharification and fermentation (SSF) of biomass to ethanol by native or genetically modified microbial strains has been studied intensively [79–83]. This approach combined the cellulase enzymes and fermenting microbes in one vessel.

A single-step process was proposed for converting lignocellulose to ethanol using mesophilically native isolated strains. The strain *Bacillus* sp. THLA0409 was identified as a dominant cellulose-degrading bacterium, while the strain *K. oxytoca* THLC0409 was determined as a dominant sugars-utilizing bacterium. The coculture of these two strains remarkably enhanced the utilization efficiency of hydrolyzates from acid-pretreated raw bamboo, *Napiergrass*, rice straw, and ethanol production [84].

During the last decade, several wild-type and genetically engineered bacteria, fungi, and yeasts have been proposed for application in a consolidated bioprocessing (CBP) system, where all the processes, including enzyme production, enzymatic saccharification, and fermentation of the resulting sugars to bioethanol or other valuable products, proceed simultaneously [25,72]. The potential microorganisms and their suitable characteristics for CBP alcohol production were comprehensively reviewed and discussed [85]. These microorganisms have combined saccharolytic and ethanologenic capabilities: they naturally degrade cellulose and ferment the resulting sugars into ethanol, lactate, acetate, carbon dioxide, and hydrogen. In particular the following strain groups were investigated:

- Bacteria (*Clostridium thermocellum*, *Clostridium phytofermentans*, *Thermoanaerobacterium* sp.)
- Fungi (*Mucor circinelloides*, *F. oxysporum*, *Fusarium verticillioides*, *Acremonium zeae*, *A. oryzae*, *Paecilomyces variotii*)

- White rot *Basidiomycetes* (*Trametes versicolor*, *Flammulina velutipes*, *Phlebia* sp., *Peniophora cinerea*, *Trametes suaveolens*)
- Yeasts (*K. marxianus*, *Clavispora*, cryophilic *Mrakia blollopis*)

In addition, two genetic engineering strategies have been extensively described [85]:

1. Engineering cellulase producers (*C. thermocellum*, *Clostridium cellulolyticum*, *T. saccharolyticum*, *Thermoanaerobacterium aotearoense*, *Thermoanaerobacter mathranii*, *Caldicellulosiruptor bescii*, *Geobacillus thermoglucosidasius*, *K. oxytoca*, *T. reesei* [59], and *F. oxysporum*) to be ethanologenic
2. Engineering ethanologens (*S. cerevisiae*, *K. marxianus*, *Hansenula polymorpha*, *Z. mobilis*, *E. coli*, *P. stipitis*, *F. velutipes*) to be cellulolytic

Hasunuma et al. [72] investigated the ethanol production from cellulosic and hemicellulosic materials with thermotolerant yeast strains in SSF or CBP at elevated temperature. In particular, they focused on the following yeast strains: *K. marxianus*, *H. polymorpha*, *Candida glabrata*, *S. cerevisiae*, *Pichia kudriavzevii*, and *Debaromyces hansenii*. Svetlitchnyi et al. [86] isolated other thermophilic bacteria suitable for a single-step conversion of lignocellulosic biomass to ethanol at temperatures >70°C.

3.2.2 Biobutanol

Biobutanol is produced from the same raw material as bioethanol, and it can be used as biofuel or fuel additive with several advantages over ethanol, because it has similar characteristics with gasoline.

A sustainable industrial-scale biobutanol production is possible by different strategies, including choice of feedstock, product toxicity to strains, multiple end products, and downstream processing of alcohol mixture [87].

Butanol can be generated as a product of anaerobic ABE (acetone—butanol—ethanol) fermentation of lignocellulosic biomasses or other feedstocks by a number of solventogenic *Clostridium* species [88], following delignification and hydrolysis pretreatments. *Clostridium* bacteria can metabolize different sugars, amino and organic acids, polyalcohols and other organic compounds to butanol, and other solvents [89]. The most intensively studied solvent-producing species is *C. acetobutylicum* [90,91] that gives a mixture of fermentation products composed of ABE at a ratio of 3:6:1. Other strains involved in ABE fermentation are *Clostridium beijerinckii* [92—100], *Clostridium pasteurianum* [92,101], *Clostridium saccharobutylicum* [94,102—106], and *Clostridium saccharoperbutylacetonicum*

[92,94,107−117]. Moreover, some strains of *C. beijerinckii* and *Clostridium aurantibutyricum* can synthesize isopropanol instead of acetone and ethanol, respectively, whereas *Clostridium tetanomorphum* synthesizes butanol and ethanol and does not synthesize any other solvents [118], as well as *Clostridium sporogenes* [119]. Some researchers investigated cocultures of *C. thermocellum* and *C. saccharoperbutylacetonicum* with the addition of cellulase [120], and of *C. acetobutylicum*, *Clostridium butylicum*, and *C. beijerinckii* with microorganisms having enzymes capable of simultaneously hydrolyzing cellulose and hemicellulose [121−125]. Another solution is the development of genetically engineered strains with activated endogenous cellulose enzyme [126].

Efficient genetic tools are crucial for the metabolic engineering of *Clostridia*, in order to enhance solvent production, to improve butanol tolerance, to increase the ratio of butanol to the solvent, and to allow the strain to grow on complex cellulosic substrates [127].

Since butanol is toxic for bacteria, it is essential to find bacterial strains tolerating higher butanol concentrations, such as the use of an "enrichment culture," with the medium containing the compound of interest. A significant number of studies has been directed toward microbial production of butanol by different metabolically engineered mesophilic and thermophilic organisms [128,129]. With respect to mesophilic microorganisms, butanol was produced by engineered strains of cyanobacterium, *Synechococcus elongatus*, [130] and bacteria, *Clostridium tyrobutyricum* [131,132], *C. beijerinckii*, and *C. acetobutylicum*, in order to enhance butanol tolerance and yield [95,118,133−144]. A number of thermophilic organisms have been engineered for butanol production: *Thermoanaerobacterium thermosaccharolyticum* [145], *T. saccharolyticum* [146], *G. thermoglucosidasius* [147], and *Pyrococcus furiosus* [148]. Moreover, other butanol-producing microorganisms have been engineered, with the aim to increase tolerance to ferulic acid and "acid crash" (*C. beijerinckii*) [149,150], to increase tolerance to lignocellulose-derived inhibitors (*Clostridium* sp. strain BOH_3) [151], and to facilitate xylose transport and metabolism (*C. tyrobutyricum*) [152].

In any case, selection of bacterial strain in ABE process depends on the raw material. For example, *C. acetobutylicum* is more suitable for starch-based medium, whereas *C. saccharobuylicum* is for molasses-based medium [92,126].

The classical ABE fermentation by *Clostridium* strains takes place under anaerobic conditions. Nevertheless, butanol can be synthesized under aerobic conditions, by introducing the corresponding genes in *Clostridia* responsible for the synthesis of butanol into *E. coli* [153–159] Moreover, different engineered strains were developed via mutagenesis based on *S. cerevisiae* [156,160,161], *Ralstonia eutropha* [162], *Lactococcus lactis*, and *Lactobacillus buchneri* [163].

Finally, even if currently biobutanol is mainly produced by *Clostridia*, they cannot directly utilize lignocellulose. Therefore metabolic construction or isolation of novel cellulolytic/hemicellulolytic and solventogenic bacteria to achieve direct butanol production from lignocellulose offers a promising alternative [164]. The present research efforts are focused on developing CBP strategies for biofuels production wherein microorganisms are used to hydrolyze and ferment inexpensive lignocellulosic materials directly into desired products without additional enzymes [85]. *C. thermocellum*, *C. cellulolyticum*, and *Clostridium thermopapyrolyticum* can directly utilize lignocellulosic biomass [165]. Several strains of *Clostridium cellulyticum*, including metabolically engineered and wild-type strains, have been reported to generate value-added products directly from cellulose, but few studies investigated wild-type or metabolically engineered strains to produce butanol directly from cellulose or xylan, which constitutes the principal hemicellulosic component of plant wastes [164,166–170].

3.3 Enzymatic hydrolysis

3.3.1 Enzymatic hydrolysis of lignocellulosic biomass

Enzymatic hydrolysis is an essential stage in the transformation of cellulose, from pretreated biomass, to glucose. The bioconversion of cellulose to glucose is being carried out by using cellulase enzymes under mild operating conditions of temperature (in the range 40°C–50°C) as well as pH ranges between 4.5 and 5.0, in order to negate the corrosion problems [171].

The efficacy of the hydrolysis process is governed by the degree of pretreatment of the biomass in terms of lignin removal and hemicelluloses solubilization including hydrolysis, and enzyme loading. Lignin and the acetyl group that are present in the hemicelluloses could produce binding with cellulase not really productive and consequently could limit the hydrolysis process [172]. The efficacy of the hydrolysis process was found to be enhanced by the addition of nonionic surfactants by which it changes the cellulose surface properties and lowering enzyme loading. Surfactants, such as polyethylene glycol (PEG), were found to intensify the enzymatic conversion of the lignocellulosic biomass from 42% to 78%

in 16 h. PEG prohibited the unproductive attachment of lignin to the surface of the enzyme [173].

Pretreatment using microwave irradiation was found to boost the enzymatic saccharification of rice straw by 31%. The enzymatic saccharification was dependent on factors such as irradiation time, substrate concentration, and microwave intensity [174]. Banerjee et al. [175] reported the utility of the using liquid fraction obtained after alkaline peroxide-assisted wet air oxidation (APAWAO) of rice husk during the process of enzymatic hydrolysis. The hydrolysis experiment was carried out with the APAWAO pretreated solid fraction (in buffer) and slurry (solid fraction in liquid fraction). It must be noted that the hydrolysis in the APAWAO pretreated liquid fraction produced extra glucose when associated to hydrolysis of the solid fraction in buffer medium. Thus recycling of APAWAO liquid portion was feasible during enzymatic hydrolysis, which caused in glucose production enhancement at 50°C hydrolysis temperature. The enzymatic cellulose conversion was up to 86 wt.% within 24 h, and the glucose yield was 21 g glucose per 100 g of the untreated rice husk. The effect of different dry matter loadings on the hydrolysis yield was also investigated and has affected the conversion [175].

The structural distinctive of biomass affects the cellulose digestibility with respect to the surface area of the particle, crystallinity of cellulose, large pore volume, presence of lignin and degree of polymerization of hemicelluloses, and acetyl group [176].

Ultrasound-assisted pretreatment is a promising and a novel process which escalates the glucose yield after the enzymatic saccharification by removing lignin and hemicelluloses [177].

Bian et al. [178] investigated the effect of ionic liquid ([Emim]Ac) pretreatment on the enzymatic hydrolysis of sugarcane bagasse sometime. It has also been observed that ionic liquid pretreatment enhanced enzymatic convertibility of cellulose and achieved 95.2% cellulose conversion after 96 h enzymatic hydrolysis. Lowering the degree of polymerization and reducing cellulose crystallinity resulted in high hydrolysis convertibility [178]. The enzymatic digestibility of cellulose was found to be more associated with lignin removal than hemicelluloses solubilization [179].

The overall bioethanol production process economics can be improved by operating hydrolysis of enzymatic at higher substrate concentration so that the glucose yields in the hydrolyzate could be added. The substrate concentration affects the rate and extent of the hydrolysis process. The hydrolysis product and substrate concentration are inversely proportional [180]. It is well acceptable that cellulolytic enzymes are completely inhibited by end product (such as cellobiose and glucose). Addition of cellobiase enzyme also alleviates the inhibitory effect of cellobiose as it also hydrolyzes the cellobiose to glucose. Operating hydrolysis reaction at low substrate concentration reduces the end product inhibition. It was observed that glucose concentration increased with increase in dry matter loading (up to 10% of dry matter). A further increase in dry matter concentration does not proportionately increase the glucose yield. This may be attributed to possible mass transfer limitations due to high substrate concentration where there is little convective mixing due to minimum water available to facilitate the mass transfer of the water soluble compounds. Sugars build up around the active sites of cellulase enzyme and inhibit further hydrolysis reaction [181]. Table 3.1 shows comparative chart for the work reported on hydrolysis of lignocellulosic biomass [182].

Table 3.1 Comparative chart for previous work reported on hydrolysis of lignocellulosic biomass [182].

Biomass	Pretreatment and operating conditions		Lignin removal (% w/w)	Cellulose potential conversion (% w/w)
Sugarcane bagasse	Steam explosion	(205°C, 40 bar, 10 min)	35	48.9
	WAO	(195°C, 12 bar, 15 min)	50	57.4
	Sodium hydroxide in liquid hot water	180°C, 20 min 1% NaOH	84.3	97.5
Energy cane bagasse	Ionic liquid ([EMIM] [OAc])	5% (w/w) 120° C 30 min	32	87
Rice husk	APAWAO	(185°C, 5 bar, 15 min)	88	86
	WAO	(185°C, 5 bar, 15 min)	–	22
Rice straw	Hydrated lime	10% (w/w) $Ca(OH)_2$ 95°C, 3 h	27	48.5
Coastal bermuda grass	Sodium hydroxide pretreatment	3% NaOH, 60 min, 121°C	86	90.43

APAWAO, Alkaline peroxide-assisted wet air oxidation; *WAO*, wet air oxidation.

3.3.2 Traditional downstream purification

Aqueous ethanol that is produced after conventional fermentation process is yet to be accepted as fuel ethanol. Ethanol and water form azeotropic mixture and make it difficult to get fuel-grade ethanol. Thus new research directs to initiate reducing high-energy demanding and industrially acceptable distillation process to break the azeotrope nature for achieving almost pure ethanol. Since many years azeotropic and extractive distillation processes are used widely for this purpose [183].

Distillation is separating liquid mixtures based on their boiling point by evaporation followed by condensation. Gill et al. designed the use of glycerol in extractive distillation processes to dehydrate ethanol. The reason is that glycerol has high availability, relatively low cost, and enhanced relative volatility, which impart a positive effect on azeotropic mixture. The study of the impact of this entrainer on feed stage, feed temperature, reflux ratio, feed molar ratio has been done to get the most appropriate design for minimizing the requirement of energy [184].

A novel approach of extractive dividing-wall columns (DWC), which combines with solvent recovery system, and preconcentration column in a single distillation unit, is efficient to separate 99.8 wt.% bioethanol. This ethanol dehydration configuration can be applied in large-scale bioethanol production anywhere in this world [185]. While using ethylene glycol and *n*-pentane as mass separating agent, DWC in extractive and azeotropic distillation can separate highly pure, that is, 99.8% ethanol, and also saves 10%–20% of overall energy [186]. Using ionic liquid as mass separating agent during extractive distillation lowers energy consumption for ethanol purification. This process can achieve 0.995 (in mass) ethanol purity and 99.9 wt.% ethanol recovery [187].

3.3.3 Possibilities of application of membrane in bioethanol production

Membrane, a permeable barrier when placed between two mediums and applying required driving force, allows the selectively transfer of one or more constituent from one phase to another. Membranes that are mostly used in bioseparation processes are generally porous in nature. The constituents that pass through the pores of the membrane and collected in other side are called permeate, whereas the components that retain are called retentate. Most useful pressure-driven membrane-based separation commonly used methods include microfiltration (MF), NF, ultrafiltration

(UF), and reverse osmosis. Most of the membrane that is used in different chemical processes is made polymer. But nowadays, the use of inorganic and hybrid membrane is common comprising inorganic nanoparticle and also polymer. The commonly used membrane configurations are hollow-fiber module that is currently being used vastly. In bioethanol industry, membrane is generally used for downstream purification to concentrate the final product. But it can be sufficiently used in cellulase enzyme recycling, cell recycling, and removal of inhibitory by-products as well.

3.3.3.1 Membrane-assisted enzymatic hydrolysis

Fermentation of lignocellulosic biomass to fuel-grade ethanol challenges two major difficulties: (1) crystalline structure of cellulose which resists enzymatic hydrolysis of cellulose and (2) lignin-cellulose association that impedes enzymatic attack on cellulose [188]. Dilute acid–catalyzed hydrolysis of cellulose uses high pressure and temperature that leads to the destruction of monomer glucose to hydroxymethylfurfural (HMF), FA, levulinic acid (LA), including side reaction that cause disruption of cellulose leading to the formation of nonreactive material [189]. On the other hand, this chemical process, enzymatic hydrolysis of cellulose, requires relatively low energy as well as normal temperature and pressure. Cellulolytic enzyme containing CBHs, EG and BGL obtained from different bacteria or fungi can sufficiently hydrolyze pretreated lignocellulosic biomass to fermentable sugar. But the cost of the enzyme constitutes 20% of the total ethanol production cost and 50% of the total cost of whole hydrolysis process [190].

One way to reduce the manufacture cost of ethanol from lignocellulosic biomass is to recycle and reuse of enzyme as well as the specific activity of enzyme should be recovered, retained. Enzyme recycling is predisposed by various reasons such as nature of biomass, process parameters and origin, and activity of enzyme [191]. After hydrolysis, cellulase can either appear in solution as free enzyme or can be adsorbed in remaining biomass [192]. Recovery and recycling of bound enzyme after initial hydrolysis require separation and desorption method [193], but the catalytic activity gradually decreases during hydrolysis steps.

Recovery of filter paper actions decreases from 91% to 22% by this method. But the use of alkali or surfactant increases enzyme activity to some point. Membrane-based process somehow adequately recovers the enzyme from hydrolysis solution, retains its catalytic activity, satisfactorily recycle, and reuses the enzyme which is further use in hydrolysis procedure and was established as best recovering procedure.

Membrane filtration, which is widely used in this process, is the most used organic membrane made of cellulose acetate (CA), polysulfone (PS), nylon (NY), polyethersulfone (PES). The membrane module that is usually adopted for this process includes hollow fiber, spiral wound, and flat sheet membrane. MF can remove remaining biomass after hydrolysis, whereas UF is used to recover soluble enzyme from hydrolysis solution.

Permeate flux for MF is higher than UF due to large pore size of MF membrane [194]. Sedimentation followed by MF and UF methods can recover and recycle 75% of the enzyme used in active catalytic form and exhibits a low-cost ethanol produced from pretreated ground yellow poplar. MF membranes were made of CA, PS, and NY, whereas UF membranes were made of PS (PSf) and PES. In combined sedimentation and cross flow UF method, sedimentation using inclined settler removes lignocelluloses particles larger than 50 μm in length, and UF transmits fermentable sugar, while retaining remaining biomass particle and cellulase enzyme. The permeate flux from UF is 64 ± 5 L/(m^2 h) when feed consists of 0.22 w/v% cellulose, and it increases up to 130 ± 20 L/(m^2 h) when feed consists of 10 wt.% lignocelluloses and settler overflow from mixture of 0.22 w/v% cellulose. As cellulose binds to the lignocellulosic particle during enzymatic hydrolysis and filtration, it prevents the enzyme from fouling the membrane, and thus increasing the permeate flux [195].

High conversion rate of cellulose and low consumption of cellulase can be achieved in a continuously or semicontinuously operated attrition bioreactor conjugated with a membrane filter which retains cellulase and remaining cellulose in reactor and allows sugar to pass through it as product [196]. Hydrothermal pretreated wheat straw for ethanol production was varied with Celluclast supplemented with BGL at different concentrations and subjected to hydrolysis followed by fermentation using yeast strain. After fermentation, cellulase was recovered by passing the liquid phase through 0.22 μm PES membrane followed by concentration and NaAc buffer exchange in a tangential UF system. Pellicon XL membrane with a 10 kDa cutoff PES membrane achieves 80% recovery of soluble enzyme, 70% recovery of activity from liquid phase providing high conversion degree. Increasing the hydrolysis temperature and enzyme loading drop the recovery of cellulose from liquid phase about 40% [197].

UF with PES10 membrane was studied to recycle 73.9% cellulase present in the hydrolyzate suspension of steam-exploded wheat straw, thereby minimizing the total hydrolysis cost. Permeate acquired from UF was concentrated by NF using NF270 membrane to concentrate glucose 3.5 times more from UF permeate, thus improving fermentation efficiency as well as lower the cost of downstream processing of fermentative product. Permeate flux for UF was 25.6 L/(m^2 h) at 1.27 bar transmembrane pressure, whereas for NF flux reduced to 13.3 L/(m^2 h) at 35 bar pressure [198].

In another study, dilute alkali-treated wheat straw showed better recycling of substantial amount of enzyme in contrast to adsorption recycling, UF using glass microfiber membrane can efficiently retain BGL, thus keeping away from the loading of BGL in successive round of hydrolysis, thereby boosting the economics of enzymatic hydrolysis [193].

Recover and recycle of cellulase and cellobiase used in hydrolysis of ammonia fiber explosion-treated corn stover can be efficiently attained by UF using PES 76 mm, 10 kDa, membrane. Electro-UF, an advancement of UF caused by intensification of UF by electric field, can efficiently lessen concentration polarization as well as increase membrane flux using cation-exchange membrane and polymeric membrane. Suitable conditions, that is, low buffer concentration, room temperature, and high current, increase the flux and sufficiently recover and recycle cellulase from hydrolizate of acid-treated wheat straw. Concentration polarization resistance decreased from 7.5 to 8×10^9 to $3-3.5 \times 10^9$ Pa s/m while increasing electric field strength from 0 to 144 V/m. Thus flux is proportional to electric field. Polymeric UF membrane can hydrolyze maximum amount of microcrystalline cellulose pretreated with ionic liquid.

This process permits maximum recovery and recycling of cellulose retaining its utmost activity for nine running cycle in semicontinuous process. Continuous process with tangential flow ceramic UF membrane shows constant permeate flux and glucose concentration at different residence time proving itself as a good practice for cellulose hydrolysis procedure [199].

Submerged vibration and stirring at membrane surface reduce fouling and concentration polarization thus enhancing permeate flux [200]. All the experimental studies reveal the importance of membranes and membrane-assisted separation process for retrieving and reuse of cellulase in an energy-efficient manner as well as reducing the cost of enzyme and enzymatic hydrolysis step in recent fuel-grade bioethanol industry. Tables 3.2 and 3.3 show the summary of membrane process assisted cellulase recovery, respectively, for UF and other membrane-based processes

Table 3.2 Summary of ultrafiltration membrane-based process assisted enzyme recovery [182].

Source of cellulose	Pretreatment method	Type of cellulase	Membrane material and module	MWCO (kDa) and rejection (*R*%)	Enzyme activity recovery
Wheat straw	Dilute acid and dilute alkali	Cellulase of *Trichoderma reesei* and β-glucosidase from *Aspergillus niger* (Novozyme 188)	Glass microfiber	10 kDa	68.6% (acid) 70.4% (alkali)
Wheat straw	Steam	Celluclast and Novozyme 188 (Novozymes A/S)	Organic (PS)	10 kDa	60%
Wheat straw	Hydrothermal	Celluclast and Novozyme 188 (Novozymes A/S)	Organic (PES)	0.2 μm	70%
Rice straw	Steam explosion	Cellulase from *T. reessei*	Organic (PS) Hollow fiber	10 kDa	50% 70% β-glucosidase
Corn stalk	Steam explosion	Cellulase from *T. reessi*	Organic (PS) Hollow fiber	10 kDa	
Sugar beet pulp		Cellulase of *T. reesei* (Celluclast Novozymes A/S0, Denmark) *Cellobiase of A. niger* (Novozyme 188, Novozymes A/S, Denmark)	Organic PES TF *Flux:* 15 L/(m^2 h)	5 kDa 4 kDa PES—*R*%: 0.48% TF—*R*%: 0.52%	
Microcrystalline cellulose (20 μm powder)	Ionic liquid (1-butyl-3-methylimidazolium chloride)	Cellulase from *T. reesei* (Celluclast 1.5 L), and cellobiase from *A. niger* (Novozyme 188)	Organic PES Inorganic: ceramic (CE) Tubular module *Average Membrane flux* 24.7 L/(m^2 h) *Fall from* 14.2 to 11.3 L/(m^2 h)	10 kDa—*R*%: 99.8 ± 0.2 5 kDa—PES—*R*%: 100 ± 0.1 5 kDa—CE—*R*%: 99.4 ± 0.3 10 k Da—*R*%: 98.4 ± 0.5 5 kDa—PES—*R*%: 99.4 ± 0.4 5 kDa—CE—*R*%: 98.1 ± 0.3	
Solka Floc BW200 microcrystalline white, delignified, pure cellulose powder, and Mavicell cellulose pellets		Celluclast (Novozyme A/S)	Organic (polymeric) flat sheet		

CE, *cellulose; PES*, Polyethersulfone ; *PS*, polysulfone.

Table 3.3 Summary of other membrane-based process-assisted enzyme recovery [182].

Source of cellulose	Pretreatment method	Type of cellulase	Membrane-based process	Membrane material and module	MWCO membrane flux and rejection *R*%	Enzyme activity recovery (%)
Knot rejects from sulfite pulping	Ammonia and deionized (DI) water	Novozymes North America (Franklinton, NC)		Tubular	10 kDa (LS) 6 kDa (PS)	
Wheat straw	Steam explosion		UF and NF	Organic (PES, NF)	UF: 10,000 g/mol—flux 25.6 L/(m^2 h) NF: 150 g/mol—flux 13.3 L/(m^2 h)	
Wheat Straw	Acid treatment	Cellulase from *Trichoderma longibrachiatum*	Electro-UF	CEM Organic (PES)	10 kDa *Flux decrease with time at all electric field*	
Corncob	Dilute sulfuric acid	Cellulase (GC220)				36.5
	Sodium hydroxide					30.4
	Aqueous ammonia	Cellobiase (Novozyme 188)		Microfiber membrane		41.2
	Dilute sulfuric acid plus aqueous ammonia					62.4
Corncob	SSA	Cellulase (Spezyme CP) β-Glucosidase (Novozyme 188)	Filtration	Glass microfiber	0.45 μm pore size	
Straw and hay mixture (1:1)	0.4 M NaOH, 20 h, room temperature, 5 g/100 g liquid			Organic polysulfone hydrophilic	10 kDa—R 83% 89, 49 L/(m^2 h bar)	55
			MMV	Composite fluoropolymer	10 kDa—R 78% 52, 35 L/(m^2 h bar)	67
			Novozyme biomass kit	Polyethersulfone	10 kDa R 83% 55, 20 L/(m^2 h bar)	30
				Composite fluoropolymer	1 kDa R 81% 27.27 L/(m^2 h bar)	100

CEM, Cation-exchange membrane; *MMV*, magnetically induced membrane vibration; *NF*, nanofiltration; *PES*, polyethersulfone; *SSA*, soaking in aqueous ammonia; *UF*, ultrafiltration.

[182]. The recovery of other cellulases by membrane processes was already reviewed elsewhere [194].

3.3.4 Final comments

1. Hydrolysis is affected by structural characteristics of cellulose and the presence of lignin and hemicellulose. The removal of lignin, solubilization of hemicelluloses, and decrystallization of cellulose by pretreatment strongly induce cellulose accessibility to enzyme.
2. The pretreatment of biomass by ionic liquid, addition of surfactant, applying microwave irradiation or ultrasound enhances cellulose conversion.
3. Glucose yield increases with increasing hydrolysis period but decreases with substrate loading above a certain limit. This is caused by end product inhibition and mass transfer limitation.
4. Most of the industries use distillation in downstream purification to dehydrate ethanol. But this technique is highly energy demanding and hence results in high power consumption. Energy consumption should be reduced in order to attain energy-efficient dehydration method.
5. Azeotropic and extractive distillations are industrially acceptable ethanol dehydration processes.
6. The use of highly available, low-cost glycerol as entrainer in extractive distillation dehydrates ethanol as well as minimizes energy consumption.
7. The use of ethylene glycol, *n*-pentane, or ionic liquid as mass separating agent minimizes energy consumption.

3.4 Simultaneous saccharification and fermentation

The fermentation of the hexoses (C6 sugars) and pentoses (C5 sugars) derived from the cellulose and hemicellulose portion of lignocellulosic biomass is performed by microorganisms, such as bacteria, yeasts, and fungi. Currently, it is possible to develop fermenting microorganisms genetically engineered by metabolic engineering approaches to convert both C5 and C6 sugars from the hydrolyzate into ethanol [201,202]. The following chemical equation summarizes the fermentation reaction of the six-carbon sugar, glucose:

$$C_6H_{12}O_6 \rightarrow 2C_2H_5OH + 6CO_2 + \text{heat}$$
$$\text{Glucose} \rightarrow \text{ethanol} + \text{carbon dioxide gas} + \text{heat}$$

where 0.51 kg of ethanol and 0.49 kg of carbon dioxide gas could be produced from 1 kg of glucose [203]. However, the fermentative microorganisms consume some of the glucose for cell growth, and consequently the actual yield is lower than 100% [204].

The fermentation process is performed by three main technologies: batch, fed-batch, and continuous fermentation [205]. In the batch process, regarded as a discontinuous fermentation technique, the starting materials and sporadically nutritive supplements are loaded in the fermenter with microorganisms only at the beginning of the fermentation. The fermentation is then carried out for a certain period of time, under optimal conditions of pH, temperature, O_2 supply, agitation, etc., without further addition of fresh culture medium. Successively, the fermenter contents are taken out for processing. In the fed-batch fermentation, a technique between batch and continuous fermentation, the substrate is progressively added at different time intervals throughout the course of fermentations. The continuous fermentation is an open system that involves the addition of the substrate in the fermenter and the removal of fermentation products continuously. Both addition and removal are done at the same rate, in order to maintain a constant working volume. Therefore the percentage of end product achieved in continuous fermentation is much higher than in batch and fed-batch processes.

During the bioethanol production from lignocellulosic biomass, pretreated lignocelluloses are converted to simple sugars, in hydrolysis reactors, by catabolic enzymes (*cellulases*); subsequently, the hydrolyzate is fermented to ethanol by ethanologenic yeasts, in separate units [205]. Currently, this method, known as separate enzymatic hydrolysis and fermentation (SHF), is still the main process configuration for the biofuels production from lignocellulose [206].

Usually, in SHF method, the hexose and pentose fermentations are carried out in independent reactors (Fig. 3.1): (1) the hydrolyzate obtained from hydrolysis reactors first enters the glucose fermentation reactor. The mixture, including unconverted xylose, is then distilled to remove the bioethanol from fermentation broth; (2) successively, xylose is fermented to bioethanol in a second reactor, and the bioethanol obtained is again separated from fermentation mixture by distillation technique [201].

The major advantage of SHF method is the ability to perform the saccharification and fermentation step at its own optimal conditions. For instance, the optimum temperature for most of the fermenting organisms is between 28°C and 37°C, while the optimum temperature for

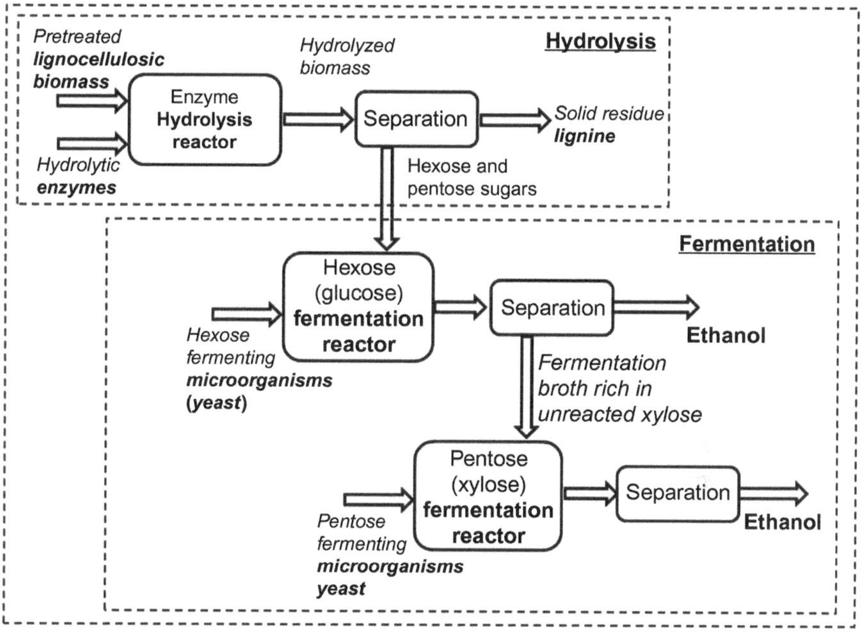

Figure 3.1 Simplified process for SHF. *SHF*, Separate enzymatic hydrolysis and fermentation.

saccharification is greater than that of fermentation, between 45°C and 50°C [205]. Moreover, the yeast could be reused after fermentation, in a SHF process [12]. However, one of the major shortcomings of SHF technique is the high production cost, due to long processing time and to great equipment costs [12]. In addition, in SHF method the possibility of microbial contaminations is high, due to long period processes, that is, 1–4 days, and the yield of ethanol is minimized by the end product inhibition of hydrolysis [207]. The cellulase activity, indeed, is inhibited by the released sugars, mainly cellobiose and glucose: a cellobiose concentration of about 6 g/L reduces the cellulase activity by 60%, while the inhibitory effect of glucose on hydrolytic enzymes is lower than that of cellobiose [205]. The enzymes could also be a possible source of contamination [208].

In order to overcome the SHF limitations, integrated conversion technologies have been developed, including SSF, simultaneous saccharification and cofermentation (SSCF), and CBP.

The SSF, SSCF, and CBP technologies combine the enzymatic hydrolysis and fermentation in one reactor, reducing the overall production

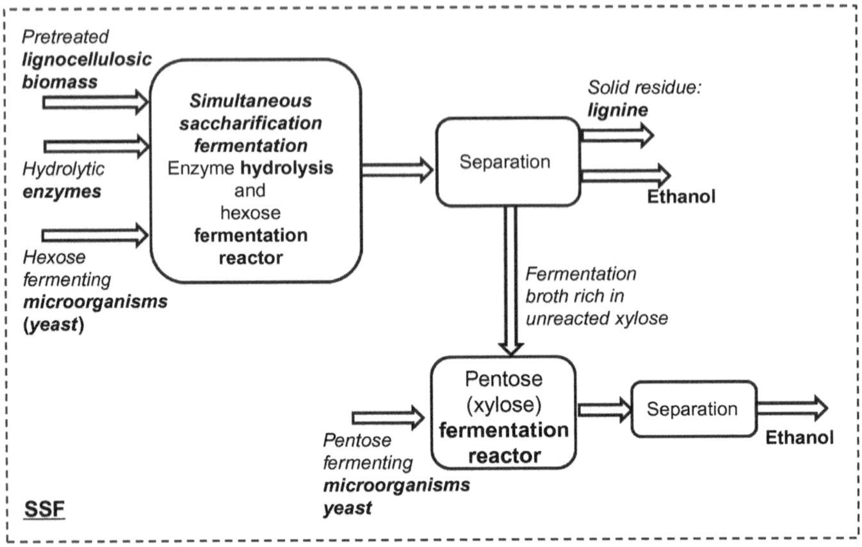

Figure 3.2 Simplified process for SSF. *SSF*, Simultaneous saccharification and fermentation.

time, operating costs, inhibitors, and increasing the hydrolysis rate [209]. However, Gauss et al. suggested to carry out the enzymatic hydrolysis and fermentation simultaneously, as early as 1976 [210]. The authors noted the low yields of glucose achieved by the fungus *T. reesei* in a traditional separate enzymatic hydrolysis, probably as a result of end product inhibition of the hydrolysis by glucose and cellobiose. Therefore they suggested that a higher overall ethanol yield was achieved during simultaneous process of enzymatic hydrolysis and fermentation, attributable to the deletion of glucose and cellobiose by the fermentation, and the consequential release of end product inhibition [211]. The main advantage to carry out saccharification and fermentation simultaneously in a single reactor (Fig. 3.2) is the possibility to rapidly convert the sugars newly formed (mainly glucose), produced by hydrolyzing enzymes, into ethanol, decreasing their buildup in the medium and alleviating feedback inhibition of cellulase [212].

A way of preventing end product inhibition by sugars cellobiose is to use commercial cellulase preparations with extra-BGL, otherwise yeasts able of fermenting cellobiose, as *Brettanomyces claussenii* [213] or recombinant *K. oxytoca* [214]. Furthermore, the several compounds contained in pretreatment hydrolyzates, which act as inhibitor of hydrolytic enzymes, may be converted by the fermenting microorganisms; for this reason an SSF process could tolerate the inhibitor formation from the pretreatment to a greater extent [215]. Therefore the rate of hydrolysis is increased, and a lower enzyme loading is required, resulting in higher ethanol yields and in a reduced risk of contamination caused by enzymes. Techno-economical assessments suggest that a 50% decrease in the enzyme loading is advantageous if the yield is less than 6%–7% and required residence time is not raised by more than 30% [216].

The risk of contamination in SSF is less than that determined in the SHF, due to the presence of ethanol in the broth that makes the reaction mixture less susceptible to the action of undesired microorganisms. Although too little attention has been paid to inhibit cellulase by produced ethanol; however, the alcohol inhibition may be a limiting factor in producing high ethanol concentration in SSF, it was indeed reported that 30 g/L ethanol reduces the enzyme activity by 25% [217]. Anyway, the SSF offers an easier operation and requires a decreased number of vessels in comparison to SHF, since no hydrolysis reactors are necessary, therefore resulting in a lower capital cost of the process [62]. The reduction of capital investment has been valued to be greater than 20% [211].

The main drawback of SSF is the different optimum temperatures of the hydrolysis and fermentation processes which means a problematic control and optimization of process parameters [218]. As discussed earlier, the optimum temperature for saccharification by cellulolytic enzymes is usually between 45°C and 50°C, whereas the fermentation step by fermenting microorganism is best done between 28°C and 37°C. For instance, the optimum temperature of *S. cerevisiae* is between 30°C and 35°C; this yeast is inactive at temperature exceeding 40°C [219]. Therefore in SHF process, the enzymatic hydrolysis temperature can be optimized regardless of the fermentation temperature, whereas a compromise is needed in SSF process [220]. Tang performed the SSF process, by using *S. cerevisiae* in the form of dry yeast, at a temperature of about 40° C, achieving a realistic compromise between the optimal temperatures for hydrolysis and fermentation [221].

Currently, hydrolysis remains the rate-limiting step in SSF. Nevertheless, given the difficulty to reduce the optimal temperature of cellulases via protein engineering, several thermotolerant bacteria and yeasts were selected for their capacity to ferment ethanol. The yeasts as *Candida acidothermophilum* and *K. marxianus* can be used in SSF with the purpose of increasing the temperature, approaching that optimal of hydrolysis [222]. In particular, many strains of *K. marxianus* appear to be very promising; indeed, these strains are able to (1) grow well at temperatures of 45°C–52°C; (2) efficiently produce ethanol at temperatures of 38°C–45°C; and (3) use several sugar substrates, such as arabinose, galactose, mannose, xylose [53].

The difficulty in implementing continuous fermentation, by recirculating and reusing the yeast, is a further obstacle of the SSF process. Indeed, owing to the presence of the lignin residues from the hydrolysis, it's proving hard to separate yeast from these solid residues after fermentation [220]. Consequently, a yield loss in an SSF process represents necessarily weakness of this method [221]. High-solid loadings are generally needed to achieve high ethanol percentages in the fermentation broths: solid loadings of pretreated biomass, greater than 30% (w/w), may be required to obtain an ethanol concentration of 4–5 wt.% that is considered a threshold level for a sustainable distillation.

The SSF process is usually carried out in a batch mode; however, the high content of solids in a bioreactor could lead to a deteriorating enzymatic activity and a viscosity increase, hampering the homogeneous and effective distribution of the enzymes in the bioreactor [220]. It was reported that the difficulty of attaining a good ethanol yields with a high loading of pretreated material contained a WIS (water insoluble solid) fraction around of 10% [211]. This problem could be overcome by a fed-batch SSF, in which a gradual or stepwise addition of hydrolyzate allows a continuous conversion of inhibitors and a gradual hydrolysis of added fibers [223].

The aim of a fermentation process is to achieve a complete assimilation by fermenting microorganisms of all sugars formerly released from lignocellulose biomass during pretreatment and hydrolysis steps. The cofermentation of hexoses and pentoses could be carried out by using of mixtures cultures of yeasts capable of assimilating both C6 and C5 sugars. However, there are significant differences with respect to temperature tolerance between the hexose-utilizing microorganisms and pentose utilizing microorganisms: usually, the first one grows faster than the latter. Consequently, the conversion efficiency of hexoses to ethanol is higher than that of pentoses [224].

The SSF process required two bioreactors and two biomass productions setup to convert C6 and C5 sugars into ethanol: in SSF bioreactor, only the hexoses fermentation is performed; pentose sugars can be fermented in another bioreactor by using different microorganisms. The

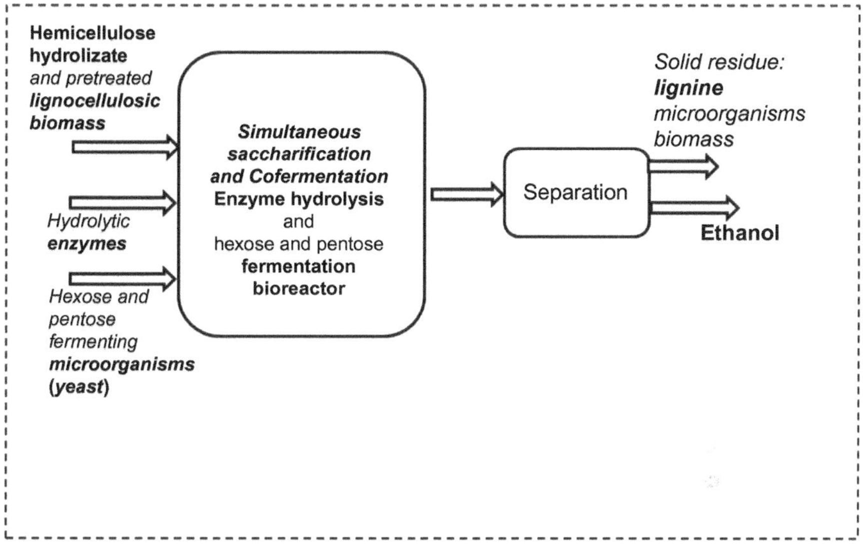

Figure 3.3 Simplified process for SSCF. *SSCF*, Simultaneous saccharification and cofermentation.

SSCF is generally superior to the SSF process, since the hexose and pentose fermentation can be achieved in a single bioreactor with a single microorganism (Fig. 3.3).

Therefore only a single fermentation step is necessary [212]. In 2016 Bondesson et al. [225] performed the ethanol production from glucose and xylose in steam-pretreated, acetic-acid-impregnated wheat straw by process design of SSCF, using a genetically modified pentose fermenting yeast strain *S. cerevisiae* [212]. In 2017 Westman et al. [226] carried out the SSCF process of steam pretreated wheat straw by using metabolically and evolutionarily engineered xylose fermenting *S. cerevisiae* strains KE6-12.A [226].

In all the previously considered processes, the enzymes are provided externally, or their production occurs in a separate unit operation. In CBP, enzymes and bioethanol are produced in a single bioreactor by a single microorganism community. Therefore in this process, also known as DMC, cellulase production, cellulose hydrolysis, and fermentation are carried out in a single step, reducing operational costs and capital investment for purchasing enzyme or its production (Fig. 3.4).

For this purpose, several thermophilic cellulolytic anaerobic bacteria are investigated, including *T. ethanolicus*, *Clostridium thermohydrosulfuricum*,

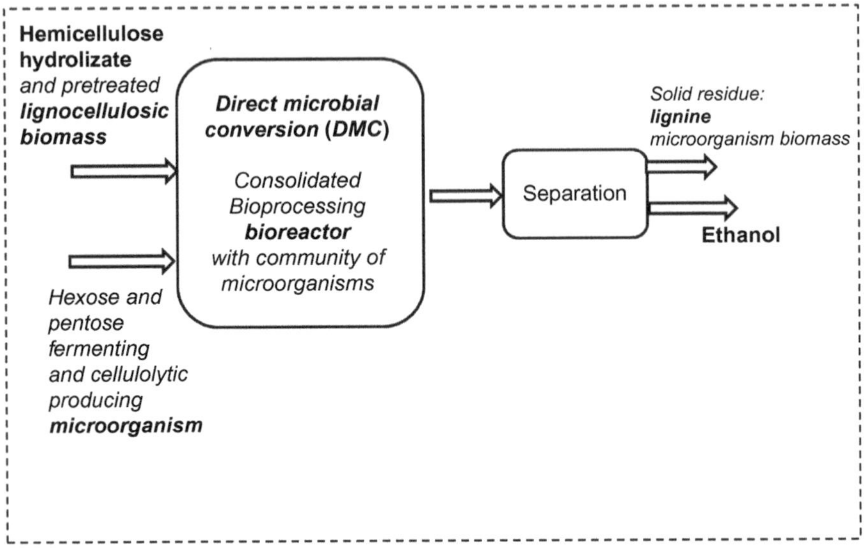

Figure 3.4 Simplified process for CBP. *CBP*, Consolidated bioprocessing.

T. mathranii, *Thermoanaerobium brockii*, and *C. thermosaccharolyticum strain*. The advantage to use thermophilic cellulolytic anaerobic bacteria compared to conventional yeasts for producing bioethanol lies in their capacity to directly employ a wide range of inexpensive biomass feedstocks and to tolerate extreme temperatures. However, one of the major drawbacks for the industrial exploitation of thermophilic cellulolytic anaerobic bacteria is their low bioethanol tolerance (<2%, v/v) [227]. Currently, several studies are focused on the development of CBP yeast for the efficient manufacture of bioethanol and on the identification and exploitation of mixed cultures able to carried out effectively the hydrolysis of lignocellulosic biomass simultaneously with fermentation step [201].

3.5 Effect of fermentation inhibitors

The rigid structure of lignocellulosic biomass hinders the enzymatic hydrolysis of biomass polysaccharides to fermentable sugars, thus preventing the biomass conversion into bioalcohols, as ethanol, and biochemicals [228]. Several different pretreatment methods (chemical, physical, electrical, biological, or a combination of those means) promote the lignocellulose breakdown, reducing recalcitrance biomass and facilitating enzymes to access their substrates [229]. However, pretreatment processes of

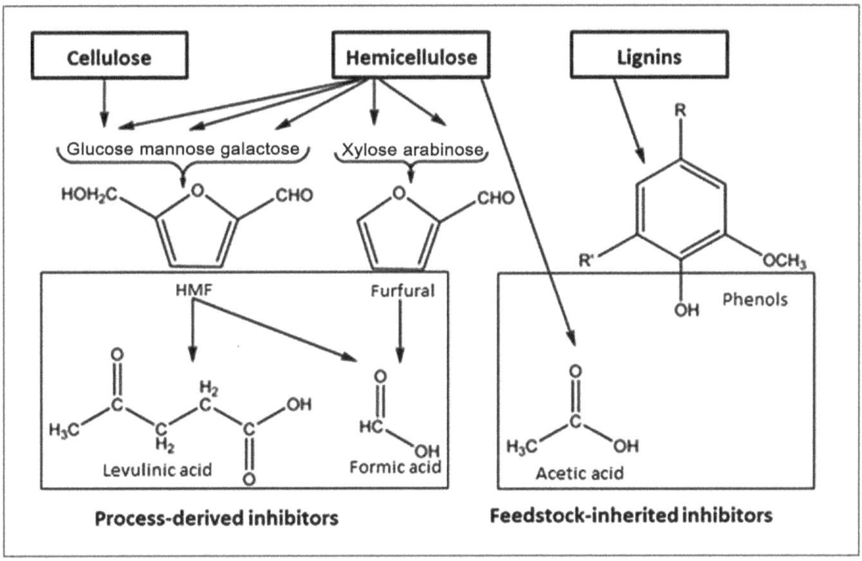

Figure 3.5 Fermentation inhibitors.

lignocellulosic materials cause increased formation of the undesirable by-products derived from lignocellulose that has the potential to inhibit microbial growth and, consequently, reduces fermentation yields [230]. The potential fermentation inhibitors, present in the liquid fraction after lignocellulosic biomass pretreatment, include furan derivatives (i.e., furfural, or 2-furaldehyde and HMF or 5-hydroxymethylfurfural), aliphatic acids (i.e., LA, FA, and acetic acid), and phenolic compounds (i.e., syringaldehyde and vanillin). These fermentation inhibitors can be grouped in two main categories (Fig. 3.5): (1) process-derived inhibitors, produced during pretreatment process (such as furan derivatives), and (2) feedstock-inherited inhibitors, in other words, naturally occurring inhibitors from the lignocellulosic biomass (primarily acetic acid and phenolic compounds) [229].

3.5.1 Furfural and 5-hydroxymethylfurfural

Pretreatment of lignocellulosic biomass is an inevitable process for the depolymerization of the holocellulosic content of biomass [231]: glucose is the only simple sugar produced by cellulose decomposition; on the other hand, the xylose is the main soluble sugar produced by hemicelluloses degradation [232]. Minor amounts of other monosaccharides, as mannose, galactose, rhamnose, and arabinose, are obtained by hemicelluloses decomposition [233].

HMF and furfural are the most process-derived inhibitors, formed by the dehydration of different hexoses and pentoses, respectively [234]. Consequently, simple sugars, such as glucose, mannose, and galactose, with the exception of rhamnose [235], can be converted to HMF, while xylose and arabinose can follow different reaction mechanisms resulting in the formation of furfural [236,237].

A study of Rasmussen et al. [238] reviewed different routes of furfural and HMF formation from xylose and glucose, respectively. The mechanism and degradation route depend on the protonation site initiating the decomposition of monosaccharides. In particular, for the degradation of xylose to furfural it was proposed one acyclic mechanism and two different types of cyclic reaction mechanisms, resulting from protonation arising from xylose protonation of either hydroxyl (1-OH or 2-OH) or the *O*-pyranose sites. For the formation of HMF from glucose they were suggested acyclic and cyclic mechanisms with or without fructose isomerization. These mechanisms are caused by 2-OH and *O*-pyranose protonation of glucose.

Furfural and HMF are commonly generated during acidic pretreatment conditions, thus at low pH, in combination with a high temperature [230,239]. Limayem [33] reports that a pretreatment of lignocellulosic material by using dilute sulfuric acid under 4 wt.% and temperatures more than 160°C determines the two furans formation in high concentrations. A furan formation of 8–25 g/kg initial lignocellulose dry weight (DW) can be reached by acid pretreatment of monocot lignocellulose, while an acid pretreatment of softwood lignocellulose can result in a furans formation of 35 g/kg initial lignocellulose DW [230]. Furthermore, as discussed by Monlau [236], thermophysical and thermochemical pretreatments release furfural and HMF at low pH (i.e., thermo-acid pretreatment); their release is instead negligible at high pH (i.e., thermo-alkaline pretreatment).

The methods of pretreatment carried out at high pH (i.e., wet oxidation, alkali hydrolysis, and AFEX) lead to a relatively low furans formation [230]. The presence of furfural and HMF is also extensively reported after the thermo-mechanicochemical pretreatment known as steam explosion [240,241], although the furan quantities produced are relatively low, about 4 g/kg initial lignocellulose DW [230]. The furans formation depends on the pretreatment method used to decompose the lignocellulose source. For instance, hemicellulose from a monocot source (grass) mainly includes xylose sugar: the furfural thus is the main furan generated in pretreated monocot; a significant amount of mannose is instead contained in the hemicellulose from softwood source: HMF thus is the most abundant furans in pretreated softwood lignocellulose [242].

According to research studies carried out by Aguilar [243] and Larsson [244], the furfural concentration in sugarcane bagasse can reach up to 5 g/L during acid hydrolysis [243], while the HMF can accumulate to 6 g/L in hydrolyzates from chipped pine wood [244]. Furfural and HMF are considered potent inhibitors of yeast cells growth and subsequent of fermentation process in a dose-dependent manner [245,246], causing the inhibition of sugar uptake rate and the decrease of ethanol production rate [247]. Anyway, they are usually toxic to many other types of organism, such as mammalian cells, fungi, and bacteria [248]. These inhibitor compounds act by blocking both different enzymes of the glycolysis pathway, that is, hexokinase, phosphofructokinase, triose-phosphate dehydrogenase [239], and nonglycolytic enzymes. Modig et al. [248] reported the furfural and HMF inhibitory effects on three dehydrogenase enzymes: alcohol dehydrogenase (ADH), aldehyde dehydrogenase (AlDH), and the pyruvate dehydrogenase (PDH) complex. The relevant reactions carried out by these different enzymes are given by the following equations:

$$\text{Acetaldehyde} + \text{NADH} + \text{H}^+ \xrightarrow{\text{ADH}} \text{NAD}^+ + \text{ethanol}$$

$$\text{Acetaldehyde} + \text{NAD}^+ \xrightarrow{\text{AlDH}} \text{acetic acid} + \text{NADH} + \text{H}^+$$

$$\text{Pyruvic acid} + \text{CoA} + \text{NAD}^+ \xrightarrow{\text{PDH}} \text{acetyl-CoA} + \text{NADH} + \text{H}^+ + \text{CO}_2$$

Furfural and HMF can act synergistically, but the microbial activity of yeast strains seems to be more sensitive to inhibition by furfural than HMF, at the same concentration [245,249]. For instance, the presence of furfural negatively affects the activities of the ADH, AlDH, and PDH enzymes; in contrast, HMF has a less inhibitory effect on AlDH and PDH enzymes compared to furfural, while the effect on ADH enzyme is similar to that of furfural [248]. However, HMF can increase the *lag phase* of microbial growth, depleting the cell growth [250]. It was reported that in cultivating *S. cerevisiae* K35 and CBS 1200, and *Z. mobilis* ATCC 10988, 5-HMF significantly inhibited cell growth and ethanol production [251].

Moreover, furan derivatives affect cell wall formation, DNA, RNA, plasmids, and/or protein synthesis, and cause the accumulation of reactive oxygen species damaging cytoskeleton, nuclear chromatin, mitochondria, and vacuole membranes [237].

Some microorganisms are capable of converting furfural and HMF to less toxic compounds [230]. For example, *S. cerevisiae* and *P. stipitis* can reduce furfural and 5-HMF, respectively, into 5-hydroxymethyl furfuryl alcohol (2,5-bis-hydroxymethylfuran, furan-2,5-dimethanol) and furfuryl alcohol (2-furanmethanol) by ADH and AlDH [252]. For most microorganisms, such as *S. cerevisiae*, 5-hydroxymethyl furfuryl- and furfuryl alcohols are less toxic than HMF and furfural, causing only slight inhibition, whereas the inhibitory effect is still acute for other microorganisms such as *Rhodosporidium toruloides* [253]. In addition, the inhibition of *P. stipitis* aerobic growth has been reported [254]. The industrially important yeasts *Tepidibacillus fermentans* and *S. cerevisiae* (primarily, under aerobic conditions), the fungi *C. ligniaria*, and the bacteria *Pseudomonas putida* and *Cupriavidus basilensis* can oxidize furfural and/or HMF to 2-furoic acid, that has a lower toxicity than furfural and furfural alcohols [230]. The HMF conversion rate is four times slower than that of furfural, probably as result of lower membrane permeability, causing a longer lag phase of the cell growth cycle, as indicated earlier [254].

3.5.2 Weak acids

Acids are categorized as either weak or strong on the basis of their acid dissociation constant, K_a. The pK_a is indeed the negative logarithmic of K_a (p$K_a = -\log 10\ K_a$) and corresponds to the pH value at which dissociated (S) and undissociated (HS) states concentrations of the acid are equivalent. The weak acids contained in lignocellulose hydrolyzates, mainly LA, FA, and acetic acid, exist in equilibrium between HS and S that highly depends on both pK_a of the particular acid and pH of the hydrolyzed [255]. In particular, HS concentration of weak acids increases when the pH value of the hydrolyzed is lower than pK_a value of weak acid, in according to the Henderson–Hasselbalch equation [254]. HS of weak acids is liposoluble and can transfer from the fermentation medium across the plasma membrane by a passive diffusion process [255] or enter into the cell through the Fps1 aquaglyceroporin channel, as observed for acetic acid [256]. In according to the uncoupling theory proposed by Russell in 1992 [257], HS of weak acids dissociates inside the microbial cells into charged anions and protons, due to their higher intracellular pH (typically well above the pK_a of organic acids). The protons release in cytoplasm causes the acidification intracellular [257,258]. In order to keep constant intracellular pH (pHi), the excessive proton accumulation is pumped out of the cells by various mechanisms, including proton translocation with the plasma membrane H^+-ATPase mediated by ATP hydrolysis. It means a depletion of ATP levels that could be used to form biomass and, consequently, inhibit a cellular growth [254]. The increase of the PHi acidity affects the purine bases integrity, causing DNA and RNA damage and denaturation of essential enzymes inside the cell, such as glycolysis enzymes and, in particular, the phosphofructokinase that is sensitive to low pH in vitro [259,260]. Although HS of weak acids is lipophilic, the polar anionic forms of weak acids are lipophobic and cannot readily across the lipid bilayer of membrane plasma. On the *anion accumulation theory*, anionic forms of the weak acids are captured within the cells where can reach toxic concentrations [254]. The buildup of anions formed by dissociation of weak acids results in growth of potassium ions transport within the cell and, consequently, in a turgor pressure increased. For the purpose of maintaining a constant osmotic pressure and cell volume, more glutamate is translocated out of the cell in the opposite direction to the potassium, disrupting the cytoplasm osmolarity, impairing important metabolic processes and reducing the cell's growth potential and viability [261]. The pHi acidification and the anions accumulation could also affect different enzymes involved in protein synthesis and decrease the aromatic amino acid uptake from culture medium through inhibition of amino acid permease [262]. However, alterations of the structure of the plasma membrane or cell wall and challenges to cellular energy balance while maintaining the pHi could also result in the cellular growth inhibition by weak organic acids [263].

Similar to furfural toxicity, low concentrations of weak acids (<100 mmol/L) were shown to enhance ethanol yield to pH value of around 5.5, whereas higher concentrations of weak acids are correlated with strong inhibitory effects on cell growth and fermentation process [262]. The weak acids toxicity depend also on their structure; for instance, the toxicity of two weak acids, with equal pK_a values, increases with their carbon chain length, affecting lipophilicity. The exact mechanisms that yeast uses to adapt to weak acids are still unknown. *S. cerevisiae* utilizes the plasma membrane H^+-ATPase Pma1p to move protons out of the cell, increasing the proton pumping capacity of the cell and induces the ATP-binding cassette anion efflux pump Pdr12p to export the remaining anions out of the cell. Moreover, *S. cerevisiae* seems to induce certain heat shock genes at low PHi and in the presence of a weak acid as sorbic acid [250].

The main weak acids contained in lignocellulosic hydrolyzates are acetic acid, FA, and LA. The formation of acetic acid (AcH) is primarily due to the hemicellulose deacetylation, while FA and LA arise as acid-catalyzed thermochemical degradation products from polysaccharides: FA is a degradation product of furfural and HMF, while LA is obtained by HMF breakdown [247]. The content of AcH, FA, and LA in hydrolyzates or slurries is very variable and depends on the feedstock and the severity of the pretreatment process. For example, feedstocks as agricultural residues and hardwood with a high percentage of acetylated xylan provide greater concentrations of weak acids than softwood [247]. Differences in toxicity between AcH, FA, and LA are correlated to differences in permeability of the membrane, or in toxicity of the acids anionic form within the cell [250].

3.5.2.1 Acetic acid

The pK_a of AcH (CH_3COOH) is 4.75 at 25°C; thus at low pH values of the medium, it is present in the undissociated form able to diffuse across the plasma membrane. Inside the cell, HS of AcH dissociates, discharging protons and anions in the cytoplasm and inhibiting cell activity and growth [264]. The AcH toxicity varies depending on the cultivation conditions employed in the fermentative process; for this reason, it is necessary to consider not only concentration of weak acid in the hydrolyzates but also the oxygen concentration and the pH of the medium, and the presence of other toxic compounds in the medium [264]. A research work by van Zyl et al. [265], in which *P. stipitis* was used to obtain ethanol from sugarcane bagasse hemicellulosic hydrolyzate, reported a drop to 50% in the ethanol production at pH 6.5 and with AcH concentrations of about 15 g/L. However, the same drop in ethanol concentration was observed at pH 5.1 when the AcH concentration was only 1 g/L [265]. According to Felipe et al., AcH concentrations more than 3 g/L were dangerous for the xylose fermentation to xylitol by *Candida guilliermondii*, while ethanol production was stimulated in medium free of other toxic compounds by AcH concentrations up to 10 g/L [266]. The weak acids as benzoic and sorbic acid cannot be metabolized by *S. cerevisiae*, causing membrane damage and severe oxidative stress under aerobic conditions. On the contrary, *S. cerevisiae* cells are characterized by the ability to grow on acetic acid medium. Under aerobic conditions the AcH is found primarily in a dissociated form as acetate anion. The acetate enters the plasma membrane by two specific transport proteins encoded by ADY2 (electroneutral proton symport transporter) and JEN1 (monocarboxylate transporter). The peroxisomal or cytosolic acetyl−CoA synthetases convert acetate to acetyl−CoA, that is, in turn oxidized in mitochondria through the Krebs cycle or consumed in the anabolic pathway, called glyoxylate cycle [267]. However, *S. cerevisiae* cells in a glucose-containing media are sensitive to AcH stress and unable to metabolize this weak acid, due to the activation of pathways for glucose repression [267]. Differently, the spoilage yeast *Zygosaccharomyces bailii* showed high resistance to weak acids in media containing glucose and, in particular, the capacity to metabolize AcH in the presence of glucose [268].

3.5.2.2 Formic acid

FA (HCOOH), contained in lignocellulosic hydrolyzates at a typical concentration of about 1.4 g/L (30 mM), can be formed when furfural and HMF are broken down at high temperatures of hydrolysis [269]. The pK_a value of FA, equal to 3.75 at 20°C, is considerably lower than those of AcH (4.75 at 25°C) and LA (4.66 at 25°C); therefore FA toxicity in yeast strains is more severe than that observed for AcH and LA [269−271]. The toxic effect on *S. cerevisiae* increases in the order AcH $<$ LA $<$ FA. The undissociated form of FA should be present in lower concentrations at the same pHi, and subsequently be less harmless to the cells. However, the increased FA toxicity seems to be related to a smaller molecule size that facilitates its diffusion across the plasma membrane, causing its higher anion toxicity [269]. Oshoma et al. [269] investigated the FA tolerance in 7 non−*S. cerevisiae* yeast strains. In particular, *S. arboricolus* exhibited a higher FA tolerance than other strains, confirmed by a series of fermentations. Then, this yeast could be used as a novel bioethanol producing strain or as a new source of gene donor to other inhibitor-sensitive strains that are able to produce high ethanol levels [269].

3.5.2.3 Levulinic acid

According to the National Renewable Energy Laboratory (Denver, CO, United States), LA $C_5H_8O_3$ represents one of the 12 keys sugar-derived platform chemicals that can be produced from biomass, in order to obtain a series of biochemicals, including solvents, biofuels, food/flavoring/fragrance components, chemical intermediates, plasticizers, polymers, herbicides, antifreeze, and pharmaceutical agents, by different chemical reaction (such as esterification, halogenation, hydrogenation, and condensation) [272]. The production of LA is mainly achieved by acid-catalyzed dehydration of hexose sugars to HMF, followed by a rehydration step of HMF to LA [271].

An often overlooked pathway in the LA production from lignocellulosic material is derived from the furfural. The furfural can be reduced to furfuryl alcohol that can in turn be converted to LA by acid hydrolysis [273,274].

In addition, the hydroxymethylation process of furfural with formaldehyde determines the HMF production. The HMF can be subsequently converted to LA [275,276].

3.5.3 Phenols

Phenolic compounds, or phenols, are the most abundant secondary metabolites in plants, biosynthesized via the PP, shikimate (SK), and phenylpropanoid (PhP) pathways in plants [277].

In the synthesis of phenolic compounds in plants the glucose is initially committed in the PP pathway, and the glucose-6-phosphate is irreversibly dehydrogenated to ribulose-5-phosphate by glucose-6-phosphate dehydrogenase (G6PD). In the last reactions of the PP pathway, ribose-5-phosphate is converted to erythrose-4-phosphate that enters in the SK pathway along with phosphoenol pyruvate obtained from glycolysis [278].

The phenylalanine produced in the SK pathway is used through the PhP pathway to generate phenolic compounds [278]. The phenols range from simple phenolic molecules to highly polymerized compounds and include various classes of compounds, such as phenolic acids, colored anthocyanins, simple, and complex flavonoids [279]; their structure comprises an aromatic ring and various functional groups such as aldehyde, ketone, acid and the side groups such as methoxy and hydroxyl [277].

The phenols make up the lignin building blocks and play a key role in plant defense systems against microbial infections, in addition to being involved in several plant physiological processes [277]. Different phenols are indeed formed as residues chiefly by lignin degradation during biomass pretreatment, depending on both the plant source and the pretreatment method [278]. In general, the phenolic compounds formed during pretreatment processes are a mix of phenolic acids, phenolic aldehydes, phenolic alcohols, and phenolic ketones [277].

Phenolic compound commonly found in a typical spruce hydrolyzates includes gallic, vanillic, syringic and ferulic acids, catechine, picein and pungenin, taxifolin, and coniferyl aldehyde [255]. All these phenolics are strong inhibitors to cellulolytic enzymes and to yeast cells, affecting negatively the hydrolysis and fermentation processes [240].

The phenols are also indicated as more toxic compounds than furan aldehydes and weak acids, even at lower concentrations. The high toxicity of phenols is related to their low molecular weight (MW) that allows them able to cross cell membranes and damage internal structures by cellular morphological changes. Phenolics indeed can perturb the cell membrane, increasing its fluidity and causing a significant drop of intracellular potassium concentrations. They can also promote the DNA breakdown and inhibition of RNA and protein synthesis [237].

The toxicity of phenolics is very variable, and it not necessarily depends on the functional groups. Adeboye et al. [277] investigated the severity of the inhibitory effects of 13 phenolic compounds (commonly present in spruce hydrolyzates) on the *S. cerevisiae* growth, biomass, and ethanol yield. They observed different toxicity limits even between phenolic compounds that share the same functional groups. For instance, toxicity limits of ferulic and *p*-coumaric acids are 1.8 and 9.7 mM, respectively. Moreover, the concentration and the nature of phenolic compounds affect the fermentability of substrates rich in phenols. The inhibition mechanisms among phenols are dissimilar and cannot be identified on the basis of main classes of phenolic compounds (aldehydes, acids, alcohols, and ketones) [277]. Li et al. [280] investigated the different inhibitor effects of three phenols, such as syringaldehyde, vanillin, and phenol, on cell growth and fermentation of glucose and xylose. The vanillin showed the strongest inhibitory effect on glucose fermentation, followed by phenol and then syringaldehyde, while phenol had the strongest inhibition on xylose fermentation among the three phenols considered. Moreover, the results obtained from this research work indicated that low MW phenols inhibit xylose fermentation more strongly than high MW phenols [280].

3.5.4 Interaction effects

In according to work of Palmqvist et al. [250], in the presence of mixtures of acetic acid and furfural in the fermentation medium, the specific growth rate decreased more than the sum of the individual effect, suggesting that acetic acid and furfural interact antagonistically on cell growth [250]. In the presence of mixtures of acetic acid, FA, LA, furfural, and HMF in the fermentation medium, the ethanol yield decrease was only slight compared to a reference fermentation [244], indicating that these compounds inhibit cell growth more than ethanol formation [250].

3.5.5 Other inhibitor compounds

Other strong inhibitory compounds, generated from lignocellulose feedstocks and during pretreatment processes, include quinones and small aliphatic aldehydes [281]. The ethanol, generated during fermentation, also shows a strong inhibitor effect on the viability, growth, glucose transport systems, and proton fluxes of *S. cerevisiae*. Nevertheless, the concentrations of ethanol tolerate from ethanologenic microbes *S. cerevisiae* and *Z. mobilis* correspond to 18% and 12%, respectively [247].

The formation of dark-colored substances, known as humins, during the thermal hydrolysis of glucose, has been reported relatively recently. The humins can be formed from reaction of HMF with glucose and/or via reaction of HMF with 2,5-dioxo-6-hydroxy-hexanal (hydrated HMF) and subsequent polymerization. However, the humins impact on cellulolytic enzymes and/or yeast during ethanol production remain at present uncertain [247].

3.5.6 Strategies for minimizing inhibitor effects

Different strategies can be employed to counteract problems with fermentation inhibitors.

1. Several natural varieties of feedstocks with low recalcitrance can be identified and used for bioconversion of lignocellulose in biofuels. The use of feedstocks with relatively low recalcitrance allows to perform pretreatment under mild conditions and without any addition of acid catalysts, in order to reduce the concentrations of furan aldehydes and phenol in lignocellulose hydrolyzates [282]. The feedstock engineering is another strategy to generate novel bioenergy crops with low recalcitrance and then obtain a reduced release of inhibitors during pretreatment processes.
2. Detoxification or conditioning of lignocellulosic hydrolyzates and slurries is one of the most powerful ways to render strongly inhibitory hydrolyzates as fermentable as reference fermentations with the same amounts of sugar but no inhibitors [283]. This process involves

different treatments that include (1) the use of chemical additives (e.g., alkali), reducing agents, and polymers; (2) enzymatic treatment; (3) liquid–liquid extraction; and (4) liquid–solid extraction (e.g., ion-exchange and treatment with activated carbon) [234]. The detoxification is selective and targets inhibitors rather than fermentable sugars [284]. However, many detoxification methods require more chemicals and frequently a separate process step [237].

3. The bioabatement, or microbial treatment, is another potential method to remove inhibitory compounds from lignocellulose hydrolyzate, improving both fermentability and enzymatic hydrolysis. Nichols et al. [285] reported the abatement of hydrolyzate by inoculation with the *C. ligniaria* NRRL30616, an Ascomycete able to metabolize furfural and HMF, allowed subsequent yeast fermentation of cellulose to ethanol [285].
4. An important approach to tackle the inhibition problem is applying various culture schemes. The strategies as SSF, fed batch, and CBP are being employed for SSF. In these methods the fermenting microorganism also contributes to supply of enzyme. However, the costs related to designing fermentation process, fermenting microorganism, enzyme loading, inoculums size, and time of fermentation under the inhibitory conditions have significant impact on the bioconversion process [234].
5. Another possible approach for counteracting problems of inhibitors in processing of lignocellulose is the screening of microorganisms collected from natural or industrial environments with high resistance to inhibitors. The strains of *S. cerevisiae*, highly resistant for aliphatic carboxylic acids and furans aldehydes, were isolated from the grape marc in a winery [286]. An important drawback of this approach is the specific productivity of the microorganism: the selection of microorganism, indeed, ought to be made primarily on the basis of their specific productivity and product yields.
6. Evolutionary engineering is a technique that permits to generate ethanol-fermenting microorganisms with enhanced inhibitor resistance [287].
7. Genetic engineering can be used to obtain recombinant microorganisms with enhanced resistance to hydrolyzates of lignocellulose [55].

Even if the various approaches proposed can be used to tackle the inhibitors problems, the chemical detoxification is currently the most effective way to achieve a fermentation level comparable with that of a medium devoid of inhibitor compounds.

3.6 Conclusion

During recent years a significant work has been done to improve efficiency and yield of the production of bioethanol. A growing interest has been demonstrated to the innovation of pretreatment methods, in order to increase sustainability and productivity with reduction of costs.

Commercial plants have been realized but both the processes innovation and intensification and the research of new no-food competitive raw matter make bioethanol production an interesting field of research.

References

[1] M.R. Ladisch, K. Dyck, Dehydration of ethanol: new approach gives positive energy balance, Science 205 (4409) (1979) 898–900.

[2] T. Neelakandan, G. Usharani, Optimisation and production of bioethanol from Cashew apple juice using yeast cells by *Saccharomyces cerevisiae*, Am. Eurasian J. Sci. Res. 4 (2) (2009) 85–88.

[3] I. Ballesteros, M.J. Negro, J.M. Oliva, A. Cabanas, P. Manzanares, M. Ballesteros, Ethanol production from steam-explosion pretreated Wheat Straw, Appl. Biochem. Biotechnol. 496 (2006) (2006) 129–132.

[4] R. Ortiz, K.D. Sayre, B. Govaerts, R. Gupta, G.V. Subbarao, T. Ban, et al., Climate change: can wheat beat the heat? Agric. Ecosyst. Environ. 126 (2008) 46–58.

[5] J.S. Van Dyk, B.I. Pletschke, A review of lignocelluloses bioconversion using enzymatic hydrolysis and synergistic cooperation between enzymes: factors affecting enzymes, conversion and synergy, Biotechnol. Adv. 30 (2012) 1458–1480.

[6] F.M. Girio, C. Fonseca, F. Carvalheiro, L.C. Duarte, S. Marques, R. Bogel-Lukasik, Hemicelluloses for fuel ethanol: a review, Bioresour. Technol. 101 (2010) 4775–4800.

[7] M. Meinita, J.Y. Kang, G.T. Jeong, H. Koo, S. Park, Y.K. Hong, Bioethanol production from the acid hydrolysate of the carrageenophyte *Kappaphycus alvarezii* (*cottonii*), J. Appl. Phycol. 24 (2012) 857–862.

[8] C.S. Jones, S.P. Mayfield, Algae biofuels: versatility for the future of bioenergy, Curr. Opin. Biotechnol. 23 (2012) 346–351.

[9] S. Kumar, R. Gupta, G. Kumar, D. Sahoo, R.C. Kuhad, Bioethanol production from *Gracilaria verrucosa*, a red alga, in a biorefinery approach, Bioresour. Technol. 135 (2013) 150–156.

[10] C.G. Boeriu, D. Bravo, R.J.A. Gosselink, J.E.G. Van D, Characterisation of structure dependent functional properties of lignin with infrared spectroscopy, Ind. Crop Prod. 20 (2004) 205–218.

[11] M. Norgren, H. Edlund, Lignin: recent advances and emerging applications, Curr. Opin. Colloid 19 (2014) 406–419.

[12] N. Sarkar, S.K. Ghosh, S. Banarjee, K. Aikat, Bioethanol production from agricultural wastes: an overview, Renew. Energy 37 (1) (2012) 19–27.

[13] H.T. Tan, K.T. Lee, Understanding the impact of ionic liquid pretreatment on biomass and enzymatic hydrolysis, Chem. Eng. J. 183 (2012) 448–458.

[14] A. Casas, M. Oliet, M.V. Alonso, F. Rodriguez, Dissolution of *Pinus radiata* and *Eucalyptus globulus* woods in ionic liquids under microwave radiation: lignin regeneration and characterization, Sep. Purif. Technol. 97 (2012) 113–122.

[15] N.L. Mai, K. Ahn, Y.-M. Koo, Methods for recovery of ionic liquids—a review, Process Biochem. 49 (2014) 872–881.
[16] T.-Q. Yuan, T.-T. You, W. Wang, F. Xu, R.-C. Sun, Synergistic benefits of ionic liquid and alkaline pretreatments of poplar wood. Part 2: Characterization of lignin and hemicelluloses, Bioresour. Technol. 136 (2013) 345–350.
[17] N. Sun, M. Rahman, Y. Qin, M.L. Maxim, H. Rodriguez, R.D. Rogers, Complete dissolution and partial delignification of wood in the ionic liquid 1-ethyl-3-methylimidazolium acetate, Green. Chem. 11 (2009) 646–655.
[18] L. Moghaddam, Z. Zhang, R.M. Wellard, J.P. Bartley, I.M. O'Hara, W.O.S. Doherty, Characterisation of lignins isolated from sugarcane bagasse pretreated with acidified ethylene glycol and ionic liquids, Biomass-Bioenergy 70 (2014) 498–512.
[19] A. Pinkert, D.F. Goeke, K.N. Marsh, S. Pang, Extracting wood lignin without dissolving or degrading cellulose: investigations on the use of food additive-derived ionic liquids, Green. Chem. 13 (2011) 3124–3136.
[20] N. Srivastava, R. Rawat, H. Singh Oberoi, P.W. Ramteke, A review on fuel ethanol production from lignocellulosic biomass, Int. J. Green Energy 12 (2015) 949–960.
[21] D. Moreno, D. Ibarra, P. Alvira, E. Tomás-Pejó, M. Ballesteros, A review of biological delignification and detoxification methods for lignocellulosic bioethanol production, Crit. Rev. Biotechnol. 35 (2015) 342–354.
[22] A.K. Chandel, C. Es, R. Rudravaram, M.L. Narasu, L.V. Rao, P. Ravindra, Economics and environmental impact of bioethanol production technologies: an appraisal, Biotechnol. Mol. Biol. Rev. 2 (2007) 14–32.
[23] R. Yamada, N. Taniguchi, T. Tanaka, C. Ogino, H. Fukuda, A. Kondo, Direct ethanol production from cellulosic materials using a diploid strain of *Saccharomyces cerevisiae* with optimized cellulase expression, Biotechnol. Biofuels 4 (2011) 1–8.
[24] T. Treebupachatsakul, K. Shioya, H. Nakazawa, T. Kawaguchi, Y. Morikawa, Y. Shida, et al., Utilization of recombinant *Trichoderma reesei* expressing *Aspergillus aculeatus* β-glucosidase I (JN11) for a more economical production of ethanol from lignocellulosic biomass, J. Biosci. Bioeng. 120 (2015) 657–665.
[25] B.G. Schuster, M.S. Chinn, Consolidated bioprocessing of lignocellulosic feedstocks for ethanol fuel production, Bioenergy Res. 6 (2013) 416–435.
[26] R.R. Singhania, A.K. Patel, R.K. Sukumaran, C. Larroche, A. Pandey, Role and significance of beta-glucosidases in the hydrolysis of cellulose for bioethanol production, Bioresour. Technol. 127 (2013) 500–507.
[27] T. Hasunuma, F. Okazaki, N. Okai, K.Y. Hara, J. Ishii, A. Kondo, A review of enzymes and microbes for lignocellulosic biorefinery and the possibility of their application to consolidated bioprocessing technology, Bioresour. Technol. 135 (2013) 513–522.
[28] M.G. Tabka, I. Herpoël-Gimbert, F. Monod, M. Asther, J.C. Sigoillot, Enzymatic saccharification of wheat straw for bioethanol production by a combined cellulase xylanase and feruloyl esterase treatment, Enzyme Microb. Technol. 39 (2006) 897–902.
[29] R. Arora, S. Behera, S. Kumar, Bioprospecting thermophilic/thermotolerant microbes for production of lignocellulosic ethanol: a future perspective, Renew. Sustain. Energy Rev. 51 (2015) 699–717.
[30] J.K. Saini, R. Saini, L. Tewari, Lignocellulosic agriculture wastes as biomass feedstocks for second-generation bioethanol production: concepts and recent developments, 3 Biotech. 5 (2015) 337–353.
[31] A. Bhalla, N. Bansal, S. Kumar, K.M. Bischoff, R.K. Sani, Improved lignocellulose conversion to biofuels with thermophilic bacteria and thermostable enzymes, Bioresour. Technol. 128 (2013) 751–759.

[32] Y. Lin, S. Tanaka, Ethanol fermentation from biomass resources: current state and prospects, Appl. Microbiol. Biotechnol. 69 (2006) 627–642.
[33] A. Limayem, S.C. Ricke, Lignocellulosic biomass for bioethanol production: current perspectives, potential issues and future prospects, Prog. Energy Combust. Sci. 38 (2012) 449–467.
[34] F.B. Pereira, A. Romaní, H.A. Ruiz, J.A. Teixeira, L. Domingues, Industrial robust yeast isolates with great potential for fermentation of lignocellulosic biomass, Bioresour. Technol. 161 (2014) 192–199.
[35] R. Lamed, J.G. Zeikus, Ethanol production by thermophilic bacteria: relationship between fermentation product yields of and catabolic enzyme activities in *Clostridium thermocellum* and *Thermoanaerobium brockii*, J. Bacteriol. 144 (1980) 569–578. http://www.pubmedcentral.nih.gov/articlerender.fcgi?artid = 294704&tool = pmcentrez&rendertype = abstract%5Cn%3CGo to ISI%3E://WOS:A1980KQ60500011.
[36] M.P. Taylor, K.L. Eley, S. Martin, M.I. Tuffin, S.G. Burton, D.A. Cowan, Thermophilic ethanologenesis: future prospects for second-generation bioethanol production, Trends Biotechnol. 27 (2009) 398–405.
[37] D.G. Olson, R. Sparling, L.R. Lynd, Ethanol production by engineered thermophiles, Curr. Opin. Biotechnol. 33 (2015) 130–141.
[38] E.A. Da Silva Filho, H.F. De Melo, D.F. Antunes, S.K.B. Dos Santos, A. Do Monte Resende, D.A. Simões, et al., Isolation by genetic and physiological characteristics of a fuel-ethanol fermentative *Saccharomyces cerevisiae* strain with potential for genetic manipulation, J. Ind. Microbiol. Biotechnol. 32 (2005) 481–486.
[39] A.J. Mattam, A. Kuila, N. Suralikerimath, N. Choudary, P.V.C. Rao, H.R. Velankar, Cellulolytic enzyme expression and simultaneous conversion of lignocellulosic sugars into ethanol and xylitol by a new *Candida tropicalis* strain, Biotechnol. Biofuels 9 (2016) 1–12.
[40] C. Laluce, A.C.G. Schenberg, J.C.M. Gallardo, L.F.C. Coradello, S.R. Pombeiro-Sponchiado, Advances and developments in strategies to improve strains of *Saccharomyces cerevisiae* and processes to obtain the lignocellulosic ethanol—a review, Appl. Biochem. Biotechnol. 166 (2012) 1908–1926.
[41] J. Zaldivar, J. Nielsen, L. Olsson, Fuel ethanol production from lignocellulose: a challenge for metabolic engineering and process integration, Appl. Microbiol. Biotechnol. 56 (2001) 17–34.
[42] A. Aristidou, M. Penttilä, Metabolic engineering applications to renewable resource utilization, Curr. Opin. Biotechnol. 11 (2000) 187–198.
[43] A. Matsushika, H. Inoue, T. Kodaki, S. Sawayama, Ethanol production from xylose in engineered *Saccharomyces cerevisiae* strains: current state and perspectives, Appl. Microbiol. Biotechnol. 84 (2009) 37–53.
[44] K. Deanda, M. Zhang, C. Eddy, S. Picataggio, Development of an arabinose-fermenting *Zymomonas mobilis* strain by metabolic pathway engineering, Microbiology 62 (1996) 4465–4470.
[45] B.S. Dien, M.A. Cotta, T.W. Jeffries, Bacteria engineered for fuel ethanol production: current status, Appl. Microbiol. Biotechnol. 63 (2003) 258–266.
[46] L.O. Ingram, T. Conway, D.P. Clark, G.W. Sewell, J.F. Preston, Genetic engineering of ethanol production in *Escherichia coli*, Appl. Environ. Microbiol. 53 (1987) 2420–2425.
[47] L.O. Ingram, P.F. Gomez, X. Lai, M. Moniruzzaman, B.E. Wood, L.P. Yomano, et al., Metabolic engineering of bacteria for ethanol production, Biotechnol. Bioeng. 58 (1998) 204–214.
[48] K. Ohta, D.S. Beall, J.P. Mejia, K.T. Shanmugam, L.O. Ingram, Genetic improvement of *Escherichia coli* for ethanol production: chromosomal integration of

Zymomonas mobilis genes encoding pyruvate decarboxylase and alcohol dehydrogenase II, Appl. Environ. Microbiol. 57 (1991) 893–900.
[49] M. Zhang, C. Eddy, K. Deanda, M. Finkelstein, S. Picataggio, Metabolic engineering of a pentose metabolism pathway in ethanologenic *Zymomonas mobilis*, Science (80-.) 267 (1995) 240–243.
[50] B.C. Saha, N.N. Nichols, M.A. Cotta, Ethanol production from wheat straw by recombinant *Escherichia coli* strain FBR5 at high solid loading, Bioresour. Technol. 102 (2011) 10892–10897.
[51] T. Jojima, R. Noburyu, M. Sasaki, T. Tajima, M. Suda, H. Yukawa, et al., Metabolic engineering for improved production of ethanol by *Corynebacterium glutamicum*, Appl. Microbiol. Biotechnol. 99 (2014) 1165–1172.
[52] B. Hahn-Hägerdal, K. Karhumaa, C. Fonseca, I. Spencer-Martins, M.F. Gorwa-Grauslund, Towards industrial pentose-fermenting yeast strains, Appl. Microbiol. Biotechnol. 74 (2007) 937–953.
[53] L. Viikari, M. Alapuranen, T. Puranen, J. Vehmaanperä, M. Siika-Aho, Biofuels, 2007.
[54] L. de Figueiredo Vilela, V.P.G. de Araujo, R. de, S. Paredes, E.P. da, S. Bon, et al., Enhanced xylose fermentation and ethanol production by engineered *Saccharomyces cerevisiae* strain, AMB Express 5 (2015) 1–7.
[55] M.M. Demeke, F. Dumortier, Y. Li, T. Broeckx, M.R. Foulquié-Moreno, J.M. Thevelein, Combining inhibitor tolerance and D-xylose fermentation in industrial *Saccharomyces cerevisiae* for efficient lignocellulose-based bioethanol production, Biotechnol. Biofuels. 6 (2013) 1–17.
[56] M.M. Demeke, H. Dietz, Y. Li, M.R. Foulquié-Moreno, S. Mutturi, S. Deprez, et al., Development of a D-xylose fermenting and inhibitor tolerant industrial *Saccharomyces cerevisiae* strain with high performance in lignocellulose hydrolysates using metabolic and evolutionary engineering, Biotechnol. Biofuels 6 (2013) 1–24.
[57] A. Eliasson, C. Christensson, C.F. Wahlbom, B. Hahn-Hägerdal, Anaerobic xylose fermentation by recombinant *Saccharomyces cerevisiae* carrying XYL1, XYL2, and XKS1 in mineral medium chemostat cultures, Appl. Environ. Microbiol. 66 (2000) 3381–3386.
[58] N.W.Y. Ho, Z. Chen, A.P. Brainard, Genetically engineered saccharomyces yeast capable of effective cofermentation of glucose and xylose, Appl. Environ. Microbiol. 64 (1998) 1852–1859. Available from: http://aem.asm.org/content/64/5/1852%5Cnhttp://aem.asm.org/content/64/5/1852.full.pdf%5Cnhttp://aem.asm.org/content/64/5/1852.short.
[59] S. Katahira, A. Mizuike, H. Fukuda, A. Kondo, Ethanol fermentation from lignocellulosic hydrolysate by a recombinant xylose- and cellooligosaccharide-assimilating yeast strain, Appl. Microbiol. Biotechnol. 72 (2006) 1136–1143.
[60] S.R. Kim, Y.-C. Park, Y.-S. Jin, J.-H. Seo, Strain engineering of *Saccharomyces cerevisiae* for enhanced xylose metabolism, Biotechnol. Adv. 31 (2013) 851–861.
[61] J.K. Ko, Y. Um, H.M. Woo, K.H. Kim, S.-M. Lee, Ethanol production from lignocellulosic hydrolysates using engineered *Saccharomyces cerevisiae* harboring xylose isomerase-based pathway, Bioresour. Technol. 209 (2016) 290–296.
[62] P. Kötter, M. Ciriacy, Xylose fermentation by *Saccharomyces cerevisiae*, Appl. Microbiol. Biotechnol. 38 (1993) 776–783.
[63] A. Matsushika, H. Inoue, K. Murakami, O. Takimura, S. Sawayama, Bioethanol production performance of five recombinant strains of laboratory and industrial xylose-fermenting *Saccharomyces cerevisiae*, Bioresour. Technol. 100 (2009) 2392–2398.
[64] A. Matsushika, H. Inoue, S. Watanabe, T. Kodaki, K. Makino, S. Sawayama, Efficient bioethanol production by a recombinant flocculent *Saccharomyces cerevisiae*

strain with a genome-integrated NADP + -dependent xylitol dehydrogenase gene, Appl. Environ. Microbiol. 75 (2009) 3818–3822.
[65] A. Matsushika, S. Watanabe, T. Kodaki, K. Makino, S. Sawayama, Bioethanol production from xylose by recombinant *Saccharomyces cerevisiae* expressing xylose reductase, $NADP^+$-dependent xylitol dehydrogenase, and xylulokinase, J. Biosci. Bioeng. 105 (2008) 296–299.
[66] B. Petschacher, B. Nidetzky, Altering the coenzyme preference of xylose reductase to favor utilization of NADH enhances ethanol yield from xylose in a metabolically engineered strain of *Saccharomyces cerevisiae*, Microb. Cell. Fact. 7 (2008) 1–12.
[67] A. Romaní, F. Pereira, B. Johansson, L. Domingues, Metabolic engineering of *Saccharomyces cerevisiae* ethanol strains PE-2 and CAT-1 for efficient lignocellulosic fermentation, Bioresour. Technol. 179 (2015) 150–158.
[68] M. Tantirunghkij, N. Nakashima, T. Seki, T. Yoshida, Construction of xylose-assimilating *Saccharomyces cerevisiae*, J. Ferment. Bioeng. 75 (1993) 83–88.
[69] J. Becker, E. Boles, A modified *Saccharomyces cerevisiae* strain that consumes L-arabinose and produces ethanol, Appl. Environ. Microbiol. 69 (2003) 4144–4150.
[70] T. Inaba, D. Watanabe, Y. Yoshiyama, K. Tanaka, J. Ogawa, H. Takagi, et al., An organic acid-tolerant HAA1-overexpression mutant of an industrial bioethanol strain of *Saccharomyces cerevisiae* and its application to the production of bioethanol from sugarcane molasses, AMB Express 3 (2013) 1–7.
[71] A. Edgardo, P. Carolina, R. Manuel, F. Juanita, B. Jaime, Selection of thermotolerant yeast strains *Saccharomyces cerevisiae* for bioethanol production, Enzyme Microb. Technol. 43 (2008) 120–123.
[72] T. Hasunuma, A. Kondo, Consolidated bioprocessing and simultaneous saccharification and fermentation of lignocellulose to ethanol with thermotolerant yeast strains, Process Biochem. 47 (2012) 1287–1294.
[73] H. Abe, Y. Fujita, Y. Takaoka, E. Kurita, S. Yano, N. Tanaka, et al., Ethanol-tolerant *Saccharomyces cerevisiae* strains isolated under selective conditions by over-expression of a proofreading-deficient DNA polymerase δ, J. Biosci. Bioeng. 108 (2009) 199–204.
[74] T. Sanda, T. Hasunuma, F. Matsuda, A. Kondo, Repeated-batch fermentation of lignocellulosic hydrolysate to ethanol using a hybrid *Saccharomyces cerevisiae* strain metabolically engineered for tolerance to acetic and formic acids, Bioresour. Technol. 102 (2011) 7917–7924.
[75] D.-Q. Zheng, X.-C. Wu, X.-L. Tao, P.-M. Wang, P. Li, X.-Q. Chi, et al., Screening and construction of *Saccharomyces cerevisiae* strains with improved multi-tolerance and bioethanol fermentation performance, Bioresour. Technol. 102 (2011) 3020–3027.
[76] S. Benjaphokee, D. Hasegawa, D. Yokota, T. Asvarak, C. Auesukaree, M. Sugiyama, et al., Highly efficient bioethanol production by a *Saccharomyces cerevisiae* strain with multiple stress tolerance to high temperature, acid and ethanol, N. Biotechnol. 29 (2012) 379–386.
[77] D. Chung, M. Cha, A.M. Guss, J. Westpheling, Direct conversion of plant biomass to ethanol by engineered *Caldicellulosiruptor bescii*, Proc. Natl. Acad. Sci. U.S.A. 111 (2014) 8931–8936.
[78] J. Huang, D. Chen, Y. Wei, Q. Wang, Z. Li, Y. Chen, et al., Direct ethanol production from lignocellulosic sugars and sugarcane bagasse by a recombinant *Trichoderma reesei* strain HJ48, Sci. World J. 2014 (2016) 1–8.
[79] M. Ballesteros, J.M. Oliva, M.J. Negro, P. Manzanares, I. Ballesteros, Ethanol from lignocellulosic materials by a simultaneous saccharification and fermentation process (SFS) with *Kluyveromyces marxianus* CECT 10875, Process Biochem. 39 (2004) 1843–1848.

[80] Y. Fujita, J. Ito, M. Ueda, H. Fukuda, A. Kondo, Synergistic saccharification, and direct fermentation to ethanol, of amorphous cellulose by use of an engineered yeast strain codisplaying three types of cellulolytic enzyme, Appl. Environ. Microbiol. 70 (2004) 1207–1212.

[81] K. Karimi, G. Emtiazi, M.J. Taherzadeh, Ethanol production from dilute-acid pretreated rice straw by simultaneous saccharification and fermentation with *Mucor indicus*, *Rhizopus oryzae*, and *Saccharomyces cerevisiae*, Enzyme Microb. Technol. 40 (2006) 138–144.

[82] A.D. Kroumov, A.N. Módenes, M.C. de, A. Tait, Development of new unstructured model for simultaneous saccharification and fermentation of starch to ethanol by recombinant strain, Biochem. Eng. J. 28 (2006) 243–255.

[83] M. Narra, J.P. James, V. Balasubramanian, Simultaneous saccharification and fermentation of delignified lignocellulosic biomass at high solid loadings by a newly isolated thermotolerant *Kluyveromyces* sp. for ethanol production, Bioresour. Technol. 179 (2015) 331–338.

[84] D.-T. Tran, Y.-P. I, C.-W. Lin, Developing co-culture system of dominant cellulolytic Bacillus sp. THLA0409 and dominant ethanolic *Klebsiella oxytoca* THLC0409 for enhancing ethanol production from lignocellulosic materials, J. Taiwan Inst. Chem. Eng. 44 (2013) 762–769.

[85] G. Salehi Jouzani, M.J. Taherzadeh, Advances in consolidated bioprocessing systems for bioethanol and butanol production from biomass: a comprehensive review, Biofuel Res. J. 5 (2015) 152–195.

[86] V.A. Svetlitchnyi, O. Kensch, D.A. Falkenhan, S.G. Korseska, N. Lippert, M. Prinz, et al., Single-step ethanol production from lignocellulose using novel extremely thermophilic bacteria, Biotechnol. Biofuels 6 (2013) 1–15.

[87] A. Amelio, B. Van der Bruggen, C.G. Lopresto, A. Verardi, V. Calabrò, P. Luis, Pervaporation membrane reactors: biomass conversion into alcoholsin: Membr. Technol. Biorefining (2016) 331–381.

[88] M.K. Mahapatra, A. Kumar, A short review on biobutanol, a second generation biofuel production from lignocellulosic biomass, J. Clean Energy Technol. 5 (2017) 27–30.

[89] S. Yadav, G. Rawat, P. Tripathi, R.K. Saxena, A novel approach for biobutanol production by Clostridium acetobutylicum using glycerol: a low cost substrate, Renew. Energy 71 (2014) 37–42.

[90] F. Monot, J.-R. Martin, H. Petitdemange, R. Gay, Acetone and butanol production by Clostridium acetobutylicum in a synthetic medium acetone and butanol production by Clostridium acetobutylicum in a synthetic medium, Appl. Environ. Microbiol. 44 (1982) 1318–1324.

[91] C.R. He, Y.Y. Kuo, S.Y. Li, Lignocellulosic butanol production from Napier grass using semi-simultaneous saccharification fermentation, Bioresour. Technol. 231 (2017) 101–108.

[92] P. Patáková, J. Lipovský, H. Čížková, J. Fořtová, M. Rychtera, K. Melzoch, Exploitation of food feedstock and waste for production of biobutanol, Czech J. Food Sci. 27 (2009) 276–283.

[93] I. Komonkiat, B. Cheirsilp, Felled oil palm trunk as a renewable source for biobutanol production by *Clostridium* spp, Bioresour. Technol. 146 (2013) 200–207.

[94] B.L. Magalhães, M.C.B. Grassi, G.A.G. Pereira, M. Brocchi, Improved *n*-butanol production from lignocellulosic hydrolysate by *Clostridium* strain screening and culture-medium optimization, Biomass Bioenergy 108 (2018) 157–166.

[95] N. Qureshi, T.C. Ezeji, J. Ebener, B.S. Dien, M.A. Cotta, H.P. Blaschek, Butanol production by *Clostridium beijerinckii*. Part I: Use of acid and enzyme hydrolyzed corn fiber, Bioresour. Technol. 99 (2008) 5915–5922.

[96] X. Wang, Y. Wang, B. Wang, H. Blaschek, H. Feng, Z. Li, Biobutanol production from fiber-enhanced DDGS pretreated with electrolyzed water, Renew. Energy 52 (2013) 16–22.
[97] S. Nanda, A.K. Dalai, J.A. Kozinski, Butanol and ethanol production from lignocellulosic feedstock: biomass pretreatment and bioconversion, Energy Sci. Eng. 2 (2014) 138–148.
[98] V. Ujor, A.K. Bharathidasan, K. Cornish, T.C. Ezeji, Feasibility of producing butanol from industrial starchy food wastes, Appl. Energy 136 (2014) 590–598.
[99] M. Hijosa-Valsero, A.I. Paniagua-García, R. Díez-Antolínez, Biobutanol production from apple pomace: the importance of pretreatment methods on the fermentability of lignocellulosic agro-food wastes, Appl. Microbiol. Biotechnol. 101 (2017) 8041–8052.
[100] Z.-Y. Liu, X.-Q. Yao, Q. Zhang, Z. Liu, Z.-J. Wang, Y.-Y. Zhang, et al., Modulation of the acetone/butanol ratio during fermentation of corn stover-derived hydrolysate by *Clostridium beijerinckii* strain NCIMB 8052, Appl. Environ. Microbiol. 83 (2017) 1–12.
[101] B. Dabrock, H. Bahl, G. Gottschalk, Parameters affecting solvent production by *Clostridium pasteurianum*, Appl. Environ. Microbiol. 58 (1992) 1233–1239.
[102] S.T. Liew, A. Arbakariya, M. Rosfarizan, A.R. Raha, Production of solvent (acetone–butanol–ethanol) in continuous fermentation by *Clostridium saccharobutylicum* DSM 13864 using gelatinised sago starch as a carbon source, Malaysian J. Microbiol. 2 (2006) 42–50. Available from: http://www.mendeley.com/research/production-solvent-acetonebutanolethanol-continuous-fermentation-clostridium-saccharobutylicum-dsm-13864-using-gelatinised-sago-starch-carbon-source/.
[103] K. Gao, S. Boiano, A. Marzocchella, L. Rehmann, Cellulosic butanol production from alkali-pretreated switchgrass (*Panicum virgatum*) and phragmites (*Phragmites australis*), Bioresour. Technol. 174 (2014) 176–181.
[104] K. Gao, L. Rehmann, ABE fermentation from enzymatic hydrolysate of NaOH-pretreated corncobs, Biomass Bioenergy 66 (2014) 110–115.
[105] N. Qureshi, a Friedl, I.S. Maddox, Butanol production from concentrated lactose/whey permeate: use of pervaporation membrane to recover and concentrate product, Appl. Microbiol. Biotechnol. 98 (2014) 9859–9867.
[106] J.J. Dong, J.C. Ding, Y. Zhang, L. Ma, G.C. Xu, R.Z. Han, et al., Simultaneous saccharification and fermentation of dilute alkaline-pretreated corn stover for enhanced butanol production by *Clostridium saccharobutylicum* DSM 13864, FEMS Microbiol. Lett. 363 (2016) 1–6.
[107] L.T. Mun, A. Ishizaki, S. Yoshino, K. Furukawa, Production of acetone, butanol and ethanol from palm oil waste by *Clostridium saccharoperbutylacetonicum* N1-4, Biotechnol. Lett. 17 (1995) 649–654.
[108] A. Ishizaki, S. Michiwaki, E. Crabbe, G. Kobayashi, K. Sonomoto, S. Yoshino, Extractive acetone–butanol–ethanol fermentation using methylated crude palm oil as extractant in batch culture of *Clostridium saccaharoperbutylacetonicum* N1-4 (ATCC 13564), J. Biosci. Bioeng. 87 (1999) 352–356.
[109] Y. Tashiro, K. Takeda, G. Kobayashi, K. Sonomoto, A. Ishizaki, S. Yoshino, High butanol production by *Clostridium saccharoperbutylacetonicum* N1-4 in fed-batch culture with pH-Stat continuous butyric acid and glucose feeding method, J. Biosci. Bioeng. 98 (2004) 263–268.
[110] V.H. Thang, K. Kanda, G. Kobayashi, Production of acetone–butanol–ethanol (ABE) in direct fermentation of cassava by *Clostridium saccharoperbutylacetonicum* N1-4, Appl. Biochem. Biotechnol. 161 (2010) 157–170.
[111] N.K.N. Al-Shorgani, M.S. Kalil, W.M.W. Yusoff, Biobutanol production from rice bran and de-oiled rice bran by *Clostridium saccharoperbutylacetonicum* N1-4, Bioprocess. Biosyst. Eng. 35 (2012) 817–826.

[112] N.K.N. Al-Shorgani, M.S. Kalil, W.M.W. Yusoff, Fermentation of sago starch to biobutanol in a batch culture using *Clostridium saccharoperbutylacetonicum* N1-4 (ATCC 13564), Ann. Microbiol. 62 (2012) 1059–1070.
[113] N.K.N. Al-Shorgani, M.S. Kalil, E. Ali, A.A. Hamid, W.M.W. Yusoff, The use of pretreated palm oil mill effluent for acetone–butanol–ethanol fermentation by *Clostridium saccharoperbutylacetonicum* N1-4, Clean Technol. Environ. Policy 14 (2012) 879–887.
[114] J.T. Ellis, N.N. Hengge, R.C. Sims, C.D. Miller, Acetone, butanol, and ethanol production from wastewater algae, Bioresour. Technol. 111 (2012) 491–495.
[115] H. Shukor, N.K.N. Al-Shorgani, P. Abdeshahian, A.A. Hamid, N. Anuar, N.A. Rahman, et al., Production of butanol by *Clostridium saccharoperbutylacetonicum* N1-4 from palm kernel cake in acetone–butanol–ethanol fermentation using an empirical model, Bioresour. Technol. 170 (2014) 565–573.
[116] J. Zheng, Y. Tashiro, Q. Wang, K. Sakai, K. Sonomoto, Feasibility of acetone–butanol–ethanol fermentation from eucalyptus hydrolysate without nutrients supplementation, Appl. Energy 140 (2015) 113–119.
[117] D. Yao, S. Dong, P. Wang, T. Chen, J. Wang, Z.B. Yue, et al., Robustness of *Clostridium saccharoperbutylacetonicum* for acetone–butanol–ethanol production: effects of lignocellulosic sugars and inhibitors, Fuel 208 (2017) 549–557.
[118] O.A. Tigunova, S.M. Shulga, Y.B. Blume, Biobutanol as an alternative type of fuel, Cytol. Genet. 47 (2013) 366–382.
[119] L.D. Gottumukkala, B. Parameswaran, S.K. Valappil, K. Mathiyazhakan, A. Pandey, R.K. Sukumaran, Biobutanol production from rice straw by a non acetone producing *Clostridium sporogenes* BE01, Bioresour. Technol. 145 (2013) 182–187.
[120] K. Kiyoshi, M. Furukawa, T. Seyama, T. Kadokura, A. Nakazato, S. Nakayama, Butanol production from alkali-pretreated rice straw by co-culture of *Clostridium thermocellum* and *Clostridium saccharoperbutylacetonicum*, Bioresour. Technol. 186 (2015) 325–328.
[121] H.T.M. Tran, B. Cheirsilp, B. Hodgson, K. Umsakul, Potential use of *Bacillus subtilis* in a co-culture with *Clostridium butylicum* for acetone–butanol–ethanol production from cassava starch, Biochem. Eng. J. 48 (2010) 260–267.
[122] M.H. Abd-Alla, A.-W. Elsadek El-Enany, Production of acetone–butanol–ethanol from spoilage date palm (*Phoenix dactylifera* L.) fruits by mixed culture of *Clostridium acetobutylicum* and *Bacillus subtilis*, Biomass Bioenergy 42 (2012) 172–178.
[123] F. Xin, J. He, Characterization of a thermostable xylanase from a newly isolated *Kluyvera* species and its application for biobutanol production, Bioresour. Technol. 135 (2013) 309–315.
[124] Z. Wen, M. Wu, Y. Lin, L. Yang, J. Lin, P. Cen, Artificial symbiosis for acetone–butanol–ethanol (ABE) fermentation from alkali extracted deshelled corn cobs by co-culture of *Clostridium beijerinckii* and *Clostridium cellulovorans*, Microb. Cell Fact. 13 (2014) 92–102.
[125] S.B. Bankar, S.A. Survase, H. Ojamo, T. Granström, Biobutanol: the outlook of an academic and industrialist, RSC Adv. 3 (2013) 24734–24757.
[126] A. Ranjan, V.S. Moholkar, Biobutanol: science, engineering, and economics, Int. J. Energy Res. 36 (2012) 277–323.
[127] J. Zheng, Y. Tashiro, Q. Wang, K. Sonomoto, Recent advances to improve fermentative butanol production: genetic engineering and fermentation technology, J. Biosci. Bioeng. 119 (2014) 1–9.
[128] H. Huang, H. Liu, Y.R. Gan, Genetic modification of critical enzymes and involved genes in butanol biosynthesis from biomass, Biotechnol. Adv. 28 (2010) 651–657.

[129] T. Lütke-Eversloh, Application of new metabolic engineering tools for *Clostridium acetobutylicum*, Appl. Microbiol. Biotechnol. 98 (2014) 5823–5837.
[130] E.I. Lan, J.C. Liao, Metabolic engineering of cyanobacteria for 1-butanol production from carbon dioxide, Metab. Eng. 13 (2011) 353–363.
[131] M. Yu, Y. Zhang, I.C. Tang, S.T. Yang, Metabolic engineering of *Clostridium tyrobutyricum* for *n*-butanol production, Metab. Eng. 13 (2011) 373–382.
[132] C. Ma, K. Kojima, N. Xu, J. Mobley, L. Zhou, S.-T. Yang, et al., Comparative proteomics analysis of high *n*-butanol producing metabolically engineered *Clostridium tyrobutyricum*, J. Biotechnol. 193 (2015) 108–119.
[133] M. Parekh, J. Formanek, H.P. Blaschek, Pilot-scale production of butanol by *Clostridium beijerinckii* BA101 using a low-cost fermentation medium based on corn steep water, Appl. Microbiol. Biotechnol. 51 (1999) 152–157.
[134] N. Qureshi, H.P. Blaschek, Butanol production using *Clostridium beijerinckii* BA101 hyper-butanol producing mutant strain and recovery by pervaporation, Appl. Biochem. Biotechnol. 84–86 (2000) 225–235.
[135] N. Qureshi, A. Lolas, H.P. Blaschek, Soy molasses as fermentation substrate for production of butanol using *Clostridium beijerinckii* BA101, J. Ind. Microbiol. Biotechnol. 26 (2001) 290–295.
[136] T.C. Ezeji, N. Qureshi, H.P. Blaschek, Butanol fermentation research: upstream and downstream manipulations, Chem. Rec. 4 (2004) 305–314.
[137] R. Sillers, A. Chow, B. Tracy, E.T. Papoutsakis, Metabolic engineering of the non-sporulating, non-solventogenic Clostridium acetobutylicum strain M5 to produce butanol without acetone demonstrate the robustness of the acid-formation pathways and the importance of the electron balance, Metab. Eng. 10 (2008) 321–332.
[138] T. Lütke-Eversloh, H. Bahl, Metabolic engineering of *Clostridium acetobutylicum*: recent advances to improve butanol production, Curr. Opin. Biotechnol. 22 (2011) 634–647.
[139] G. Jurgens, S. Survase, O. Berezina, E. Sklavounos, J. Linnekoski, A. Kurkijärvi, et al., Butanol production from lignocellulosics, Biotechnol. Lett. 34 (2012) 1415–1434.
[140] D. Cai, T. Zhang, J. Zheng, Z. Chang, Z. Wang, P.Y. Qin, et al., Biobutanol from sweet sorghum bagasse hydrolysate by a hybrid pervaporation process, Bioresour. Technol. 145 (2013) 97–102.
[141] Y.-S. Jang, A. Malaviya, S.Y. Lee, Acetone–butanol–ethanol production with high productivity using *Clostridium acetobutylicum* BKM19, Biotechnol. Bioeng. 110 (2013) 1646–1653.
[142] C. Lu, J. Dong, S.T. Yang, Butanol production from wood pulping hydrolysate in an integrated fermentation-gas stripping process, Bioresour. Technol. 143 (2013) 467–475.
[143] W. Jiang, J. Zhao, Z. Wang, S.T. Yang, Stable high-titer *n*-butanol production from sucrose and sugarcane juice by *Clostridium acetobutylicum* JB200 in repeated batch fermentations, Bioresour. Technol. 163 (2014) 172–179.
[144] H.G. Li, W. Luo, Q. Wang, X. Bin Yu, Direct fermentation of gelatinized cassava starch to acetone, butanol, and ethanol using *Clostridium acetobutylicum* mutant obtained by atmospheric and room temperature plasma, Appl. Biochem. Biotechnol. 172 (2014) 3330–3341.
[145] A. Bhandiwad, A. Guseva, L. Lynd, Metabolic engineering of *Thermoanaerobacterium thermosaccharolyticum* for increased *n*-butanol production, Adv. Microbiol. 3 (2013) 46–51.
[146] A. Bhandiwad, A.J. Shaw, A. Guss, A. Guseva, H. Bahl, L.R. Lynd, Metabolic engineering of *Thermoanaerobacterium saccharolyticum* for *n*-butanol production, Metab. Eng. 21 (2014) 17–25.

[147] P.P. Lin, K.S. Rabe, J.L. Takasumi, M. Kadisch, F.H. Arnold, J.C. Liao, Isobutanol production at elevated temperatures in thermophilic *Geobacillus thermoglucosidasius*, Metab. Eng. 24 (2014) 1–8.
[148] M.W. Keller, G.L. Lipscomb, A.J. Loder, G.J. Schut, R.M. Kelly, M.W.W. Adams, A hybrid synthetic pathway for butanol production by a hyperthermophilic microbe, Metab. Eng. 27 (2015) 101–106.
[149] J. Liu, T. Guo, D. Wang, J. Xu, H. Ying, Butanol production by a *Clostridium beijerinckii* mutant with high ferulic acid tolerance, Biotechnol. Appl. Biochem. 63 (2016) 727–733.
[150] C. Lu, L. Yu, S. Varghese, M. Yu, S.T. Yang, Enhanced robustness in acetone–butanol–ethanol fermentation with engineered *Clostridium beijerinckii* overexpressing *adh*E2 and *ctf*AB, Bioresour. Technol. 243 (2017) 1000–1008.
[151] Y. Yan, J. He, *Clostridium* species strain BOH_3 tolerates and transforms inhibitors from horticulture waste hydrolysates, Appl. Microbiol. Biotechnol. 101 (2017) 6289–6297.
[152] L. Yu, M. Xu, I.-C. Tang, S.-T. Yang, Metabolic engineering of *Clostridium tyrobutyricum* for *n*-butanol production through co-utilization of glucose and xylose, Biotechnol. Bioeng. 112 (2015) 2134–2141.
[153] S. Atsumi, J.C. Liao, Metabolic engineering for advanced biofuels production from *Escherichia coli*, Curr. Opin. Biotechnol. 19 (2008) 414–419.
[154] S. Atsumi, A.F. Cann, M.R. Connor, C.R. Shen, K.M. Smith, M.P. Brynildsen, et al., Metabolic engineering of *Escherichia coli* for 1-butanol production, Metab. Eng. 10 (2008) 305–311.
[155] C.R. Shen, J.C. Liao, Metabolic engineering of *Escherichia coli* for 1-butanol and 1-propanol production via the keto-acid pathways, Metab. Eng. 10 (2008) 312–320.
[156] D.R. Nielsen, E. Leonard, S.H. Yoon, H.C. Tseng, C. Yuan, K.L.J. Prather, Engineering alternative butanol production platforms in heterologous bacteria, Metab. Eng. 11 (2009) 262–273.
[157] C.R. Shen, E.I. Lan, Y. Dekishima, A. Baez, K.M. Cho, J.C. Liao, Driving forces enable high-titer anaerobic 1-butanol synthesis in *Escherichia coli*, Appl. Environ. Microbiol. 77 (2011) 2905–2915.
[158] Q. Wang, Y. Ding, L. Liu, J. Shi, J. Sun, Y. Xue, Engineering *Escherichia coli* for autoinducible production of *n*-butanol, Electron. J. Biotechnol. 25113 (2015) 1–5.
[159] M. Saini, L.J. Lin, C.J. Chiang, Y.P. Chao, Synthetic consortium of *Escherichia coli* for *n*-butanol production by fermentation of the glucose–xylose mixture, J. Agric. Food. Chem. 65 (2017) 10040–10047.
[160] E.J. Steen, R. Chan, N. Prasad, S. Myers, C.J. Petzold, A. Redding, et al., Metabolic engineering of *Saccharomyces cerevisiae* for the production of *n*-butanol, Microb. Cell Fact. 7 (2008) 36–43.
[161] Z. Dai, Y. Liu, X. Zhang, M. Shi, B. Wang, D. Wang, et al., Metabolic engineering of *Saccharomyces cerevisiae* for production of ginsenosides, Metab. Eng. 20 (2013) 146–156.
[162] H. Li, P.H. Opgenorth, D.G. Wernick, S. Rogers, T.-Y. Wu, W. Higashide, et al., Integrated electromicrobial conversion of CO_2 to higher alcohols, Science (80-.) 335 (2012) 1596.
[163] S. Liu, K.M. Bischoff, N. Qureshi, S.R. Hughes, J.O. Rich, Functional expression of the thiolase gene thl from *Clostridium beijerinskii* p260 in *Lactococcus lactis* and *Lactobacillus buchneri*, N. Biotechnol. 27 (2010) 283–284.
[164] Y. Jiang, D. Guo, J. Lu, P. Dürre, W. Dong, W. Yan, et al., Consolidated bioprocessing of butanol production from xylan by a thermophilic and butanologenic Thermoanaerobacterium sp. M5, Biotechnol. Biofuels 11 (2018) 89.
[165] S.H. Lee, E.J. Yun, J. Kim, S.J. Lee, Y. Um, K.H. Kim, Biomass, strain engineering, and fermentation processes for butanol production by solventogenic clostridia, Appl. Microbiol. Biotechnol. 100 (2016) 8255–8271.

[166] Y. Yan, A. Basu, T. Li, J. He, Direct conversion of xylan to butanol by a wild-type *Clostridium* species strain G117, Biotechnol. Bioeng. 113 (2016) 1702–1710.
[167] S.M. Gaida, A. Liedtke, A.H.W. Jentges, B. Engels, S. Jennewein, Metabolic engineering of *Clostridium cellulolyticum* for production of *n*-butanol from crystalline cellulose, Microb. Cell Fact. 15 (2016) 1–11.
[168] T. Li, J. He, Simultaneous saccharification and fermentation of hemicellulose to butanol by a non-sporulating *Clostridium* species, Bioresour. Technol. 219 (2016) 430–438.
[169] F. Xin, T. Chen, Y. Jiang, W. Dong, W. Zhang, M. Zhang, et al., Strategies for improved isopropanol-butanol production by a *Clostridium* strain from glucose and hemicellulose through consolidated bioprocessing, Biotechnol. Biofuels 10 (2017) 1–13.
[170] Y. Jiang, T. Chen, W. Dong, M. Zhang, W. Zhang, H. Wu, et al., The draft genome sequence of *Clostridium beijerinckii* NJP7, a unique bacterium capable of producing isopropanol–butanol from hemicellulose through consolidated bioprocessing, Curr. Microbiol. 75 (2018) 305–308.
[171] M.J. Taherzadeh, K. Karimi, Acid based hydrolysis processes for ethanol from lignocellulosic materials: a review, Bioresources 2 (3) (2007) 472–499.
[172] S. Banerjee, R. Sen, R.A. Pandey, T. Chakrabarti, D. Satpute, B.S. Giri, Commercializing lignocellulosic bioethanol: technology bottlenecks and possible remedies, Biofuels Bioprod. Biorefin. 4 (2010) 77–93.
[173] J. Borjesson, R. Peterson, F. Tjerneld, Enhanced enzymatic conversion of softwood lignocelluloses by poly(ethylene glycol) addition, Enzym Microb Technol. 40 (2007) 754–762.
[174] H. Ma, W. Liu, X. Chen, Y.J. Wua, Z.L. Yu, Enhanced enzymatic saccharification of rice straw by microwave pretreatment, Bioresour. Technol. 100 (2009) 1279–1284.
[175] S. Banerjee, R. Sen, S. Mudliar, R.A. Pandey, T. Chakrabarti, D. Satpute, Alkaline peroxide assisted wet air oxidation pretreatment approach to enhance enzymatic convertibility of rice husk, Biotechnol. Prog. 27 (3) (2011) 691–697.
[176] L. Zhu, J.P. Dwyer, V.S. Chang, C.B. Granda, M.T. Holtzapple, Structural features affecting biomass enzymatic digestibility, Bioresour. Technol. 99 (2008) 3817–3828.
[177] F.C. Lunelli, P. Sfalcin, et al., Ultrasound-assisted enzymatic hydrolysis of sugarcane bagasse for the production of fermentable sugars, Biosyst. Eng. 124 (2014) 24–28.
[178] J. Bian, F. Peng, X.P. Peng, X. Xiao, P. Peng, F. Xu, et al., Effect of [Emim]Ac pretreatment on the structure and enzymatic hydrolysis of sugarcane bagasse cellulose, Carbohydr. Polym. 100 (2014) 211–217.
[179] Y. Gao, J. Xu, Y. Zhang, Q. Yu, Z. Yuan, Y. Liu, Effects of different pretreatment methods on chemical composition of sugarcane bagasse and enzymatic hydrolysis, Bioresour. Technol. 144 (2013) 396–400.
[180] N. Ortega, D. Busto, M. Mateos, Kinetics of cellulose saccharification by *Trichoderma reesei* cellulases, Int. Biodeterior. Biodegrad. 47 (1) (2001) 7–14.
[181] D.B. Hodge, M.N. Karim, D.J. Schell, J.D. McMillan, Soluble and insoluble solids contributions to high-solids enzymatic hydrolysis of lignocelluloses, Bioresour. Technol. 99 (18) (2008) 8940–8948.
[182] K. Saha, U. Maheswari R, J. Sikder, S. Chakraborty, S.S. da Silva, J.C. dos Santos, Membranes as a tool to support biorefineries: applications in enzymatic hydrolysis, fermentation and dehydration for bioethanol production, Renew. Sustain. Energy Rev. 74 (2017) 873–890.
[183] V. Gomis, R. Pedraza, M.D. Saquete, A. Font, J. García-Cano, Ethanol dehydration via azeotropic distillation with gasoline fractions as entrainers: a pilot-scale

study of the manufacture of an ethanol–hydrocarbon fuel blend, Fuel 139 (2015) 568–574.

[184] I.D. Gil, J.M. Gomez, J.M.G. Rodríguez, Control of an extractive distillation process to dehydrate ethanol using glycerolas entrainer, Comput. Chem. Eng. 39 (2012) 29–42.

[185] A.A. Kiss, R.M. Ignat, Innovative single step bioethanol dehydration in an extractive dividing-wall column, Sep. Purif. Technol. 98 (2012) 290–297.

[186] A.A. Kiss, D.J.-P.C. Suszwalak, Enhanced bioethanol dehydration by extractive and azeotropic distillation individing-wall columns, Sep. Purif. Technol. 86 (2012) 70–78.

[187] A.J.J. Figueroa, B.H. Lunelli, R.M. Filho, M.R.W. Maciel, Improvements on anhydrous ethanol production by extractive distillation using ionic liquid as solvent, Procedia Eng. 42 (2012) 1016–1026.

[188] R.N. Maeda, C.A. Barcelos, L.M.M.S. Anna, N. Pereira Jr, Cellulase production by Penicillium funiculosum and its application in the hydrolysis of sugar cane bagasse for second generation ethanol production by fed batch operation, J. Biotechnol. 163 (2013) 38–44.

[189] S.E. Jacobsen, C.E. Wyman, Cellulose and hemicellulose hydrolysis models for application to current and novel pretreatment processes, Appl. Biochem. Biotechnol. 84–86 (2000) 81–96.

[190] Wooley R., Ruth M., Sheehan J., Ibsen K., Majdeski H., Galvez A. Lignocellulosic biomass to ethanol process design and economics utilizing co-current dilute acid prehydrolysis and enzymatic hydrolysis current and futuristic scenarios, in: Technical Report. NREL/TP-580-26157, National Renewable Energy Laboratory, Golden, CO, 1990.

[191] J. Lindedam, M.O. Haven, P. Chylenski, H. Jørgensen, C. Felby, Recycling cellulases for cellulosic ethanol production at industrial relevant conditions: potential and temperature dependency at high solid processes, Bioresour. Technol. 148 (2013) 180–188.

[192] M. Tu, R.P. Chandra, J.N. Saddler, Evaluating the distribution of cellulases and the recycling of free cellulases during the hydrolysis of lignocellulosic substrates, Biotechnol. Prog. 23 (2) (2007) 398–406.

[193] B. Qi, X. Chen, Y. Su, Y. Wan, Enzyme adsorption and recycling during hydrolysis of wheat straw lignocelluloses, Bioresour. Technol. 102 (2011) 2881–2889.

[194] Y. He, D.M. Bagley, K.T. Leung, S.N. Liss, B.-Q. Liao, Recent advances in membrane technologies for biorefining and bioenergy production, Biotechnol. Adv. 30 (2012) 817–858.

[195] J.S. Knutsen, R.H. Davis, Combined sedimentation and filtration process for cellulase recovery during hydrolysis of lignocellulosic biomass, Appl. Biochem. Biotechnol. 98–100 (2002) 1161–1172.

[196] J.M. Lee, J.H. Wolf, Continuous attrition bioreactor with enzyme recycling for the bioconversion of cellulose, Appl. Biochem. Biotechnol. 18 (1988) 203–215.

[197] A.C. Rodrigues, C. Felby, M. Gama, Cellulase stability, adsorption/desorption profiles and recycling during successive cycles of hydrolysis and fermentation of wheat straw, Bioresour. Technol. 156 (2014) 163–169.

[198] B. Qi, J. Luo, G. Chen, X. Chen, Y. Wan, Application of ultrafiltration and nanofiltration for recycling cellulase and concentrating glucose from enzymatic hydrolyzate of steam exploded wheat straw, Bioresour. Technol. 104 (2012) 466–472.

[199] P. Lozano, B. Bernal, A.G. Jara, M.-P. Belleville, Enzymatic membrane reactor for full saccharification of ionic liquid-pretreated microcrystalline cellulose, Bioresour. Technol. 151 (2014) 159–165.

[200] S. Szelpal, O. Poser, M. Abel, Enzyme recovery by membrane separation method from waste products of the food industry, in: Acta Tech Corviniensis Bull Eng Tome, VI, 2013.

[201] M. Vohra, J. Manwar, R. Manmode, S. Padgilwar, S. Patil, Bioethanol production: feedstock and current technologies, J. Environ. Chem. Eng. 2 (1) (2014) 573–584.

[202] S. Xiu, B. Zhang, A. Shahbazi, Biorefinery processes for biomass conversion to liquid fuel, Biofuel's Eng. Proces Technol. (2010) 167–190.

[203] H.S. Hafid, N.A.A. Rahman, U.K.M. Shah, A.S. Baharuddin, A.B. Ariff, Feasibility of using kitchen waste as future substrate for bioethanol production: a review, Renew. Sustain. Energy Rev. 74 (2017) 671–686.

[204] A. Demirbaş, Bioethanol from cellulosic materials: a renewable motor fuel from biomass, Energy Sources 27 (4) (2005) 327–337.

[205] M.J. Taherzadeh, K. Karimi, Enzyme-based hydrolysis processes for ethanol from lignocellulosic materials: a review, Bioresources 2 (4) (2007) 707–738.

[206] V. Balan, Current challenges in commercially producing biofuels from lignocellulosic biomass, ISRN Biotechnol. 2014 (2014) 1–31. Article ID 463074.

[207] M. Yang, H. Ji, J.Y. Zhu, Batch fermentation options for high titer bioethanol production from a SPORL pretreated Douglas-fir forest residue without detoxification, Fermentation 2 (3) (2016) 10.

[208] S. Macrelli, J. Mogensen, G. Zacchi, Techno-economic evaluation of 2nd generation bioethanol production from sugar cane bagasse and leaves integrated with the sugar-based ethanol process, Biotechnol. Biofuels 5 (22) (2012) 2–18.

[209] T.D. Foust, A. Aden, A. Dutta, S. Phillips, An economic and environmental comparison of a biochemical and a thermochemical lignocellulosic ethanol conversion processes, Cellulose 16 (4) (2009) 547–565.

[210] W. Gauss, S. Suzuki, M. Takagi, Manufacture of Alcohol From Cellulosic Materials Using Plural Ferments, US Pat. 3,990,944, 39909, No. Issue 610731, 1976.

[211] K. Olofsson, M. Bertilsson, G. Lidén, A short review on SSF – an interesting process option for ethanol production from lignocellulosic feedstocks, Biotechnol. Biofuels 1 (7) (2008) 1–14.

[212] S.H. Mohd Azhar, R. Abdulla, S.A. Jambo, H. Marbawi, J.A. Gansau, A.A. Mohd Faik, et al., Yeasts in sustainable bioethanol production: a review, Biochem. Biophys. Rep. 10 (2017) 52–61.

[213] D.D. Spindler, C.E. Wyman, K. Grohmann, G.P. Philippidis, Evaluation of the cellobiose-fermenting yeast *Brettanomyces custersii* in the simultaneous saccharification and fermentation of cellulose, Biotechnol. Lett. 14 (5) (1992) 403–407.

[214] S. Zhou, L.O. Ingram, Simultaneous saccharification and fermentation of amorphous cellulose to ethanol by recombinant *Klebsiella oxytoca* SZ21 without supplemental cellulase, Biotechnol. Lett. 23 (18) (2001) 1455–1462.

[215] C. Tengborg, M. Galbe, G. Zacchi, Reduced inhibition of enzymatic hydrolysis of steam-pretreated softwood, Enzyme Microb. Technol. 28 (9–10) (2001) 835–844.

[216] P. Sassner, M. Galbe, G. Zacchi, Techno-economic evaluation of bioethanol production from three different lignocellulosic materials, Biomass Bioenergy 32 (5) (2008) 422–430.

[217] C.E. Wyman, Handbook on Bioethanol: Production and Utilization, in: Charles Wyman (Ed.), Taylor and Francis group, vol. 51, 1996.

[218] C. Ricardo Soccol, V. Faraco, S. Karp, L.P.S. Vandenberghe, V. Thomaz-Soccol, A. Woiciechowski, et al., Lignocellulosic Bioethanol: Current Status and Future Perspectives, first ed, Elsevier Inc, 2011.

[219] M.P. García-Aparicio, I. Ballesteros, A. González, et al., Effect of inhibitors released during steam-explosion pretreatment of barley straw on enzymatic hydrolysis, Apll. Biochem. Biotechnol. 129 (2006) 278–288.

[220] A. Verardi, I. De Bari, E. Ricca, V. Calabrò, Hydrolysis of lignocellulosic biomass: current status of processes and technologies and future perspectives, in: A.P.P. Natalense (Ed.), Bioethanol, InTech, 2012.

[221] Y. Tang, L. Zhu, W. Zhang, X. Shang, J. Jiang, Integrated process of starch ethanol and cellulosic lactic acid for ethanol and lactic acid production, Appl. Microbiol. Biotechnol. 97 (2013) 1923–1932.
[222] J. Choudhary, S. Singh, L. Nain, Thermotolerant fermenting yeasts for simultaneous saccharification fermentation of lignocellulosic biomass, Electron. J. Biotechnol. 21 (2016) 82–92.
[223] K. Olofsson, A. Rudolf, G. Lidén, Designing simultaneous saccharification and fermentation for improved xylose conversion by a recombinant strain of *Saccharomyces cerevisiae*, J. Biotechnol. 134 (1–2) (2008) 112–120.
[224] Ó.J. Sánchez, C.A. Cardona, Trends in biotechnological production of fuel ethanol from different feedstocks, Bioresour. Technol. 99 (2008) 5270–5295.
[225] P.M. Bondesson, M. Galbe, Process design of SSCF for ethanol production from steam-pretreated, acetic-acid-impregnated wheat straw, Biotechnol. Biofuels 9 (1) (2016).
[226] J.O. Westman, R. Wang, V. Novy, C.J. Franzén, Sustaining fermentation in high-gravity ethanol production by feeding yeast to a temperature-profiled multifeed simultaneous saccharification and co-fermentation of wheat straw, Biotechnol. Biofuels 10 (1) (2017).
[227] C.R. Carere, R. Sparling, N. Cicek, D.B. Levin, Third generation biofuels via direct cellulose fermentation, Int. J. Mol. Sci. 9 (7) (2008) 1342–1360.
[228] M.Á.B. Alcántara, J. Dobruchowska, P. Azadi, B.D. García, F.P. Molina-Heredia, F.M. Reyes-Sosa, Recalcitrant carbohydrates after enzymatic hydrolysis of pretreated lignocellulosic biomass, Biotechnol. Biofuels 9 (1) (2016) 207.
[229] A.V. Suko, R. Bura, Enhanced xylitol and ethanol yields by fermentation inhibitors in steam-pretreated lignocellulosic biomass, Ind. Biotechnol. 12 (3) (2016) 187–194.
[230] E.C. van der Pol, R.R. Bakker, P. Baets, G. Eggink, By-products resulting from lignocellulose pretreatment and their inhibitory effect on fermentations for (bio)chemicals and fuels, Appl. Microbiol. Biotechnol. 98 (2014) 9579–9593.
[231] A.M. Socha, R. Parthasarathi, J. Shi, S. Pattathil, D. Whyte, M. Bergeron, et al., Efficient biomass pretreatment using ionic liquids derived from lignin and hemicellulose, Proc. Natl. Acad. Sci. U.S.A. 111 (35) (2014) E3587–E3595.
[232] C. Carrasco, H.M. Baudel, J. Sendelius, T. Modig, C. Roslander, M. Galbe, et al., SO_2-catalyzed steam pretreatment and fermentation of enzymatically hydrolyzed sugarcane bagasse, Enzyme Microb. Technol. 46 (2010) 64–73.
[233] A. Verardi, A. Blasi, A. Molino, L. Albo, V. Calabrò, Improving the enzymatic hydrolysis of *Saccharum officinarum* L. bagasse by optimizing mixing in a stirred tank reactor: quantitative analysis of biomass conversion, Fuel Process. Technol. 149 (2016) 15–22.
[234] L.J. Jönsson, C. Martín, Pretreatment of lignocellulose: formation of inhibitory by-products and strategies for minimizing their effects, Bioresour. Technol. 199 (2016) 103–112.
[235] A.T. Smit, J.W. Van Hal, L. Lanting, K.J. Damen, Effective fractionation of lignocellulose using a mild acetone-based organosolv process, Green Chem. 19 (2017) 5505–5514.
[236] F. Monlau, C. Sambusiti, A. Barakat, M. Quéméneur, E. Trably, J. Steyer, et al., Do furanic and phenolic compounds of lignocellulosic and algae biomass hydrolyzate inhibit anaerobic mixed cultures? A comprehensive review, Biotechnol. Adv. 32 (5) (2014) 934–951.
[237] D. Kim, Physico-chemical conversion of lignocellulose: inhibitor effects and detoxification strategies: a mini review, Molecules 23 (2) (2018) 309.

[238] H. Rasmussen, H.R. Sørensen, A.S. Meyer, Formation of degradation compounds from lignocellulosic biomass in the biorefinery: sugar reaction mechanisms, Carbohydr. Res. 385 (2014) 45–57.

[239] M.K. Malav, S. Prasad, S.K. Kharia, Furfural and 5-HMF: potent fermentation inhibitors and their removal techniques, Int. J. Curr. Microbiol. Appl. Sci. 6 (3) (2017) 2060–2066.

[240] A. Verardi, A. Blasi, T. Marino, A. Molino, V. Calabrò, Effect of steam-pretreatment combined with hydrogen peroxide on lignocellulosic agricultural wastes for bioethanol production: analysis of derived sugars and other by-products, J. Energy Chem. 27 (2) (2018) 535–543.

[241] A. Verardi, A. Blasi, I. De Bari, V. Calabrò, Steam pretreatment of *Saccharum officinarum* L. bagasse by adding of impregnating agents for advanced bioethanol production, Ecotoxicol. Environ. Saf. 134 (Pt 2) (2015) 1–8.

[242] M. Shafiei, K. Karimi, M.J. Taherzadeh, Pretreatment of spruce and oak by *N*-methylmorpholine-*N*-oxide (NMMO) for efficient conversion of their cellulose to ethanol, Bioresour. Technol. 101 (13) (2010) 4914–4918.

[243] R. Aguilar, J.A. Ramírez, G. Garrote, M. Vázquez, Kinetic study of the acid hydrolysis of sugar cane bagasse, J. Food Eng. 55 (4) (2002) 309–318.

[244] S. Larsson, E. Palmqvist, B. Hahn-Hägerdal, C. Tengborg, K. Stenberg, G. Zacchi, et al., The generation of fermentation inhibitors during dilute acid hydrolysis of softwood, Enzyme Microb. Technol. 24 (3–4) (1999) 151–159.

[245] A. Iwaki, T. Kawai, Y. Yamamoto, S. Izawa, Biomass conversion inhibitors furfural and 5-hydroxymethylfurfural induce formation of messenger RNP granules and attenuate translation activity in *Saccharomyces cerevisiae*, Appl. Environ. Microbiol. 79 (5) (2013) 1661–1667.

[246] J.R.M. Almeida, T. Modig, A. Petersson, B. Hähn-Hägerdal, G. Lidén, M.F. Gorwa-Grauslund, Increased tolerance and conversion of inhibitors in lignocellulosic hydrolysates by *Saccharomyces cerevisiae*, J. Chem. Technol. Biotechnol. 82 (4) (2007) 340–349.

[247] L.J. Jönsson, B. Alriksson, N.-O. Nilvebrant, Bioconversion of lignocellulose: inhibitors and detoxification, Biotechnol. Biofuels 6 (1) (2013) 16.

[248] T. Modig, G. Lidén, M.J. Taherzadeh, Inhibition effects of furfural on alcohol dehydrogenase, aldehyde dehydrogenase and pyruvate dehydrogenase, Biochem. J. 363 (3) (2002) 769–776.

[249] Z.L. Liu, P.J. Slininger, B.S. Dien, M.A. Berhow, C.P. Kurtzman, S.W. Gorsich, Adaptive response of yeasts to furfural and 5-hydroxymethylfurfural and new chemical evidence for HMF conversion to 2,5-bis-hydroxymethylfuran, J. Ind. Microbiol. Biotechnol. 31 (8) (2004) 345–352.

[250] E. Palmqvist, B. Hahn-Hägerdal, Fermentation of lignocellulosic hydrolysates. II: Inhibitors and mechanisms of inhibition, Bioresour. Technol. 74 (1) (2000) 25–33.

[251] S.K. Kim, D.H. Park, S.H. Song, Y.J. Wee, G.T. Jeong, Effect of fermentation inhibitors in the presence and absence of activated charcoal on the growth of *Saccharomyces cerevisiae*, Bioprocess. Biosyst. Eng. 36 (6) (2013) 659–666.

[252] Z. Lewis Liu, M. Ma, M. Song, Evolutionarily engineered ethanologenic yeast detoxifies lignocellulosic biomass conversion inhibitors by reprogrammed pathways, Mol. Genet. Genomics 282 (3) (2009) 233–244.

[253] C. Hu, X. Zhao, J. Zhao, S. Wu, Z.K. Zhao, Effects of biomass hydrolysis by-products on oleaginous yeast *Rhodosporidium toruloides*, Bioresour. Technol. 100 (20) (2009) 4843–4847.

[254] E. Palmqvist, B. Hahn-Hägerdal, Fermentation of lignocellulosic hydrolysates. I: Inhibition and detoxification, Bioresour. Technol. 74 (1) (2000) 17–24.

[255] Z. Guo, L. Olsson, Physiological response of *Saccharomyces cerevisiae* to weak acids present in lignocellulosic hydrolysate, FEMS Yeast Res. 14 (2014) 1234–1248.
[256] M. Mollapour, P.W. Piper, W. Bank, S. Sheffield, Hog1 mitogen-activated protein kinase phosphorylation targets the yeast Fps1 aquaglyceroporin for endocytosis, thereby rendering cells resistant to acetic acid, Mol. Cell. Biol. 27 (18) (2007) 6446–6456.
[257] J.B. Russell, Another explanation for the toxicity of fermentation acids at low pH: anion accumulation versus uncoupling, J. Appl. Bacteriol. 73 (5) (1992) 363–370.
[258] L.M. Maurer, E. Yohannes, S.S. Bondurant, M. Radmacher, J.L. Slonczewski, pH regulates genes for flagellar motility, catabolism, and oxidative stress in *Escherichia coli* K-12, J. Bacteriol. 187 (1) (2005) 304–319.
[259] H.A. Krebs, D. Wiggins, M. Stubbs, A. Sols, F. Bedoya, Studies on the mechanism of the antifungal action of benzoate, Biochem. J. 214 (3) (1983) 657–663.
[260] M. Stratford, P.A. Anslow, Evidence that sorbic acid does not inhibit yeast as a classic 'weak acid preservative', Lett. Appl. Microbiol. 27 (1998) 203–206.
[261] T. Warnecke, R.T. Gill, Organic acid toxicity, tolerance, and production in *Escherichia coli* biorefining applications, Microb. Cell Fact. 4 (2005) 1–8.
[262] M. Huang, G. Peabody, K.C. Kao, Tolerance of microbial biocatalysts to feedstocks, products, and environmental conditions, Metab. Eng. Bioprocess Commer. (2016) 73–100.
[263] D.B. Flagfeldt, V. Siewers, L. Huang, J. Nielsen, Characterization of chromosomal integration sites for heterologous gene expression in *Saccharomyces cerevisiae*, Yeast 26 (10) (2009) 545–551.
[264] S.I. Mussatto, I.C. Roberto, Alternatives for detoxification of diluted-acid lignocellulosic hydrolyzates for use in fermentative processes: a review, Bioresour. Technol. 93 (1) (2004) 1–10.
[265] C. van Zyl, B.A. Prior, J.C. du Preez, Acetic acid inhibition of D-xylose fermentation by *Pichia stipitis*, Enzyme Microb. Technol. 13 (1) (1991) 82–86.
[266] M.G.A. Felipe, D.C. Vieira, M. Vitolo, S.S. Silva, I.C. Roberto, I.M. Manchilha, Effect of acetic acid on xylose fermentation to xylitol by *Candida guilliermondii*, J. Basic Microbiol. 35 (3) (1995) 171–177.
[267] S. Giannattasio, N. Guaragnella, M. Ždralević, E. Marra, Molecular mechanisms of *Saccharomyces cerevisiae* stress adaptation and programmed cell death in response to acetic acid, Front. Microbiol. 4 (2013) 1–7.
[268] F. Rodrigues, M.J. Sousa, P. Ludovico, H. Santos, M. Côrte-Real, C. Leão, The fate of acetic acid during glucose co-metabolism by the spoilage yeast *Zygosaccharomyces bailii*, PLoS One 7 (12) (2012) 1–7.
[269] C.E. Oshoma, D. Greetham, E.J. Louis, K.A. Smart, T.G. Phister, C. Powell, et al., Screening of non-*Saccharomyces cerevisiae* strains for tolerance to formic acid in bioethanol fermentation, PLoS One 10 (8) (2015) 1–17.
[270] L. Axelsson, M. Franzén, M. Ostwald, G. Berndes, G. Lakshmi, N.H. Ravindranath, Perspective: Jatropha cultivation in southern India: assessing farmers' experiences, Biofuels Bioprod. Biorefin. 6 (3) (2012) 246–256.
[271] J. Su, F. Shen, M. Qiu, X. Qi, T.J. Farmer, J.H. Clark, High-yield production of levulinic acid from pretreated cow dung in dilute acid aqueous solution, Molecules 22 (2) (2017) 285.
[272] C. Liu, Q. Feng, J. Yang, X. Qi, Catalytic production of levulinic acid and ethyl levulinate from uniconazole-induced duckweed (Lemna minor), Bioresour. Technol. 255 (2018) 50–57.
[273] D.W. Rackemann, W.O. Doherty, The conversion of lignocellulosics to levulinic acid, Biofuels Bioprod. Biorefin. 5 (2) (2011) 198–214.

[274] J.Q. Li, The chemistry and technology of furfural and its many by-products, Chem. Eng. J. 81 (1–3) (2001) 338–339.
[275] J. Lecomte, A. Finiels, C. Moreau, A new selective route to 5-hydroxymethylfurfural from furfural and furfural derivatives over microporous solid acidic catalysts, Ind. Crops Prod. 9 (3) (1999) 235–241.
[276] R. Weingarten, J. Cho, W.C. Conner Jr., G.W. Huber, Kinetics of furfural production by dehydration of xylose in a biphasic reactor with microwave heating, Green Chem. 12 (8) (2010) 1423.
[277] P.T. Adeboye, M. Bettiga, L. Olsson, The chemical nature of phenolic compounds determines their toxicity and induces distinct physiological responses in *Saccharomyces cerevisiae* in lignocellulose hydrolysates, AMB Exp. 4 (1) (2014) 1–10.
[278] D. Lin, M. Xiao, J. Zhao, Z. Li, B. Xing, X. Li, et al., An overview of plant phenolic compounds and their importance in human nutrition and management of type 2 diabetes, Molecules 21 (10) (2016) 1374.
[279] N. Babbar, H.S. Oberoi, S.K. Sandhu, V.K. Bhargav, Influence of different solvents in extraction of phenolic compounds from vegetable residues and their evaluation as natural sources of antioxidants, J. Food Sci. Technol. 51 (10) (2014) 2568–2575.
[280] Y.C. Li, Z.X. Gou, Y. Zhang, Z.Y. Xia, Y.Q. Tang, K. Kida, Inhibitor tolerance of a recombinant flocculating industrial *Saccharomyces cerevisiae* strain during glucose and xylose co-fermentation, Braz. J. Microbiol. 48 (4) (2017) 791–800.
[281] S. Stagge, A. Cavka, L.J. Jönsson, Identification of benzoquinones in pretreated lignocellulosic feedstocks and inhibitory effects on yeast, AMB Exp. 5 (1) (2015) 62.
[282] D. Chiaramonti, M. Prussi, S. Ferrero, L. Oriani, P. Ottonello, P. Torre, et al., Review of pretreatment processes for lignocellulosic ethanol production, and development of an innovative method, Biomass Bioenergy 46 (2012) 25–35.
[283] A. Cavka, L.J. Jönsson, Detoxification of lignocellulosic hydrolysates using sodium borohydride, Bioresour. Technol. 136 (2013) 368–376.
[284] A. Cavka, A. Wallenius, B. Alriksson, N.O. Nilvebrant, L.J. Jönsson, Ozone detoxification of steam-pretreated Norway spruce, Biotechnol. Biofuels 8 (1) (2015) 196.
[285] N.N. Nichols, B.S. Dien, G.M. Guisado, M.J. López, Bioabatement to remove inhibitors from biomass-derived sugar hydrolysates, Appl. Biochem. Biotechnol. 121–124 (2005) 379–390.
[286] L. Favaro, M. Basaglia, A. Trento, E. Van Rensburg, M. García-Aparicio, W.H. Van Zyl, et al., Exploring grape marc as trove for new thermotolerant and inhibitor-tolerant *Saccharomyces cerevisiae* strains for second-generation bioethanol production, Biotechnol. Biofuels 6 (1) (2013) 168.
[287] J. Smith, E. van Rensburg, J.F. Görgens, Simultaneously improving xylose fermentation and tolerance to lignocellulosic inhibitors through evolutionary engineering of recombinant *Saccharomyces cerevisiae* harbouring xylose isomerase, BMC Biotechnol. 14 (2014) 41.

Further reading

C. Bro, B. Regenberg, J. Förster, J. Nielsen, In silico aided metabolic engineering of *Saccharomyces cerevisiae* for improved bioethanol production, Metab. Eng. 8 (2006) 102–111.
H.H. Cheng, L.M. Whang, K.C. Chan, M.C. Chung, S.H. Wu, C.P. Liu, et al., Biological butanol production from microalgae-based biodiesel residues by *Clostridium acetobutylicum*, Bioresour. Technol. 184 (2015) 379–385.

I. Dolejš, V. Krasňan, R. Stloukal, M. Rosenberg, M. Rebroš, Butanol production by immobilised *Clostridium acetobutylicum* in repeated batch, fed-batch, and continuous modes of fermentation, Bioresour. Technol. 169 (2014) 723–730.

P. Dürre, Fermentative butanol production. Bulk chemical and biofuel, Ann. N.Y. Acad. Sci. 1125 (2008) 353–362.

M.I. Foda, H. Dong, Y. Li, Study the suitability of cheese whey for bio-butanol production by clostridia, J. Am. Sci. 6 (2010) 39–46.

L. Frank, Membrane process opportunities and challenges in the bioethanol industry, Desalination 250 (2010) 1067–1069.

R.L. Kudahettige-Nilsson, J. Helmerius, R.T. Nilsson, M. Sjöblom, D.B. Hodge, U. Rova, Biobutanol production by *Clostridium acetobutylicum* using xylose recovered from birch Kraft black liquor, Bioresour. Technol. 176 (2015) 71–79.

W.D. Mores, J.S. Knutsen, R. Davis, Cellulase recovery via membrane filtration, Appl. Biochem. Biotechnol. 91–93 (2001) 297–309.

F. Raganati, A. Procentese, F. Montagnaro, G. Olivieri, A. Marzocchella, Butanol production from leftover beverages and sport drinks, BioEnergy Res. 8 (2014) 369–379.

A. Ranjan, S. Khanna, V.S. Moholkar, Feasibility of rice straw as alternate substrate for biobutanol production, Appl. Energy 103 (2013) 32–38.

J. Roth, N. Tippkotter, Evaluation of lignocellulosic material for butanol production using enzymatic hydrolysate medium, Cellul. Chem. Technol. 50 (2016) 405–410. Available from: https://www.scopus.com/inward/record.uri?eid = 2-s2.0-84986587086& partnerID = 40&md5 = ad23340c7e91690928bb8fe22f3720f6.

R. Sindhu, M. Kuttiraja, V.E. Preeti, S. Vani, R.K. Sukumaran, P. Binod, A novel surfactant-assisted ultrasound pretreatment of sugarcane tops for improved enzymatic release of sugars, Bioresour. Technol. 135 (2012) 67–72.

M. Yang, S. Kuittinen, J. Vepsäläinen, J. Zhang, A. Pappinen, Enhanced acetone–butanol–ethanol production from lignocellulosic hydrolysates by using starchy slurry as supplement, Bioresour. Technol. 243 (2017) 126–134.

J. Yu, T. Zhang, J. Zhong, X. Zhang, T. Tan, Biorefinery of sweet sorghum stem, Biotechnol. Adv. 30 (2012) 811–816.

CHAPTER 4

Lignocellulosic biomass to biodiesel

Gaetano Zuccaro[1,2], Domenico Pirozzi[2] and Abu Yousuf[3]
[1]Laboratory of Environmental Biotechnology, National Institute of Agronomic Research,University of Montpellier, Narbonne, France
[2]Department of Chemical, Materials and Production Engineering, University of Naples "Federico II", Naples, Italy
[3]Department of Chemical Engineering & Polymer Science, Shahjalal University of Science and Technology, Sylhet, Bangladesh

Contents

4.1 Introduction

The application of the "circular economy" concept requires a growing effort to recycle the wastes of each human activity [1,2], which should be transformed into secondary raw materials offering equivalent functionality. In particular, technical products should be reused or recycled after their

Lignocellulosic Biomass to Liquid Biofuels
DOI: https://doi.org/10.1016/B978-0-12-815936-1.00004-6

life cycle, while organic products could be decomposed through a biological cycle. Moreover, all primary raw materials and the energy needed should be derived from renewable sources and efficiently use to improve potential savings.

A very deep change is obviously needed in the structure of industrial systems to extend this logic across the economy. In this view the biorefinery approach embodies the circular economy concept. Biorefinery is the sustainable processing of biomass into a wide spectrum of marketable bioproducts, such as food, feed, biomaterials, and chemicals, and bioenergy, such as fuels, power, and/or heat [3]. The development of a sustainable concept of biorefinery needs to solve the problem related to the competition with food chain, the impact on water use, the use of land to biomass supply, the influence on the balance of soil carbon stock and on its fertility, the balance of gas emissions, the impact on biodiversity, the potential toxicological risks, and the energy efficiency. Moreover, the concept of biorefinery should take into account the impacts on international and regional dynamics. For this reason the sustainability assessment is not an absolute number but should be compared to conventional systems that provide the same products and services [4].

However, new biorefinery concepts are still mostly in the R&D (Research and Development) phase or in pilot or small-scale demonstration state, and their industrial application is still far away, but these new concepts will be implemented in different countries in the medium term (2015–25) [5], although current economic conditions could be a reason for a delay in the application of these concepts.

The technologies to implement these processes still need to become commercial, but they require an important effort due to the complexity in terms of feedstocks used (e.g., algae, energy crops, and wood chips from short rotation), as well as the spectrum of bioproducts. Among these, particular attention is paid to microbial cultures that can be established to convert lignocellulosic sugars or low-value hydrophobic substrates into biodiesel and biochemicals.

4.2 Potentiality of lignocellulosic biomass as a source of biodiesel

Biomass and biomass-derived materials have been considered one of the most promising alternatives to fossil resources to fuel production, since they are a sustainable source of organic carbon with net zero carbon emission, as well as they are generated from available atmospheric CO_2, water and sunlight, and obtained by using biological processes, such as photosynthesis [6,7]. The biomass contributes to around 10% of global primary energy, about 55 EJ in 2013, and the technical potential derived from available annual supplies of residues and wastes has been estimated in energy terms above 100 EJ/year at delivered costs in the range of US $2–3/GJ [8]. However, in many regions because of limiting supplies of these biomass feedstocks, the growth of vegetative grasses or short rotation crops may be necessary to meet the demand for biofuels. Local field trials to determine the levels of productivity under a range of growth conditions is required to identify potential energy crops able to grow in marginal and degraded lands and avoid the direct competition with food chain and fiber crops, which require better quality of arable lands.

In recent decades, many countries are implementing biofuel policies to reduce the dependence on fossil fuels, to improve energy and food security or to create job opportunities and to develop the rural regions. Around 4.4 EJ were produced as liquid biofuels in 2014 [9]; ethanol and biodiesel currently supply about 3% of world's energy demand in road transport, with forecasts to contribute up to 30% in 2050, when it is expected that 1.7–2.1 billion cars will be running in our planet, about 2.6 times the global fleet in 2010 [8,10]. The renewable alternatives to the increasing consumption of liquid fuels are essentially liquid biofuels (Fig. 4.1).

Global biodiesel production, around 30 billion liters, is based largely on vegetable oils, mostly from rapeseed and sunflower (Europe), soybeans (United States, Brazil, and Argentina), palm (Indonesia), and other sources such as jatropha and coconut, cooking oils (the main feedstock in China), and animal fats [11]. The top producers were the United States, which accounted for 16% of the total production, followed by Brazil and Germany (both with 11%), Indonesia (10%), and Argentina (9.7%). Europe accounted for 39% of global biodiesel production in 2014 [8].

Different raw materials used as feedstock for biodiesel, including agricultural residues and lignocellulosic biomasses, have recalcitrant properties that represent a critical factor for their use and require highly efficient, environmental friendly, and cost-effective pretreatments for subsequent hydrolysis and fermentation processes [12,13].

Nevertheless, lignocellulosic biomass, which is the most abundant and biorenewable biomass on earth is the feedstock offering the most promising perspectives [14,15]. It has been estimated that 3700×10^6 t of

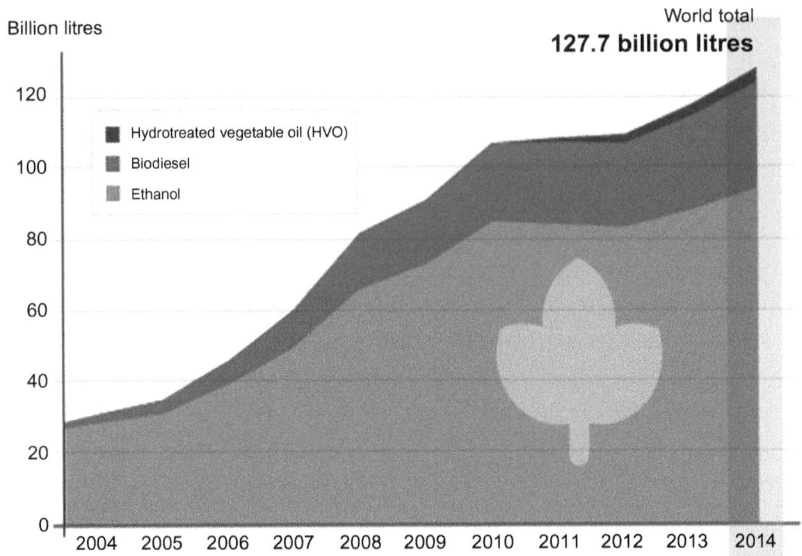

Figure 4.1 Ethanol, biodiesel, and HVO (Hydrotreated Vegetable Oil) Global Production, 2004–14 [9].

agricultural residues are produced world-wide annually as agro-forestry by-products and that 1376×10^6 t cellulose and 848×10^6 t hemicellulose are produced globally every year. Furthermore, it has been estimated that 1300×10^6 t of food are wasted every year, corresponding to one-third of the food globally produced for human consumption [16,17]. Lignocellulosic biomass can be considered the most abundant carbon-neutral renewable source, which can decrease CO_2 emissions and atmospheric pollution. Furthermore, its major component, the cellulose, is the strongest potential candidate for the substitution of petroleum-based polymers because of its ecofriendly properties such as renewability, biocompatibility, and biodegradability [18].

Table 4.1 shows the comparison of various biodiesel lignocellulosic sources grown worldwide in terms of oil yield, oil content, land area use, productivity, and price.

The development of lignocellulosic biomass conversion to biodiesel and polymers still remains a big challenge due to the technical and economic obstacles [12,15,19] mostly related to the biomass recalcitrance stemming from the barriers naturally developed to avoid the degradation from natural infections. The future of second-generation biofuel production will be critically affected by the price of the oil, as well. It is

Table 4.1 Comparison of various biodiesel lignocellulosic sources.

Plant	Oil yield (L/ha year)	Oil content (% oil by wt. in biomass)	Land area use (m^2 year/kg biodiesel)	Biodiesel productivity (kg biodiesel/ha year)	Price (US $/L)
Corn (*Zea mays* L.)	172	44	66	152	0.8
Soybean (*Glycine max* L.)	446–636	18	18	562	0.8
Camelina (*Camelina sativa* L.)	915	42	12	809	1.1
Sunflower (*Helianthus annuus* L.)	952–1070	40	11	946	1.4
Canola/Rapeseed (*Brassica napus* L.)	974–1190	41	12	862	1.1
Jatropha (*Jatropha curcas* L.)	1892	28	15	656	0.9
Oil palm (*Elaeis guineensis*)	5366–5950	36	2	4747	0.9

estimated that to produce a cost-competitive lignocellulosic biodiesel, the price of the oil required is in the range of \$70–\$85/bl [20]. Some studies estimated the value that can be potentially generated by the lignocellulosic industry. Hertel et al. [21] gave a value to second-generation industry of \$64 billion under baseline conditions.

A number of novel technologies are being developed to refine lignocellulosic biomass for the production of renewable oil and green monomers [22]. In addition, the number of biorefinery-related pilot and demonstration plants is increasing [23], though only few companies, such as Lignol, Verenium, and Mascoma, have so far focused their attention on the development of biorefining technologies for the production of advanced biodiesel, biochemicals, and biomaterials from nonfood cellulosic biomass feedstocks [18].

Lignocellulosic biomasses consist of micro- and macrofibrils organized in crystalline structures, that need to be pretreated in order to promote access to their cellulose, hemicellulose, lignin and small amounts of pectin, protein, and ash fractions in the subsequent hydrolysis step [24]. Cellulose, hemicellulose, and lignin fractions depend by species, age, and growing conditions of lignocellulosic biomasses [25,26].

The cellulose is the most abundant constituent of plant cells. It is a polysaccharide characterized by a linear sequence of D-glucose molecules linked by β-1,4-glycosidic bonds with a high degree of polymerization equal to 10,000 or even higher. It consists of intra- and intermolecular hydrogen bonds able to influence the specific crystallinity. It is chemically stable and resistant to microbial degradation [27,28]. The cellulose fibrils are responsible for the great tensile strength of the cell wall [29]. This structural and inherent integrity of cellulose plays an important role in the recalcitrance of lignocellulosic biomass [30]. The hydrophobic surface involves the formation of water layer that prevents the diffusion of enzymes and the degradation of the products on the surface [31].

Hemicellulose is the second most abundant polymer of plant cells. Unlike cellulose, hemicellulose has a random and amorphous structure, which is composed by several monomers including D-glucose, D-galactose, D-mannose, D-xylose, L-arabinose, D-glucuronic acid, and 4-O-methyl-D-glucuronic acid. It is characterized by a degree of polymerization lower than 200. Hemicelluloses are embedded in the plant cell walls to form a complex network of bonds that provides structural strength by linking cellulose fibers into microfibrils and cross-linking with lignin.

The lignin is a complex network of phenyl propane units. It is the most important nonpolysaccharide fraction of lignocellulosic biomass. The three characteristic lignin's monomers are *p*-cumarilic alcohol, sinapyl alcohol, and conferilic alcohol related by different ether bonds. The lignin also provides protection against chemical and microbial degradation [26] (Fig. 4.2).

4.3 Pathway: lignocellulosic biomass to lipids

The biochemical pathway to lipid production from lignocellulosic biomass typically provides the following steps (Fig. 4.3):

- Biomass pretreatment
- Enzymatic hydrolysis of pretreated biomass to generate fermentable sugars
- Fermentation of sugars to lipids

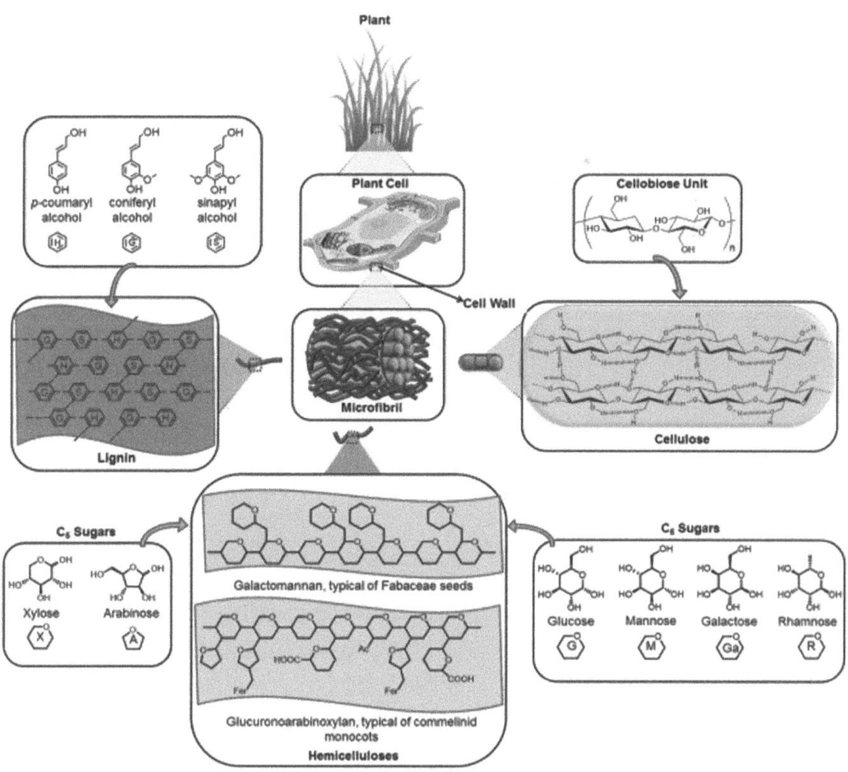

Figure 4.2 The main components and structure of lignocellulose [18].

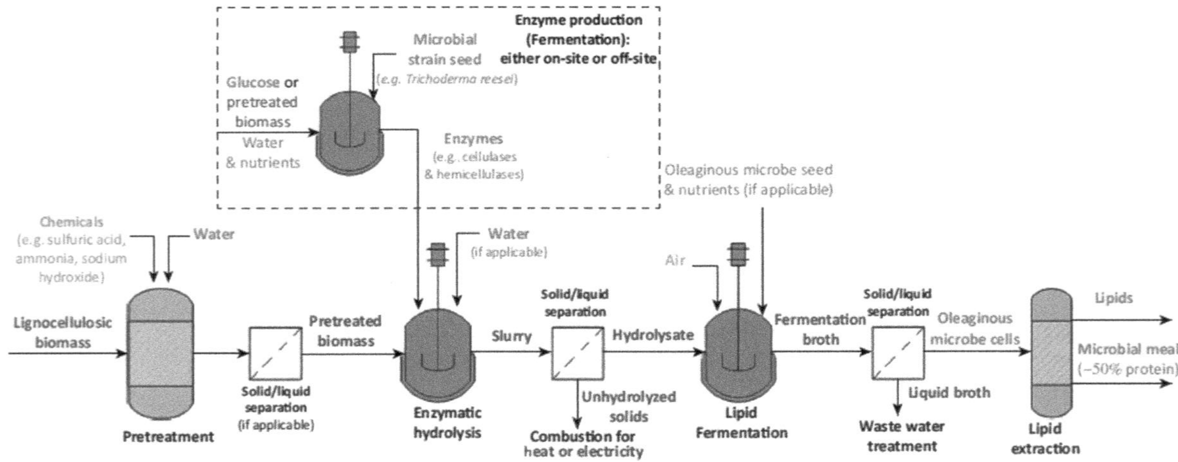

Figure 4.3 Biochemical route for lipid production from lignocellulosic biomass [32].

- Lipid extraction

  Lipid production is mainly related to several technical challenges including (1) the control of C:N ratio, exceeding 65 and near 100 [33,34]; (2) sugar concentrations (almost 100 g/L); and (3) the presence of inhibitors, as side-products formed during the pretreatments, which can include furfural, hydroxymethylfurfural, acetic acid, phenolic acids, and other compounds [32].

4.4 Preprocessing of lignocellulosic biomass (mechanical, chemical, and biological)

The lignocellulosic biomass is characterized by a complex structure that requires pretreatment processes in order to make the structure accessible to hydrolytic enzymes [24] and, in the same time, to prevent carbohydrate degradation since the carbohydrates represent the most essential substrate for the microorganisms. The pretreatment of the lignocellulosic biomass is also required to prevent the generation of inhibitors directly related to lignin fraction. Thus pretreatment processes are of crucial importance, due to their large impact on all subsequent steps of lipid production. However, these processes are usually energy-expensive and significantly affect process costs [35,36]. For this reason, one of the objectives is to optimize and reduce the energy demand and to improve the processing of lignocellulosic biomasses. The pretreatment methodologies can be classified in physical, chemical–physical, chemical, and biological. A list of the most promising and commonly employed pretreatment processes is reported in Table 4.2. A single method cannot represent the universal choice due to the variability of lignocellulosic biomass composition. Usually, extra costs are necessary to solve the negative effects of pretreatments on subsequent steps.

4.5 Hydrolysis of lignocellulosic biomass

4.5.1 Enzymatic hydrolysis

The pretreatment of lignocellulosic biomasses is important to facilitate the access of hydrolytic enzymes to the crystalline structure Enzymatic hydrolysis of cellulose and hemicellulose is carried out under mild conditions (pH 4.5–5.0 and temperature 40°C–50°C), which ensure reducing corrosion problems, low energy consumption, and reduced generation of toxic by-products [80].

Table 4.2 Summary of various processes used for the pretreatment of biomass.

Pretreatment process	Feedstock	Procedure	Composition changed	Advantages	Disadvantages	References
Physical						
Mechanical size reduction	Rice straw, corn stover, wheat straw, cane and sweet sorghum bagasse	Chipping, grinding, milling	Cellulose and hemicellulose conversion	Moderate temperature Cellulose crystallinity reduction Adaptability for many process modifications Less hazardous process Good sugar yield	Partially sugar degradation Generation of inhibitor compounds High power consumption	[36,37]
Mechanical extrusion	Corn stover, switchgrass, soybean	Heating process under shear mixing, $T > 300°C$	Cellulose and hemicellulose crystallinity disruption	Low by-product formation	High energy demand Difficult in scaling up for industrial purposes	[38,39]
Microwave	Switchgrass, corn stover, sugarcane bagasse, rape straw, *Miscanthus*	1%–5%NaOH, H_3PO_4 4–30 min 300–600 W	Degradation of structural organization of cellulose fractionReduction in lignin composition	Easy operation Low energy requirements Minimum generation of inhibitors	No special advantage over conventional heat processing	[40–42]
Ultrasound	Sugarcane bagasse, corn stover, switchgrass, *Sorghum*, poplar	Sonication 10–100 kHz 50°C	Rupture of the cellulose and hemicellulose fractions	High accessibility to cellulose degrading enzymes for effective breakdown into simpler reducing sugars	Costly process	[43,44]

Pyrolysis	Wood, waste cotton, corn stover, switchgrass	$T > 300°C$, then cooling and condensing	Thermal decomposition of cellulose Separation of the main portions of hemicellulose and residual lignin from cellulose	Gas and liquid production	Production of volatile products and char High temperature requirement Ash production Particle size reduction can be costly	[45,46]
Pulsed electric field	Wood chip, switchgrass	High voltage (5–20 kV/cm) for short duration (nano to milliseconds)	Lignin content decrease	Very low energy requirement Ambient conditions Simple equipment	Process requires more research	[47,48]
Chemical						
Acid hydrolysis	Poplar wood bagasse, corn stover, wheat straw, rye straw, rice hulls, switchgrass	(Dilute) 0.75%–5% H_2SO_4, HCl, HNO_3 $T = 160°C–200°C$ $P = 1$ MPa Continuous process for low solids loads (5%–10% substrate/mixture) Batch process for high loads (10%–40% substrate/mixture) (Concentrated) 10%–30% H_2SO_4 $T = 170°C–190°C$ 1:2 Solid/Liquid ratio	Effective hydrolyze of hemicelluloses with high sugar yieldLignin is not solubilized	Low thermal energyPractical and simple technique	Generation of toxic inhibitors Equipment corrosion Request recovery steps pH neutralization	[36,49–56]
Alkaline hydrolysis	Hardwood, bagasse, corn stover, cane leaves	Dilute NaOH, 60°C, 24 h $Ca(OH)_2$, 120°C, 4 h Adding of H_2O_2 (0.5–2.15% v/v), 35°C	High hemicellulose fraction hydrolysis High cellulose conversion Lignin removal	Reactor cost lower if compared to acid hydrolysis Low inhibitor formation Hemicellulose and lignin removal	Long residence time request Presence of salts incorporated into biomass	[36,49,50,57,58]

(Continued)

Table 4.2 (Continued)

Pretreatment process	Feedstock	Procedure	Composition changed	Advantages	Disadvantages	References
Ozonolysis	Wheat straw, bagasse, peanut, poplar, sawdust, green hay, pine	Ozone, room temperature, and pressure	Effectively remove lignin	No toxic compound generation	Costly process Request of large amount of ozone	[36]
Organosolv	Wheat straw, sugarcane bagasse, poplar, spruce, pine	Organic solvents (methanol, ethanol, acetone, ethylene glycol, triethylene glycol) or mixture with 1% of H_2SO_4 or HCl $T = 185°C$, 30–60 min, pH = 2.0–3.4	Total hydrolysis of hemicellulose Lignin solubilization	Pure lignin removal as by-product	Formation of toxic inhibitors High capital investment Organic solvents require recycling	[59–61]
Ionic liquids	Wheat straw, bagasse, peanut, poplar and corn stover	Ionic liquid/biomass ratio equal to 20:1 $T = 120°C$, 30 min	Cellulose conversion	Efficiently dissolution of cellulose	Costly processSolutions viscous and difficult to handle	[62,63]
Natural deep eutectic solvents	Rice straw	Choline, urea, sugars, amino acids, and several other organic acids	Lignin solubilization	Green solvent at room temperature Cost effective Synthesis of nontoxic compounds Biocompatible Highly biodegradable	High viscosity The dilution with water decreases the interactions	[64,65]

Physical–chemical

Steam explosion	Wheat straw, corn stalk, poplar, aspen, eucalyptus softwood, rice straw, barley straw, sweet sorghum bagasse, olive stones, common reed	Saturated steam ($T = 160°C–290°C$, $P = .69–4.85$ atm, 1–10 min), then decompression until atmosphere pressure	High hemicellulose fractions removal	Good sugar recovery Cost-effective Less hazardous process Lignin transformation and hemicellulose solubilization	Generation of inhibitory compoundsIncomplete destruction of lignin matrix	[36,49,50,66–69]
LHW	Sugarcane bagasse, corn stover, wheat straw, sunflower stalk, bagasse, corn stover, olive pulp	Pressurized hot water ($T = 170°C–230°C$, $P > 5$ MPa, 1–46 min)	The majority of hemicellulose can be dissolved	High recovery of pure hemicellulose No addition of catalyst or chemicals Hydrolysis of hemicellulose Little formation of inhibitors	High energy demand	[50,70–73]
AFEX	Wheat straw, corn stover, rice straw, aspen wood chips bagasse, rice hulls, corn stover, switchgrass, coastal Bermuda grass	$T = 90°C$, $P = 1.12–1.36$ MPa, 30 min 1–2 kg ammonia/kg dry biomass	Not suitable for high-lignin materials High cellulose conversion	Efficient removal of lignin Low formation of inhibitor compounds and sugar degradation Moderate process conditions	Costly process Not efficient for biomass with high lignin content	[36,50,74,75]

(Continued)

Table 4.2 (Continued)

Pretreatment process	Feedstock	Procedure	Composition changed	Advantages	Disadvantages	References
CO_2 explosion	Wheat straw, sugarcane bagasse, recycled paper	4 kg CO_2/kg fiber $P = 5.6$ MPa	Cellulose conversion	Moderate process conditions Increasing of accessible surface area No toxic compound generation	No lignin and hemicellulose modification	[36,76]
Wet oxidation	Common reed, rice husk	Air/oxygen along with water or hydrogen peroxide at high temperature (120°C–170°C, 30 min)	Hemicellulose is broken down into smaller pentose monomers and the lignin undergoes oxidation, while the cellulose is least affected	The combination of oxygen, water, high temperature, and alkali reduce toxic inhibitors High delignification and solubilization of cellulosic material	High costs of hydrogen peroxide and oxygenLimitation in large-scale processes	[77,78]
Biological						
Biological	Wheat straw, rice straw, corn stover	Cellulase, hemicellulase and lignin degrading enzyme production by fungi using solid state fermentation of biomass	Hemicellulose conversion	Low energy requirement No toxic compound generation Cost effective environmentally friendly	Slow bioconversion	[36,79]

AFEX, Ammonia fiber explosion; *LHW*, liquid hot water.

The insoluble portion of the biomass is initially broken as a result of synergistic action of different enzymes because of multistep and heterogeneous reactions [81,82].

The hydrolytic enzymes can be divided into the following species:
- Endoglucanases, which hydrolyze randomly internal β-1,4-D glucosidic linkages in the cellulose chain
- Cellobiohydrolases (CBHs, also known as exoglucanases), which progress along the cellulose line and cleave off cellobiose units from the ends
- β-Glucosidases (BG), which hydrolyze cellobiose to glucose and also cleave off glucose units from cello-oligosaccharides

The synergistic action of these enzymes is able to increase glucose conversion starting from lignocellulosic biomass. In addition, the introduction of hemicellulosic enzymes is required, such as:
- Endo-1,4-β-D-xylanases to make xylan chain
- 1,4-β-D-xylosidases to release xylose
- Endo-1,4-β-D-mannanases to break internal bonds to obtain mannans
- 1,4-β-D-mannosidase that cleave molecules of manno-oligosaccharide in mannose.

Cellulases and hemicellulases described above can be produced by bacteria, such as *Clostridium*, *Ruminococcus*, *Streptomyces*, *Cellulomonas*, *Bacillus* and *Erwinia*, and fungi such as *Trichoderma*, *Penicillium*, *Fusarium*, and *Humicola* [36,83]. Among the cellulases produced by various microorganisms, those derived from *Trichoderma reesei* or from *Trichoderma viride* have been widely studied and better characterized. The stability of their enzymatic activity and the resistance to the presence of inhibitors are positive properties, though these enzymes show a reduced BG activity. On the other hand, cellulases from *Aspergillus* are able to overcome the inefficiencies related to *Trichoderma*. In many studies, it has been observed a synergy between enzymes derived from two microbial strains to improve the efficiency of the hydrolysis process [84–88] (Fig. 4.4).

The obstacles still preventing the implementation of this process on an industrial scale are associated to the cost of the enzymes. In the last years the intense research activity of different companies, such as Novozymes and Genencor, that focused on the process operating-cost minimization, has increased the use of these enzymes [89].

Since industrial processes require high product concentrations, the inhibition phenomena significantly reduce the reaction rate and the efficiency of enzymatic hydrolysis reactions, both in batch and in continuous

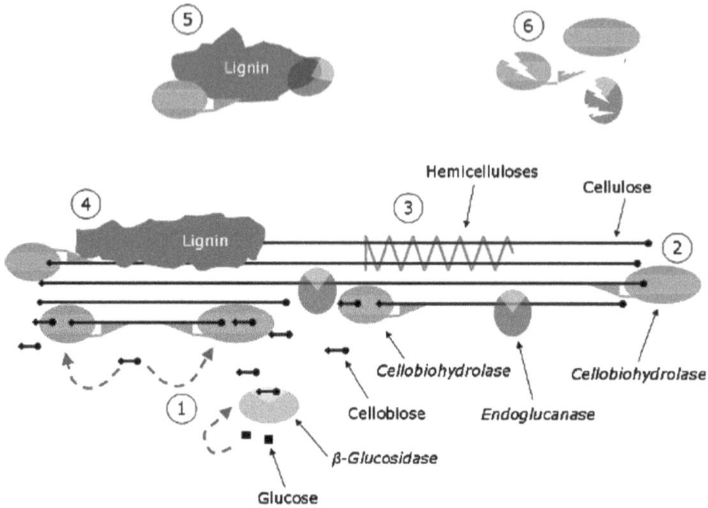

Figure 4.4 Simplified diagram of lignocellulose hydrolysis showing synergism and limiting factors. Cellulose is symbolized straight lines. (1) Product inhibition of BG and CBH by glucose and cellobiose. (2) CBH hydrolyzing from the end of a cellulose chain. (3 and 4) Hemicelluloses and lignin associated with or covering the microfibrils prevent the cellulases from accessing the cellulose surface. (5) Enzymes (both cellulases and hemicellulases) can be unspecifically adsorbed onto lignin particles or surfaces. (6) Denaturation or loss of enzyme activity due to mechanical shear, proteolytic activity or low thermostability [88]. *BG*, β-Glucosidase; *CBH*, cellobiohydrolase.

reactors [90]. The increase of the enzyme/substrate ratio is a possible solution to overcome this problem, though it is in contrast with the objective to minimizing the operating costs. Alternative solutions to reduce inhibition phenomena are the addition of surfactants to change surface properties of cellulose. The economical balance of the process can be further on improved by adopting suitable techniques to recover the enzymes, once immobilized, through recycling mechanisms [91]. Unfortunately, all of these techniques for recycling and reducing enzyme adsorption have been so far only tested at laboratory scale. Furthermore, most of the studies do not include the costs associated to the recycling of the enzymes and to the reduction of the enzyme. The ability to scale-up the techniques, as well as their robustness and feasibility, still need to be demonstrated [88].

Finally, in order to reduce operating costs and the enzyme dosage, innovative methodologies are being developed to carry out the cellulose hydrolysis and the subsequent fermentation in a single reactor. This type of process is commonly called simultaneous saccharification and fermentation

(SSF). Unfortunately, hydrolysis and fermentation require different operating conditions: the optimal temperature for the growth of yeasts is 30°C–37°C, whereas the activity of hydrolytic enzymes is maximum at 40°C–50°C [92–94].

SSF of lignocellulosic biomass is commonly studied for bioethanol production [95,96]. SSF can potentially be developed to obtain oleaginous microorganisms, able to produce intracellular lipids, though this field is still substantially unexplored [97]. The reason is attributable to the lower metabolic activity of oleaginous microorganisms and to the difficulty of oxygen supply due to the increase of culture medium viscosity especially in the case of high concentrations of processed biomass [33,97,98].

4.6 Oleaginous strains and their productivity

The oleaginous microorganisms have the ability to produce and accumulate a large amount of lipids if compared with their dry mass [99]. They consist of different families, such as microalgae, bacillus, and fungi (molds and yeasts) [100].

4.6.1 Yeasts

The increasing interest to the applications directly connected to the lipid production by oleaginous fermentation, in the nutraceutical and pharmaceutical fields, as well as the possibility to use them also to produce biodiesel, has created scientific appeal aimed to obtain alternative ways to those offered by oleaginous yeasts.

In comparison to oleaginous plants, oleaginous yeasts are not influenced by climate variability and can grow in the presence of different carbon sources, for example, hexose and pentose sugars with high growth rates [101]. Generally, they have the ability to accumulate microbial oils, commonly called single cell oils (SCOs). These oils are usually more than 20%–25% of their total dry weight [98,102] reaching 65% of their dry weight in specific growth conditions [103].

When cultured at low concentrations of nitrogen, the oleaginous yeasts have the ability to trigger a cascade of reactions leading to intermediate compounds formation such as acetyl-CoA [104], enabling oil accumulation mechanisms related to the tricarboxylic acid cycle. In eukaryotic microorganisms, these mechanisms take place in the mitochondria (Fig. 4.5) [32].

Though oleaginous and non-oleaginous yeasts share the same biosynthetic pathways, there is a fundamental difference concerning their

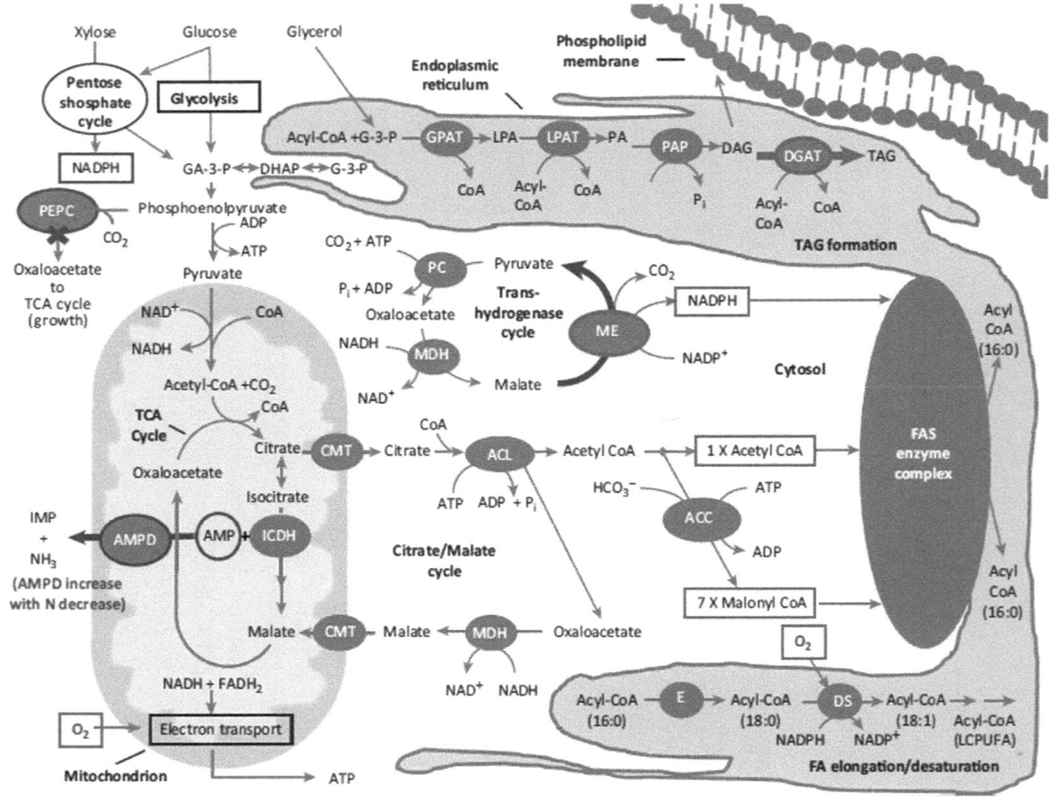

Figure 4.5 Major cellular pathways, enzymes, control points, and organelles involved in the conversion of carbohydrates to lipids [32].

behavior in presence of a carbon source excess. In fact, in a medium with abundant carbon source and limiting amount of nitrogen, if all nitrogen source has been consumed, oleaginous microorganisms utilize the remaining carbon source for the synthesis of lipids, mainly triacylglycerols (TAGs); on the contrary, non-oleaginous microorganisms convert carbon sources into polysaccharides (glycogen, glucans, mannans, etc.) and are not predisposed to accumulate lipids (maximum up to 10%–20%). Although, the biosynthetic pathway to fatty acids substantially the same, in oleaginous yeasts, the reason of this difference is related to the production of acetyl-CoA in the cytosol as a precursor for FAS (fatty acids synthetase) and to NADPH (nicotinamide adenine dinucleotide phosphate hydrogen) production, which is used as a reducing agent in the synthesis of fatty acids [105]. In oleaginous yeasts, the most abundantly produced fatty acids are oleic acid (C18:1, n-9), linoleic acid (C18:2, n-6), palmitic acid (C16:0), and palmitoleic acid (C16:1), as well as by C18:3 or alpha-linolenic acid that, in general, represent less than 10% of the total [106].

In the recent years, it has been proven that yeasts such as *Rhodotorula glutinis* [107], *Rhorosporidium toruloides* [104], *Trichosporon fermentans* [108], and *Lipomyces starkeyi* [103,109] show a potential to microbial oil production also because they can be cultivated in simple media containing low cost substrates [110] (Table 4.3).

TAGs and steryl-esters (SE) are the lipids most abundantly accumulated in the cells during stationary growth phase. TAGs have a distribution of acyl groups very similar to the plant oils, in particular, the central position is occupied almost exclusively by an unsaturated group. Since these molecules are without positive or negative charge, they cannot be part of cell membranes. However, they are sequestered in hydrophobic particles called lipid particles or lipid bodies and are accumulated in the form of micro droplets. Though the lipid fraction of the cells is composed almost entirely of TAGs, the latter are extracted together with other lipid fractions (phospholipids, sterols, SE, and others) associated to the cell membrane and free fatty acids related to an uncontrolled lipolysis which can occur during the extraction process.

4.6.2 Molds

Some oleaginous molds can store up to 80% of their biomass as lipids. As in most other microorganisms, the increase in lipid yield is related to C/N ratio, temperature, and pH. Table 4.4 shows the values in terms of lipid

Table 4.3 Oleaginous yeast species for lipid production.

Oleaginous yeast species	Substrate	Culture mode	Lipid production	References
Cryptococcus curvatus	Corn cob hydrolysate, sweet sorghum bagasse enzymatic hydrolysate	Flask	10.83 g/L[a] 61%–73.26%[b]	[111,112]
Lipomyces starkeyi	Corn cob acid hydrolysate, sugarcane bagasse acid hydrolysate	250 mL flask	8.1 g/L[a] 26.9%–55%[b]	[113,114]
Rhodosporidium toruloides	Jerusalem artichoke, cassava starch enzymatic hydrolysate	2 L fermenter	14–39.6 g/L[a] 43.3%–63.4%[b]	[115,116]
Rhodotorula glutinis	Corn cob acid hydrolysate, wheat straw acid hydrolysate	5 L fermenter	1.4–5.5 g/L[a] 11.86%–36.4%[b]	[117]
Trichosporon cutaneum	Corn cob acid and enzymatic hydrolysates, diluted acid pretreated and biodetoxified corn stover	250 mL flask	9.8–12.3 g/L[a] 32.1%–40%[b]	[97,118]
Yarrowia lipolytica	Sugarcane bagasse and rice bran hydrolysate	250 mL flask	5.2–6.68 g/L[a] 48%–58.5%[b]	[119]

[a]Lipid yield.
[b]Lipid content (% w/w).

Table 4.4 Lipid yield and content of oleaginous molds in different fermentation conditions.

Mold species	Substrate	Culture media	Lipid production	References
Mortierella isabellina	Corn stover enzymatic hydrolysate, sweet sorghum	Liquid solid	0.016–0.11 mg/gds[a] 29.47%–38.36%[b]	[120,121]
Microsphaeropsis sp.	Wheat straw	Solid	80 mg/gds[a] 10.2%[b]	[122]
Aspergillus oryzae	Wheat straw + bran	Solid	62.9 mg/gds[a] N.A.%[b]	[123]
Colletotrichum sp.	Rice straw + wheat bran	Solid	68.2 mg/gds[a] N.A.%[b]	[124]
Alternaria sp.	Rice straw + wheat bran	Solid	60.3 mg/gds[a] N.A.%[b]	[124]

[a]Lipid yield (mg/g dry substrate).
[b]Lipid content (% w/w).

yield and content of oleaginous molds in different fermentation conditions [100,125].

4.6.3 Bacteria

Oleaginous bacteria show a high cell growth rate and simpler cultivation requirements. In fermenters operated in batch mode the growth of bacteria can be affected by micro- and macronutrient limitation. The excess of micronutrients promotes a very high biomass concentration. On the other hand, when the macronutrients are progressively consumed, they represent a cause of slowing down and finally of halt in biomass growth [126]. Bacterial species, such as *Mycobacterium*, *Streptomyces*, *Rhodococcus*, and *Nocardia*, can accumulate TAGs at high concentrations. The actinomycete group could be used for their remarkable fatty acid accumulation (up to 70% of the cell dry weight) [127].

Bacterial strains, such as *Dietzia maris* sp., *Stappia* sp., *Nocardioides* sp., *Sphingomonas* sp., *Oceanicaulis alexandrii* sp., *O. alexandrii* sp., and *Micrococcus* sp., isolated from marine living cells, contain a total fatty acid content of 0.3% to 4% dry weight [128].

Generally, most bacteria are not oil producers. Only few of them can accumulate lipids, mostly in the form of polyhydroxyalkanoates [129]. For this reason, there is no industrial significance to produce biodiesel by using oleaginous bacteria [100].

4.6.4 Microalgae

Microalgae have been widely used for decades as feedstock for traditional applications in cosmetic, pharmaceutical, and nutrition sectors. In addition, a variety of bioactive substances, such as carotenoids, polysaccharides, and β-carotene, can be produced. The forms of microalgal products include tablets, capsules, liquids, as well as pure molecules with high added value such as fatty acids, pigments, stable isotope biochemicals, face and skin care products, antiaging creams, refreshing or regenerant care products, emollient and antiirritant in peelers [130–132].

Microalgae are prokaryotic or eukaryotic photosynthetic microorganisms that can grow rapidly even in harsh conditions due to their unicellular or simple multicellular structure. Examples of prokaryotic and eukaryotic microorganisms are Cyanobacteria (*Cyanophyceae*) and green algae (*Chlorophyta*). Microalgae assume many types of metabolisms, such as photoautotrophic, heterotrophic, mixotrophic, and photoheterotrophic

[133,134], offering the highest yields of lipids [135]. Microalgal lipids are composed of saturated and unsaturated fatty acids with 12–22 carbon atoms, some of them belonging to ω-3 and ω-6 families.

Recently, the research activity has focused the attention on the microalgal oil production in photoautotrophic regime using light as energy source, though there are significant drawbacks associated to this cultivation regime. First of all, it is difficult to combine biomass accumulation and lipid synthesis during the microalgal life cycle [136]. Second, light attenuation is unavoidable for photoautotrophic cultures from lab to pilot scale, thus significantly reducing the productivity [137].

Many microalgae strains such as *Chlorella sorokiniana* [138], *Chlorella saccarophila* [139], *Nannochloropsis* sp. [140–142], *Neochloris oleoabundans* [143], *Cladophora fracta* [144], *Chlorella protothecoides* [144,145], *Chlorella vulgaris* [146], and *Chlamydomonas* sp. [147,148] could supply SCOs.

As above explained, microalgae cultures can have different types of metabolisms that are distinguished in autotrophic, heterotrophic, and mixotrophic. In autotrophic conditions, microorganisms absorb light energy to reduce CO_2 and release O_2. Most of algae belong to this category and require minimal amounts of organic compounds for growth, such as vitamins.

In autotrophic regime, they require organic carbon sources to promote the growth and, in the same time, it is possible to solve technical and physiological problems related to the presence and distribution of light and CO_2 associated to the autotrophic growth. Therefore heterotrophic regime offers the possibility to increase cell concentration and productivity.

The mixotrophic growth is defined as a growth where autotrophic autotrophic metabolism is integrated. In this case, CO_2 and organic carbon source are used as substrate through photosynthetic metabolism and cell respiration. The development of mixotrophic cultures may open new and interesting perspectives; in this, normally the cell growth rate is approximately the sum of growth rate in heterotrophic and autotrophic mode and the advantages of heterotrophic growth, such as high concentration and productivity, are also applicable to mixotrophic growth.

In addition to carbon (usually supplied in the form of CO_2 and organic sources) microalgal metabolism also requires vitamins, salts, and other nutrients, in particular nitrogen and phosphorus. Parameters, such as oxygen, carbon dioxide, pH, temperature, and light intensity, have to be balanced [149].

The pathway of triglyceride synthesis in microalgae consists of three steps [150]:

1. *Formation of acetyl-CoA in the cytoplasm*—Acetyl-CoA is an initiator and its formation takes place in the chloroplast, where it is formed as an intermediate, glyceraldehyde phosphate (GAP), that is transferred to the cytoplasm and subsequently consumed. After the export of GAP from chloroplast to cytoplasm, the carbon source is directed to the sugar synthesis (that represent the main storage products in the cytoplasm of plant cells) or oxidation through the glycolytic pathway to pyruvate. Therefore a part of the exogenous glucose is directly converted into starch and the rest is oxidized via glycolysis.
2. *Elongation and desaturation of carbon chains of fatty acids*—Elongation depends mainly on the reaction of two enzyme systems that include acetyl-CoA and requires the presence of malonyl-CoA.
3. *Biosynthesis of triglycerides*—Generally, L-α-phosphoglycerol and acetyl-coA are two major primers.

So far, few works have been reported about the direct effects of lignocellulosic hydrolysates on microalgae cultures (Table 4.5). Lignocellulosic hydrolysates contain organic carbon, such as fermentable sugars, volatile organic acids, and inhibitors, which should be considered to evaluate microalgae cultivation growth.

4.7 Fermentation process

The fermentation processes carried out for lipid production need a high C/N ratio, where nitrogen is not limiting. The residual carbon-to-nitrogen ratio increases during the cultivation time, and the control of this parameter is essential to prevent citric acid production instead of lipid accumulation. Continuous reactors are almost always operated under steady-state conditions, and the C/N ratio and the concentrations of C and N can be kept constant at a given dilution rate. The optimization of the process, therefore, involves determining the optimal dilution rate to obtain an optimal intermediate C/N ratio.

In fed-batch culture, nitrogen and carbon flows have to be accurately monitored to control the specific growth rate and residual C/N ratio [157].

Aeration is another important factor, affecting yeast cell growth as well as total lipid level. The amount of dissolved oxygen in liquid cultures can also influence the fatty acids composition of cell lipids. In an aerated

Table 4.5 Effect of hydrolysates on microalgae growth under different cultivation conditions.

Microalgae species	Substrate	Culture mode	Light	Lipid production	References
Chlorella pyrenoidosa	Rice straw enzymatic hydrolysate	250 mL flask	Yes	1.55 g/L[a] 53.6%[b]	[151]
Chlorella protothecoides	Cassava starch enzymatic hydrolysate, corn powder enzymatic hydrolysate	250 mL flask 5 L fermenter	No Yes	2.14 g/L[a] 22%–55.2%[b]	[152–154]
Chlorella vulgaris	Wheat bran enzymatic hydrolysate	500 mL flask	Yes	0.6%–0.9%[b]	[155]
Schizochytrium limacinum	Sweet sorghum juice (squeezed by a mill)	250 mL flask	No	2.15–4.95 g/L[a] 55.3%–70.5%[b]	[112]
Schizochytrium mangrovei	Food waste hydrolysate	2 L fermenter	No	3.52 g/L[a] 16.4%[b]	[156]

[a]Lipid yield.
[b]Lipid content (% w/w).

culture, cellular viability increases and free fatty acids are oxidized to unsaturated fatty acids since yeast cells need unsaturated fatty acids to continue their growth.

When lignocellulosic materials are used as feedstock, the cultivation strategies commonly adopted are based on the separate execution of hydrolysis and fermentation (separate hydrolysis and fermentation, SHF). The major advantage of SHF method is to carry out hydrolysis (40°C–50°C) and fermentation (30°C–37°C) in separate stages at their own optimum conditions. The main drawback of SHF is the inhibition of hydrolytic enzymes activity by released sugars. A new perspective is offered by combining enzymatic hydrolysis of pretreated lignocelluloses and fermentation in a single step or in a single reactor, also called SSF. A main advantage of this method is that the sugars produced by hydrolysis are immediately consumed by fermenting microorganisms, limiting substrate inhibition of the enzymes during the hydrolysis [94]. This process has been studied mainly for bioethanol production. Possible limitations to the applicability of the SSF processes in single reactors are the different requirements of hydrolytic enzymes and oleaginous yeasts in terms of operating temperatures and pH. Therefore for application at industrial scale, new improvements in enzyme technology (e.g., thermostable cellulases and higher inhibitor tolerance) are required [158].

Another interesting perspective, though not adequately deepened by the scientific community, is represented by mixed cultures of microorganisms, very common in natural ecological system. When a mixed culture is used, two or more microorganisms are synchronously cultivated within the same medium, so that these microorganisms can mutually exploit complementary metabolic activities to survive, grow, and reproduce [159].

Different systems have been developed for microalgae cultivation, such as open ponds or closed-up photobioreactors or bioreactors [160].

Currently, commercial cultivations of microalgae are mostly carried out in open ponds, where algae are directly exposed to sun irradiation, because of their economic feasibility and simplicity of maintenance. However, in open systems, the light provided to microalgae growth is a limiting factor, since it is correlated to the seasonal changes. In these systems the biomass density is not higher than 0.5 g/L [160], and the simultaneous presence of bacteria can be considered a competitive factor, since they are able to consume the organic source supplied to microalgae, in particular when lignocellulosic hydrolysates are used as feedstock. This problem could be solved by the addition of organic solvents such as methanol or ethanol, though not all microalgae strains are capable to metabolize these solvents. In addition, and at higher concentration, these solvents could inhibit microalgae growth [161,162].

The attention of research community is also focused about closed-up photobioreactors, where light energy can be supplied to by solar concentrators, or alternatively by artificial lamps. Closed-up photobioreactors are systems where parameters such as pH, temperature, dissolved oxygen, concentration of CO_2 or nutrient availability can be strictly controlled during cultivation.

In photobioreactor the light irradiation is more efficient, resulting in increased biomass concentration, though in some cases the photoinhibition makes difficult the scale-up of these systems [163].

As an alternative to photobioreactors, bioreactors with no direct light energy supply can be used. However, the lack of a light source makes these systems only suitable for the cultivation of microalgae that can use organic compounds as a source of energy.

Currently, increasing efforts are being made to develop efficient systems for microalgae cultivation with higher production efficiency and reduced energy input and maintenance costs. Lignocellulosic hydrolysates can be added to the bioreactor or photobioreactor to promote microalgal growth in a heterotrophic regime, as they contain organic carbon, mainly in terms of sugars and acetate, though the turbidity may cause photoinhibition phenomena. On the other hand, lignocellulosic hydrolysates are a low-cost feedstock and can significantly increase the economic sustainability of the process.

4.8 Extraction of microbial lipids

The lack of cost effective and efficient methods for the extraction of microbial lipids from cells remains a bottleneck for the commercial deployment of biodiesel technology. Solvent extraction methods are most commonly used for lipid extraction since they provide the highest lipid recovery. Environmental and friendly methods based on the use of green solvents are not yet a practicable option because of the poor lipid recovery, the possible degradation of lipids, and difficult large-scale commercialization. Cell disruption methods such as microwave, ultrasonication, bead beating, and supercritical fluid extraction may negatively affect the lipid extraction yields (Table 4.6).

Table 4.6 Comparison of different microbial lipid extraction methods.

Method	Efficiency rating	Cost involved	Advantages	Limitations
Use of organic solvents	Moderate	High cost due to the use of solvents. Reuse may help save some costs	• Regarded as "golden standard" for lipid extraction • Solvent combination successfully disrupts cell membranes • Use of polar solvents decrease interference of nonlipid compounds • Efficiently extracts polar lipids • Nonlipid compounds do not interfere, due to strong polar solvents used • Buffered solvent system prevents ionic salt adsorption effects	• Requires high volumes of solvents • Carcinogenicity • Polar lipids have been found in aqueous phase • Time-consuming • Difficulty in extracting organic layer without contamination occurring
Supercritical CO_2	High	High cost	• Rapid analysis • CO_2 gas is nontoxic and have low critical values • Reduced solvent volumes required	• Energy-intensive due to use of high pressure • CO_2 not enough polar to the separation

(Continued)

Table 4.6 (Continued)

Method	Efficiency rating	Cost involved	Advantages	Limitations
Bead beating	Moderate	Cost-effective	• Easily and relatively effective	• Direct mechanical damage to cells based on high-speed spinning with fine beads • Difficult to scale-up • High heat generation • Incomplete cell lysis • High energy demand
Microwave	Very high	High maintenance costs	• Easy scaling up (but not yet at a commercial level) • Production of good-quality extracts	
Sonication	High	High maintenance costs	• Faster extraction • Suitable for all cell type	• Damage chemical structure during the process
Osmotic shock	Moderate	Low cost	• Low energy consumption • Easier scale-up	• Long treatment time • Generation of waste salt
Electroporation	High	High investment and maintenance costs	• Low operation cost • Lipid extraction efficiency in terms of time	• Difficult to scale-up

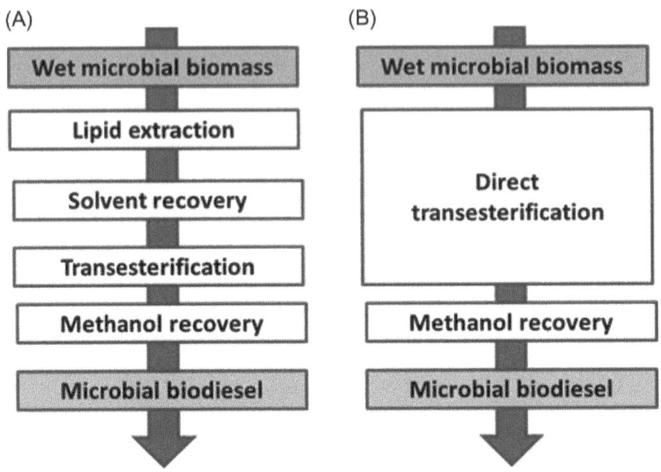

Figure 4.6 Conventional process (A) and direct transesterification process (B) for microbial biodiesel production. *Adapted from J.Y. Park, M.S. Park, Y.C. Lee, J.W. Yang, Advances in direct transesterification of algal oils from wet biomass, Bioresour. Technol. 184 (2015) 267–275 [164].*

Biodiesel production through conventional process demands high energy consumption during lipid extraction and biodiesel conversion process. Recently, the combination of lipid extraction and biodiesel conversion in one step, called direct or in situ transesterification, has been investigated. Direct transesterification could be an alternative to reduce the amount of necessary equipment. A comparison of the direct transesterification and the conventional process is shown in Fig. 4.6.

4.9 Catalysts for biodiesel synthesis

There are three categories of catalysts used for the transesterification: alkali, acids, and enzymes. The limited use of enzyme catalysts is related to their long reaction time and high cost, though they are attractive as they avoid soap formation and promote simpler processes of purification.

Homogeneous alkaline catalysts are the more traditionally used for the synthesis of biodiesel, as they offer higher reaction rates in comparison to the acid catalysts. Hydroxides and methoxides from sodium and potassium are the most commonly used alkaline catalysts.

Homogeneous acid catalysts offer a lower reaction rate in comparison to their alkaline counterpart, though they are often used when the triglyceride feedstock contains higher levels of free fatty acids, as in the case of lignocellulosic hydrolysates. As a matter of facts, acid catalysts are able to promote simultaneously the triglyceride transesterification and the fatty acid esterification.

Currently, heterogeneous alkaline catalysts are considered very promising, as they ensure relatively high reaction rates, reusability, easier separation from the products of reaction by filtration, lower cost, and require less energy as compared to acid catalysts. The main limitation of heterogeneous catalysts stems from diffusional resistances, leading to slower reaction rates [165].

Heterogenous acid catalysts offer a lower reaction rate in comparison to their alkaline counterpart, though they are specifically used with triglyceride feedstock containing higher concentrations of free fatty acids.

4.10 Genetic and metabolic engineering of microbes

Due to the large interest raised form the use of microalgae for biodiesel production, an increasing research activity is currently concerned at the metabolic and genetic engineering of microalgae. The green alga *Chlamydomonas reinhardtii* has been the most studied, being its genome the best known. The current research is mainly aimed at overcoming the bottlenecks that limit the two main pathways leading to the production of TAGs. These pathways is the acyl CoA-dependent pathway allowing the acylation of diacylglycerol, and the acyl CoA-independent pathway allowing the recycle of membrane lipids. Recent studies have elucidated the role of the enzymes involved in these pathways. In particular the enzymes acetyl CoA carboxylase (involved in the synthesis of malonyl-CoA) and type-II fatty acid synthase (involved in the fatty-acid chain elongation) have been found to be rate-limiting in fatty acid synthesis [158]. Consequently, metabolic engineering strategies are being developed to increase the activity of these enzymes [163].

Another potential target of the research is the improvement of the distribution of fatty acid residues in cellular lipids. As a matter of facts, a balanced proportion between saturated and unsaturated fatty acids encourages the use of the biodiesel as automotive fuel, leading to a moderate tendency to autoxidation and to a reduced tendency to gelification, thus improving the performance of biofuel at cold temperatures. Nevertheless, the results so far obtained in this field are not yet satisfactory, and much study is still required. Consequently, it can be said that in the near future, the production of microalgal biodiesel is not likely to be significantly affected by the improvements offered by the metabolic and genetic engineering.

4.11 Future prospects and conclusions

The biorefinery concept is becoming more attractive, taking into account that the present economic model based on the nonrenewable fossil resources for energy and industrial production is the reason of energy instability, climate changes and cannot be considered sustainable. An important output of biorefineries is the biodiesel, a biofuel increasingly used as it reduces the net greenhouse generation. The cost of biodiesel is primarily affected by the cost of the feedstock used, and secondarily by the process adopted. Consequently, lignocellulosic biomasses are increasingly studied as a feedstock for biodiesel production, since they are the most abundant carbon-neutral renewable source and can be used as a sustainable source of organic carbon with net zero carbon emission.

The hydrolysis of cellulose and hemicellulose (the main components of lignocellulosic materials) is required to obtain fermentable sugars. In this view a preliminary pretreatment of lignocellulosic biomasses is of crucial importance to facilitate the access of hydrolytic enzymes to the crystalline structure. The correct research is devoted to solve the problems still affecting this process, stemming from the loss and the inhibition of the hydrolytic enzymes.

Different microorganisms can be used to obtain SCOs using the mixtures of fermentable sugars as feedstock. Microalgae are the most promising, due to the high flexibility of their metabolism, allowing higher growth rates under operating conditions of industrial interest, as well as higher cellular lipid fractions. Lignocellulosic hydrolysates can be used to promote microalgal growth in a heterotrophic regime. In particular, mixotrophic cultures of microalgae are of great interest, as they sum the advantages of heterotrophic and autotrophic cultures.

Significant advantages in biodiesel production can be achieved by suitable choices of the reactor configuration. SSF offers very interesting perspectives, as it prevents the inhibition phenomena and reduces the investment costs.

Other factors affecting the economical balance of microalgal production of biodiesel are the extraction of cellular lipids, and the choice of the catalyst for esterification/transesterification reaction. These aspects require a research effort to envisage satisfactory solutions.

Though there are still priorities to be met for its full exploitation, the development of microalgal biodiesel deserve a major effort from the international community, as it offers the way to exploit sustainable, cost-effective, and clean alternative energy resources and can significantly contribute to solve the energetic and environmental issues still unsolved.

References

[1] D. Ness, Sustainable urban infrastructure in China: towards a factor 10 improvement in resource productivity through integrated infrastructure system, Int. J. Sustain. Dev. World Ecol. 15 (2008) 288–301.

[2] M.S. Andersen, An introductory note on the environmental economics of the circular economy, Sustain. Sci. 2 (2006) 133–140.

[3] F. Cherubini, The biorefinery concept: using biomass instead of oil for producing energy and chemicals, Energy Convers. Manage. 51 (2010) 1412–1421.

[4] E. de Jong, G. Jungmeier, Biorefinery concepts in comparison to petrochemical refineries, Ind. Biorefin. White Biotechnol. 15 (2015) 1–31.

[5] K. Wagemann, T. Benzing, T. Böhland, U.R. Fritsche, M. Fröhling, A. Gröngröft, et al., Biorefineries Roadmap, Agency for Renewable Resources (FNR), Gülzow, 2012.

[6] D. Keegan, B. Kretschemer, B. Elbersen, C. Panoutsou, Cascading use: asystematic approach to biomass beyond the energy sector, Biofuel Bioprod. Bioref. 7 (2013) 193–206.

[7] C.H. Zhou, J.N. Beltramini, Y.X. Fan, G.Q. Lu, Chemoselective catalytic conversion of glycerol as a biorenewable source to valuable commodity chemicals, Chem. Soc. Rev. 37 (2008) 527–549.

[8] J. Gerfried, The biorefinery fact sheet. In *IEA Bioenergy Task*, 42. 2014.

[9] REN21 Renewables, Academy, Bonn, Germany, 10 November 2015. <http://www.ren21.net>.

[10] IRENA, International Renewable Energy Agency, 2014. <https://www.irena.org>.

[11] Observ'ER, Biofuels Barometer, 2014. <http://www.energies-renouvelables.org>. Other Major European Biodiesel Producers in 2014 Were Finnish Neste Oil and Swiss Biopetrol Industries.

[12] M.E. Himmel, S.Y. Ding, D.K. Johnson, W.S. Adney, M.R. Nimlos, J.W. Brady, et al., Biomass recalcitrance: engineering plants and enzymes for biofuels production, Science 315 (2007) 804–807.

[13] M. Valdivia, J.L. Galan, J. Laffarga, J.L. Ramos, Biofuels 2020: biorefineries based on lignocellulosic biomass, Microb. Biotechnol. 9 (2016) 585–594.

[14] C. Somerville, H. Youngs, C. Taylor, S. Davis, S. Long, Feedstocks for lignocellulosic biofuels, Science 329 (2010) 790–792.

[15] C.H. Zhou, X. Xia, X. Lin, S. Tong, J. Beltramini, Catalytic conversion of lignocellulosic biomass to fine chemicals and fuels, Chem. Soc. Rev. 40 (2011) 5588–5617.

[16] J. Gustavsson, C. Cederberg, U. Sonesson, R. van Otterdijk, A. Meybeck, Food Losses and Food Waste—Extent, Causes and Prevention, Food and Agriculture Organization of the United Nations, 2011, p. 38.

[17] J. Gustavsson, C. Cederberg, U. Sonesson, A. Emanuelsson, The Methodology of the FOA Study: Global Food Losses and Food Waste—Extent, Causes and Prevention, 70, The Swedish Institute for Food and Biotechnology, 2013.

[18] F.H. Isikgor, R. Becer, Lignocellulosic biomass: a sustainable platform for the production of bio-based chemicals and polymers, Polym. Chem. 6 (2015) 4497–4559.
[19] G.W. Huber, Breaking the Chemical and Engineering Barriers to Lignocellulosic Biofuels: Next Generation Hydrocarbon Biorefineries, National Science Foundation, 2008.
[20] R.E.H. Sims, W. Mabee, J. Saddler, M. Taylor, An overview of second generation biofuel technologies, Bioresour. Technol. 101 (2009) 1570–1580.
[21] T.W. Hertel, J. Steinbuks, W.E. Tyner, What is the social value of second generation biofuels? Appl. Econ. Perspect. Policy (2015) 1–19.
[22] M. Stöcker, Biofuels and biomass-to-liquid fuels in the biorefinery: catalytic conversion of lignocellulosic biomass using porous materials, Angew. Chem. 47 (2008) 9200–9211.
[23] F. Cherubini, A.H. Strømman, Chemicals from lignocellulosic biomass: opportunities, perspectives, and potential of biorefinery systems, Biofuels Bioprod. Bioref. 5 (2011) 548–561.
[24] M. Galbe, G. Zacchi, A review of the production of ethanol from softwood, Appl. Microbiol. Biotechnol. 59 (2002) 618–628.
[25] N.C. Carpita, M. Tierney, M. Campbell, Molecular biology of the plant cell wall: searching for the genes that define structure, architecture and dynamics, Plant Mol. Biol. 47 (2001) 1–5.
[26] R.C. Kuhad, A. Singh, K.E. Eriksson, Microorganisms and enzymes involved in the degradation of plant fiber cell walls, Adv. Biochem. Eng. Biotechnol. 57 (1997) 45–125.
[27] S.Y. Ding, M.E. Himmel, The maize primary cell wall microfibril: a new model derived from direct visualization, J. Agric. Food Chem. 54 (2006) 597–606.
[28] O.P. Ward, M. Moo-Young, Enzymatic degradation of cell wall and related plant polysaccharides, CRC Crit. Rev. Biotechnol., 8, 1989, pp. 237–274.
[29] N.C. Carpita, D.M. Gibeaut, Structural models of primary cell walls in flowering plants: consistency of molecular structure with the physical properties of the walls during growth, Plant J. 3 (1993) 1–30.
[30] S.Y. Ding, M.E. Himmel, Anatomy and ultrastructure of maize cell walls: an example of energy plants, in: Biomass Recalcitrance: Deconstructing the Plant Cell Wall for Bioenergy, 2008, pp. 38–60.
[31] J.F. Matthews, C.E. Skopec, P.E. Mason, P. Zuccato, R.W. Torget, J. Sugiyama, et al., Computer simulation studies of microcrystalline cellulose Iβ, Carbohydr. Res. 34 (2006) 138–152.
[32] M. Jin, P.J. Slininger, B.S. Dien, S. Waghmode, B.R. Moser, A. Orjuela, et al., Microbial lipid-based lignocellulosic biorefinery: feasibility and challenges, Trends Biotechnol., 33, 2015, pp. 42–54.
[33] J.M. Ageitos, J.A. Vallejo, P. Veiga-Crespo, T.G. Villa, Oily yeasts as oleaginous cell factories, Appl. Microbiol. Biotechnol. 90 (2011) 1219–1227.
[34] J.D. Weete, D.J. Weber, Lipid Biochemistry of Fungi and Other Organisms, Plenum Press, 1980.
[35] V.B. Agbor, N. Cicek, R. Sparling, A. Berlin, D.B. Levin, Biomass pretreatment: fundamentals toward application, Biotechnol. Adv. 29 (2011) 675–685.
[36] Y. Sun, J. Cheng, Hydrolysis of lignocellulosic materials for ethanol production: a review, Bioresour. Technol. 83 (2002) 1–11.
[37] J. Yoo, S. Alavi, P. Vadlani, V. Amanor-Boadu, Thermo-mechanical extrusion pretreatment for conversion of soybean hulls to fermentable sugars, Bioresour. Technol. 102 (2011) 7583–7590.
[38] J. Zheng, L. Rehmann, Extrusion pretreatment of lignocellulosic biomass: a review, Int. J. Mol. Sci. 15 (2014) 18967–18984.

[39] C. Karunanithy, K. Muthukumarappan, W.R. Gibbons, Effect of extruder screw speed, temperature, and enzyme levels on sugar recovery from different biomasses, ISRN Biotechnol. (2013) 1–13.
[40] S. Ethaib, R. Omar, S.M.M. Kamal, D.R.A. Biak, Microwave assisted pretreatment of lignocellulosic biomass: a review, J. Eng. Sci. Technol. (2015) 97–109.
[41] Z. Zhu, D.J. Macquarrie, R. Simister, L.D. Gomez, S.J. McQueen-Mason, Microwave assisted chemical pretreatment of *Miscanthus* under different temperature regimes, Sustain. Chem. Process 3 (2015) 15–27.
[42] J. Zhu, C.A. Rezende, R. Simister, S.J. McQueen-Mason, D.J. Macquarrie, I. Polikarpov, et al., Efficient sugar production from sugarcane bagasse by microwave assisted acid and alkali pretreatment, Biomass Bioenergy 93 (2016) 269–278.
[43] P.R. Gogate, V.S. Sutkar, A.B. Pandit, Sonochemical reactors: important design and scale up considerations with a special emphasis on heterogeneous systems, Chem. Eng. J. 166 (2011) 1066–1082.
[44] W. Chen, H. Yu, Y. Liu, P. Chen, M. Zhang, Y. Hai, Individualization of cellulose nanofibers from wood using high-intensity ultrasonication combined with chemical pretreatments, Carbohydr. Polym. 83 (2011) 1804–1811.
[45] M.A. Khiyami, A.L. Pometto, R.C. Brown, Detoxification of corn stover and corn starch pyrolysis liquors by *Pseudomonas putida* and *Streptomyces setonii* suspended cells and plastic compost support biofilms, J. Agric. Food Chem. 53 (2005) 2978–2987.
[46] T. Kan, V. Strezov, T.J. Evans, Lignocellulosic biomass pyrolysis: a review of product properties and effects of pyrolysis parameters, Renew. Sustain. Energy Rev. 57 (2016) 1126–1140.
[47] R. Kumar, C.E. Wyman, Effects of cellulase and xylanase enzymes on the deconstruction of solids from pretreatment of poplar by leading technologies, Biotechnol. Prog. 25 (2009) 302–314.
[48] A. Golberg, M. Sack, J. Teissie, G. Pataro, U. Pliquett, G. Saulis, et al., Energy efficient biomass processing with pulsed electric fields for bioeconomy and sustainable development, Biotechnol. Biofuels 9 (2016) 94–116.
[49] C.N. Hamelinck, G.V. Hooijdonk, A.P.C. Faaij, Ethanol from lignocellulosic biomass: techno-economic performance in short-, middle- and long-term, Biomass Bioenergy 28 (2005) 384–410.
[50] L.R. Lynd, P.J. Weimer, W.H. van Zyl, I.S. Pretorius, Microbial cellulose utilization: fundamentals and biotechnology, Microbiol. Mol. Biol. Rev. 66 (2002) 506–577.
[51] R. Wooley, M. Ruth, J. Sheehan, K. Ibsen, H. Majdeski, A. Galvez, Lignocellulosic biomass to ethanol process design and economics utilizing co-current dilute acid prehydrolysis and enzymatic hydrolysis, Proceedings of Current and Futuristic Scenarios. Technical report NREL/TP-580-26157. Golden, CO, National Renewable Energy Laboratory, vol. 123, 1999.
[52] J.C. Cuzens, J.R. Miller, Acid hydrolysis of bagasse for ethanol production, Renew. Energy 10 (1997) 285–290.
[53] A.K. Chandel, E. Chan, R. Rudravaram, M.L. Narasu, L.V. Rao, P. Ravindra, Economics and environmental impact of bioethanol production technologies: an appraisal, Biotechnol. Mol. Biol. Rev. 2 (2007) 14–32.
[54] C.E. Wyman, B.E. Dale, R.T. Elander, M. Holtzapple, M.R. Ladisch, Y.Y. Lee, Coordinated development of leading biomass pretreatment technologies, Bioresour. Technol. 96 (2005) 1959–1966.
[55] S. Gamez, J.A. Ramirez, G. Garrote, M. Vazquez, Manufacture of fermentable sugar solutions from sugar cane bagasse hydrolyzed with phosphoric acid at atmospheric pressure, J. Agric. Food Chem. 52 (2004) 4172–4177.

[56] T.A. Lloyd, C.E. Wyman, Combined sugar yields for dilute sulfuric acid pretreatment of corn stover followed by enzymatic hydrolysis of the remaining solids, Bioresour. Technol. 96 (2005) 1967–1977.
[57] W.E. Kaar, M.T. Holtzapple, Using lime pretreatment to facilitate the enzymic hydrolysis of corn stover, Biomass Bioenergy 18 (2000) 189–199.
[58] B.C. Saha, M.A. Cotta, Lime pretreatment, enzymatic saccharification and fermentation of rice hulls to ethanol, Biomass Bioenergy 32 (2008) 971–977.
[59] S. Monavari, M. Galbe, G. Zacchi, Impact of impregnation time and chip size on sugar yield in pretreatment of softwood for ethanol production, Bioresour. Technol. 100 (2009) 6312–6316.
[60] L. Mesa, E. González, C. Cara, M. González, E. Castro, S.I. Mussatto, The effect of organosolv pretreatment variables on enzymatic hydrolysis of sugarcane bagasse, Chem. Eng. J. 168 (2011) 1157–1162.
[61] X.J. Pan, C. Arato, N. Gilkes, D. Gregg, W. Mabee, K. Pye, et al., Biorefining of softwoods using ethanol organosolv pulping: preliminary evaluation of process streams for manufacture of fuel-grade ethanol and co-products, Biotechnol. Bioeng. 90 (2005) 473–481.
[62] Q.P. Liu, X.D. Hou, N. Li, M.H. Zong, Ionic liquids from renewable biomaterials: synthesis, characterization and application in the pretreatment of biomass, Green. Chem. 14 (2012) 304–307.
[63] Z. Qiu, G.M. Aita, M.S. Walker, Effect of ionic liquid pretreatment on the chemical composition, structure and enzymatic hydrolysis of energy cane bagasse, Bioresour. Technol. 117 (2012) 251–256.
[64] Y. Dai, G.J. Witkamp, R. Verpoorte, Y.H. Choi, Tailoring properties of natural deep eutectic solvents with water to facilitate their applications, Food Chem. 187 (2015) 14–19.
[65] A.K. Kumar, B.S. Parikh, M. Pravakar, Natural deep eutectic solvent mediated pretreatment of rice straw: bioanalytical characterization of lignin extract and enzymatic hydrolysis of pretreated biomass residue, Environ. Sci. Pollut. Res. Int. 23 (2016) 9265–9275.
[66] E. Ruiz, C. Cara, P. Manzanares, M. Ballesteros, E. Castro, Evaluation of steam explosion pre-treatment for enzymatic hydrolysis of sunflower stalks, Enzym. Microb. Technol. 42 (2008) 160–166.
[67] S. Sharma, R. Kumar, R. Gaur, R. Agrawal, R.P. Gupta, D.K. Tuli, et al., Pilot scale study on steam explosion and mass balance for higher sugar recovery from rice straw, Bioresour. Technol. 175 (2015) 350–357 (2015).
[68] M. Ballesteros, J.M. Oliva, M.J. Negro, P. Manzanares, I. Ballesteros, Ethanol from lignocellulosic materials by a simultaneous saccharification and fermentation process (SFS) with *Kluyveromyces marxianus* CECT 10875, Process Biochem. 39 (2004) 1843–1848.
[69] J. Soderstrom, M. Galbe, G. Zacchi, Separate versus simultaneous saccharification and fermentation of two-step steam pretreated softwood for ethanol production, J. Wood Chem. Technol. 25 (2005) 187–202.
[70] I. Ballesteros, J.M. Oliva, F. Sanchez, M. Ballesteros, Ethanol production from lignocellulosic byproducts of olive oil extraction, Appl. Biochem. Biotechnol. 91 (2001) 237–252.
[71] T. Rogalinski, T. Ingram, G. Brunner, Hydrolysis of lignocellulosic biomass in water under elevated temperatures and pressures, J. Supercrit. Fluids 47 (2008) 54–63.
[72] J.A. Perez, I. Ballesteros, M. Ballesteros, F. Saez, M.J. Negro, P. Manzanares, Optimizing liquid hot water pretreatment conditions to enhance sugar recovery from wheat straw for fuel-ethanol production, Fuel 87 (2008) 3640–3647.
[73] R.G. Koegel, H.K. Sreenath, R.J. Straub, Alfalfa fiber as a feedstock for ethanol and organic acids, Appl. Biochem. Biotechnol. 92 (1999) 105–115.

[74] V. Balan, B. Bals, S.P. Chundawat, D. Marshall, B.E. Dale, Lignocellulosic biomass pretreatment using AFEX, Biofuels Methods Protoc. (2009) 61–77.
[75] T.C. Bradshaw, H. Alizadeh, F. Teymouri, V. Balan, B.E. Dale, Ammonia fiber expansion pretreatment and enzymatic hydrolysis on two different growth stages of reed canary grass, Appl. Biochem. Biotechnol. 140 (2007) 395–405.
[76] H.D. Zhang, S.B. Wu, Enhanced enzymatic cellulose hydrolysis by subcritical carbon dioxide pretreatment of sugarcane bagasse, Bioresour. Technol. 158 (2014) 161–165.
[77] E. Varga, A.S. Schmidt, K. Réczey, A.B. Thomsen, Pretreatment of corn stover using wet oxidation to enhance enzymatic digestibility, Appl. Biochem. Biotechnol. 104 (2003) 37–50.
[78] H.B. Klinke, A.B. Thomsen, B.K. Ahring, Inhibition of ethanol-producing yeast and bacteria by degradation products produced during pre-treatment of biomass, Appl. Microbiol. Biotechnol. 66 (2004) 10–26.
[79] H.K. Sharma, C. Xu, W. Qion, Biological pretreatment of lignocellulosic biomass for biofuels and bioproducts: an overview, Waste Biomass Valor. 10 (2017) 1–17.
[80] M.J. Taherzadeh, K. Karimi, Enzyme-based hydrolysis processes for ethanol from lignocellulosic materials: a review, BioResources 2 (2007) 707–738.
[81] T. Eriksson, J. Karlsson, F. Tjerneld, A model explaining declining rate in hydrolysis of lignocellulose substrates with cellobiohydrolase I (Cel7A) and endoglucanase I (cel7B) of *Trichoderma reesei*, Appl. Biochem. Biotechnol. Bioeng. 101 (2002) 41–60.
[82] P. Väljamäe, K. Kipper, G. Petterson, G. Johansson, Synergistic cellulose hydrolysis can be described in terms of fractal-like kinetics, Biotechnol. Bioeng. 84 (2003) 254–257.
[83] M.L. Rabinovich, M.S. Melnik, A.V. Boloboba, Microbial cellulose (review), Appl. Biochem. Microbiol. 38 (2002) 305–321.
[84] H. Itoh, M. Wada, Y. Honda, M. Kuwahara, T. Watanabe, Bioorganosolve pretreatments for simultaneous saccharification and fermentation of beech wood by ethanolysis and white rot fungi, J. Biotechnol. 103 (2003) 273–280.
[85] N. Ortega, M.D. Busto, M. Perez-Mateos, Kinetics of cellulose saccharification by *Trichoderma reesei* cellulases, Int. Biodeterior. Biodegrad. 47 (2001) 7–14.
[86] C. Tengborg, M. Galbe, G. Zacchi, Influence of enzyme loading and physical parameters on the enzymatic hydrolysis of steam-pretreated softwood, Biotechnol. Prog. 17 (2001) 110–117.
[87] C.E. Wyman, Handbook on Bioethanol: Production and Utilization, Taylor & Francis, Washington, DC, 1996.
[88] H. Jørgensen, J.B. Kristensen, C. Felby, Enzymatic conversion of lignocellulose into fermentable sugars: challenges and opportunities, Biofuels Bioprod. Bioref. 1 (2007) 119–134.
[89] Y.H.P. Zhang, M.E. Himmel, J.R. Mielenz, Outlook for cellulose improvement: screening and selection strategies, Biotechnol. Adv. 24 (2006) 452–481.
[90] P. Andric, A.S. Meyer, P.A. Jensen, K. Dam-Johansen, Reactor design for minimizing product inhibition during enzymatic lignocellulose hydrolysis: I. Significance and mechanism of cellobiose and glucose inhibition on cellulolytic enzymes, Biotechnol. Adv. 28 (2010) 308–324.
[91] M. Tu, X. Zhang, A. Kurabi, N. Gilkes, W. Mabee, J. Saddler, Immobilization of beta-glucosidase on Eupergit C for lignocellulose hydrolysis, Biotechnol. Lett. 28 (2006) 151–156.
[92] L. Olsson, H.R. Soerensen, B.P. Dam, H. Christensen, K.M. Krogh, A.S. Meyer, Separate and simultaneous enzymatic hydrolysis and fermentation of wheat hemicellulose with recombinant xylose utilizing *Saccharomyces cerevisiae*, Twenty-Seventh Symposium on Biotechnology for Fuels and Chemicals (2006) 117–129. Humana Press.

[93] B.C. Saha, L.B. Iten, M.A. Cotta, Y.V. Wu, Dilute acid pretreatment enzymatic saccharification and fermentation of rice hulls to ethanol, Biotechnol. Prog. 21 (2005) 816–822.
[94] A. Wingren, M. Galbe, G. Zacchi, Techno-economic evaluation of producing ethanol from softwood: comparison of SSF and SHF and identification of bottlenecks, Biotechnol. Prog. 19 (2003) 1109–1117.
[95] O. Kim, B. Magnus, L. Gunnar, A short review on SSF – an interesting process option for ethanol production from lignocellulosic feedstocks, Biotechnol. Biofuels 1 (2008) 1–14.
[96] L.R. Lynd, M.S. Laser, D. Bransby, B.E. Dale, B. Davison, R. Hamilton, et al., How biotech can transform biofuels, Nat. Biotechnol. 26 (2008) 169–172.
[97] W. Liu, Y. Wang, Z. Yu, J. Bao, Simultaneous saccharification and microbial lipid fermentation of corn stover by oleaginous yeast *Trichosporon cutaneum*, Bioresour. Technol. 118 (2012) 13–18.
[98] A. Beopoulos, J.M. Nicaud, Yeast: a new oil producer? OCL 19 (2012) 22–28.
[99] G. Knothe, J. Krahl, J. Van Gerpen, The Biodiesel Handbook, Elsevier, Champaign, IL, 2005.
[100] R. Subramaniam, S. Dufreche, M. Zappi, R. Bajpai, Microbial lipids from renewable resources: production and characterization, J. Ind. Microbiol. Biotechnol. 37 (2010) 1271–1287.
[101] Z. Chi, Y. Zheng, B. Lucker, S. Chen, Integrated System for Production of Biofuel Feedstock, Patents, 2010.
[102] C. Ratledge, J.P. Wynn, The biochemistry and molecular biology of lipid accumulation in oleaginous microorganisms, Adv. Appl. Microbiol. 51 (2002) 1–51.
[103] C. Angerbauer, M. Siebenhofer, M. Mittelbach, G.M. Guebitz, Conversion of sewage sludge into lipids by *Lipomyces starkeyi* for biodiesel production, Bioresour. Technol. 99 (2008) 3051–3056.
[104] Y.H. Li, Z.B. Zhao, F.W. Bai, High-density cultivation of oleaginous yeast *Rhodosporidium toruloides* Y4 in fed-batch culture, Enzyme Microb. Technol. 41 (2007) 312–317.
[105] C. Ratledge, Fatty acid biosynthesis in microorganisms being used for Single Cell Oil production, Biochimie 86 (2004) 807–815.
[106] J. Xu, W. Du, X. Zhao, G. Zhang, D. Liu, Microbial oil production from various carbon sources and its use for biodiesel preparation: review, Biofuel Bioprod. Bioref. 7 (2013) 65–77.
[107] E.R. Easterling, W.T. French, R. Hernandez, M. Licha, The effect of glycerol as a sole and secondary substrate on the growth and fatty acid composition of *Rhodotorula glutinis*, Bioresour. Technol. 100 (2009) 356–361.
[108] L.Y. Zhu, M.H. Zong, H. Wu, Efficient lipid production with *Trichosporon fermentans* and its use for biodiesel preparation, Bioresour. Technol. 99 (2008) 7881–7885.
[109] X. Zhao, X.L. Kong, Y.Y. Hua, B. Feng, Z.B. Zhao, Medium optimization for lipid production through co-fermentation of glucose and xylose by the oleaginous yeast *Lipomyces starkeyi*, Eur. J. Lipid Sci. Technol. 110 (2008) 405–412 (2008).
[110] C. Huang, X.F. Chen, L. Xiong, X.D. Chen, L.L. Ma, Y. Chen, Single cell oil production from low-cost substrates: the possibility and potential of its industrialization, Biotechnol. Adv. 31 (2013) 129–139.
[111] Y.-H. Chang, K.-S. Chang, C.-L. Hsu, L.-T. Chuang, C.-Y. Chen, F.-Y. Huang, A comparative study on batch and fed-batch cultures of oleaginous yeast *Cryptococcus* sp. in glucose-based media and corncob hydrolysate for microbial oil production, Fuel 105 (2013) 711–717.

[112] Y. Liang, T. Tang, T. Siddaramu, R. Choudhary, A.L. Umagiliyage, Lipid production from sweet sorghum bagasse through yeast fermentation, Renew. Energy 40 (2012) 130–136.
[113] C. Huang, X.-F. Chen, L. Yang, X.Q. Xiong, J. Lin, J. Yang, Bioconversion of corncob acid hydrolysate into microbial oil by the oleaginous yeast *Lipomyces starkeyi*, Appl. Biochem. Biotechnol. 172 (2014) 2197–2204.
[114] M.C.A. Xavier, A.L.V. Coradini, A.C. Deckmann, T.T. Franco, Lipid production from hemicellulose hydrolysate and acetic acid by *Lipomyces starkeyi* and the ability of yeast to metabolize inhibitors, Biochem. Eng. J. 118 (2017) 11–19.
[115] X. Zhao, S. Wu, C. Hu, Q. Wang, Y. Hua, Z. Zhao, Lipid production from Jerusalem artichoke by *Rhodosporidium toruloides* Y4, J. Ind. Microbiol. Biotechnol. 37 (2010) 581–585.
[116] Q. Wang, F.-J. Guo, Y.-J. Rong, Z.-M. Chi, Lipid production from hydrolysate of cassava starch by *Rhodosporidium toruloides* 21167 for biodiesel making, Renew. Energy 46 (2012) 164–168.
[117] Y. Liu, Y. Wang, H. Liu, J. Zhang, Enhanced lipid production with undetoxified corncob hydrolysate by *Rhodotorula glutinis* using a high cell density culture strategy, Bioresour. Technol. 180 (2015) 32–39.
[118] Q. Gao, Z. Cui, J. Zhang, J. Bao, Lipid fermentation of corncob residues hydrolysate by oleaginous yeast *Trichosporon cutaneum*, Bioresour. Technol. 152 (2014) 552–556.
[119] Y.A. Tsigie, L.H. Huynh, I.N. Ahmed, Y.-H. Ju, Maximizing biodiesel production from *Yarrowia lipolytica* Po1g biomass using subcritical water pretreatment, Bioresour. Technol. 111 (2012) 201–207.
[120] Z. Ruan, M. Zanotti, X. Wang, C. Ducey, Y. Liu, Evaluation of lipid accumulation from lignocellulosic sugars by *Mortierella isabellina* for biodiesel production, Bioresour. Technol. 10 (2012) 198–205.
[121] C.N. Economou, A. Makri, G. Aggelis, S. Pavlou, D.V. Vayenas, Semi-solid state fermentation of sweet sorghum for the biotechnological production of single cell oil, Bioresour. Technol. 101 (2010) 1385–1388.
[122] X.W. Peng, H.Z. Chen, Single Cell Oil production in solid-state fermentation by *Microsphaeropsis* sp. from steam-exploded wheat straw mixed with wheat bran, Bioresour. Technol. 99 (2008) 3885–3889.
[123] H. Lin, W. Cheng, H.T. Ding, X.J. Chen, Q.F. Zhou, Y.H. Zhao, Direct microbial conversion of wheat straw into lipid by a cellulolytic fungus of *Aspergillus oryzae* A-4 in solid-state fermentation, Bioresour. Technol. 101 (2010) 7556–7562.
[124] P. Dey, J. Banerjee, M.K. Maiti, Comparative lipid profiling of two endophytic fungal isolates—*Colletotrichum* sp. and *Alternaria* sp. having potential utilities as biodiesel feedstock, Bioresour. Technol. 102 (2011) 5815–5823.
[125] D.J. Murphy, Storage lipid bodies in plants and other organisms, Prog. Lipid Res. 29 (1991) 299–324.
[126] H. Koku, I. Eroglu, U. Gunduz, M. Yucel, L. Turker, Kinetics of biological hydrogen production by the photosynthetic bacterium *Rhodobacter sphaeroides* O.U. 001, Int. J. Hydrogen Energy 28 (2003) 381–388.
[127] R. Kalscheuer, T. Torsten Stolting, A. Steinbuchel, Microdiesel: *Escherichia coli* engineered for fuel production, Microbiology 152 (2006) 2529–2536.
[128] N. Zabeti, P. Bonin, J.K. Volkman, S. Guasco, J.F. Rontani, Fatty acid composition of bacterial strains associated with living cells of the haptophyte *Emiliania huxleyi*, Org. Geochem. 41 (2010) 627–636.
[129] A. Steinbüchel, B. Füchtenbusch, Bacterial and other biological systems for polyester production, Trends Biotechnol. 16 (1998) 419–427.

[130] M. Koller, A. Salerno, P. Tuffner, M. Koinigg, H. Böchzelt, S. Schober, et al., Characteristics and potential of micro algal cultivation strategies: a review, J. Cleaner Prod. 37 (2012) 377–388.
[131] K. Samarakoon, Y.J. Jeon, Bio-functionalities of proteins derived from marine algae—a review, Food Res. Int. 48 (2012) 948–960.
[132] L. Zhu, Biorefinery as a promising approach to promote microalgae industry: an innovative framework, Renew. Sustan. Energy Rev. 41 (2015) 1376–1384.
[133] T.M. Mata, A.A. Martins, N.S. Caetano, Microalgae for biodiesel production and other applications: a review, Renew. Sustain. Energy Rev. 14 (2010) 217–232.
[134] G.L. Zhao, J.Y. Yu, F.F. Jiang, X. Zhang, T.W. Tan, The effect of different trophic modes on lipid accumulation of *Scenedesmus quadricauda*, Bioresour. Technol. 114 (2012) 466–471.
[135] X. Chen, Z.H. Li, X. Zhang, F.X. Hu, D.D.Y. Ryu, J. Bao, Screening of oleaginous yeast strains tolerant to lignocellulose degradation compounds, Appl. Biochem. Biotechnol. 159 (2009) 591–604.
[136] J. Liu, J. Huang, Z. Sun, Y. Zhong, Y. Jiang, F. Chen, Differential lipid and fatty acid profiles of photoautotrophic and heterotrophic *Chlorella zofingiensis*: assessment of algal oils for biodiesel production, Bioresour. Technol. 102 (2011) 106–110.
[137] C. Wilhelm, T. Jakob, From photons to biomass and biofuels: evaluation of different strategies for the improvement of algal biotechnology based on comparative energy balances, Appl. Microbiol. Biotechnol. 92 (2011) 909–919.
[138] C.Y. Chen, J.S. Chang, H.Y. Chang, T.Y. Chen, J.H. Wu, W.L. Lee, Enhancing microalgal oil/lipid production from *Chlorella sorokiniana* CY1 using deep-sea water supplemented cultivation medium, Biochem. Eng. J. 77 (2013) 74–81.
[139] V.A. Herrera-Valencia, P.Y. Contreras-Pool, S.J. López-Adrián, S. Peraza-Echeverría, L.F. Barahona-Pérez, The green microalga *Chlorella saccharophila* as a suitable source of oil for biodiesel production, Curr. Microbiol. 63 (2011) 151–157.
[140] L. Jiang, S. Luo, X. Fan, Z. Yang, R. Guo, Biomass and lipid production of marine microalgae using municipal wastewater and high concentration of CO_2, Appl. Energy 88 (2011) 3336–3341.
[141] S. Wahidin, A. Idris, S.R.M. Shaleh, The influence of light intensity and photoperiod on the growth and lipid content of microalgae *Nannochloropsis* sp., Bioresour. Technol. 129 (2013) 7–11.
[142] S.Y. Chiu, C.Y. Kao, M.T. Tsai, S.C. Ong, C.H. Chen, C.S. Lin, Lipid accumulation and CO_2 utilization of *Nannochloropsis oculata* in response to CO_2 aeration, Bioresour. Technol. 100 (2009) 833–838.
[143] Y.Q. Li, M. Horsman, B. Wang, N. Wu, C.Q. Lan, Effects of nitrogen sources on cell growth and lipid accumulation of green alga *Neochloris oleoabundans*, Appl. Microbiol. Biotechnol. 81 (2008) 629–636.
[144] A. Demirbas, Production of biodiesel from algae oils, Energy Sourc., A—Recovery Util. Environ. Eff. 31 (2009) 163–168.
[145] W. Xiong, X.F. Li, J.Y. Xiang, Q.Y. Wu, High-density fermentation of microalga *Chlorella protothecoides* in bioreactor for microbio-diesel production, Appl. Microbiol. Biotechnol. 78 (2008) 29–36.
[146] Z.Y. Liu, G.C. Wang, B.C. Zhou, Effect of iron on growth and lipid accumulation in *Chlorella vulgaris*, Bioresour. Technol. 99 (2008) 4717–4722.
[147] S.H. Ho, A. Nakanishi, X. Ye, J.S. Chang, K. Hara, T. Hasunuma, et al., Optimizing biodiesel production in marine *Chlamydomonas* sp. JSC4 through metabolic profiling and an innovative salinity-gradient strategy, Biotechnol. Biofuels 7 (2014) 97.
[148] A. Nakanishi, S. Aikawa, S.H. Ho, C.Y. Chen, J.S. Chang, T. Hasunuma, et al., Development of lipid productivities under different CO_2 conditions of marine microalgae *Chlamydomonas* sp. JSC4, Bioresour. Technol. 152 (2014) 247–252.

[149] K. Chojnacka, F.J. Marquez-Roche, Kinetic stoichiometric relationship of the energy and carbon metabolism in the culture of microalgae, Biotechnology 3 (2004) 21–34.
[150] G. Huang, F. Chen, D. Wei, X. Zhang, G. Chen, Biodiesel production by microalgal biotechnology, Appl. Energy 87 (2010) 38–46.
[151] P. Li, X. Miao, R. Li, J. Zhong, *In situ* biodiesel production from fast-growing and high oil content *Chlorella pyrenoidosa* in rice straw hydrolysate, J. Biomed. Biotechnol. 2011 (2001) 1–8.
[152] A. Wei, X. Zhang, D. Wei, G.U. Chen, Q. Wu, S.T. Yang, Effects of cassava starch hydrolysate on cell growth and lipid accumulation of the heterotrophic microalgae *Chlorella protothecoides*, J. Ind. Microbiol. Biotechnol. 36 (2009) 1383–1389.
[153] H. Xu, X. Miao, Q. Wu, High quality biodiesel production from a microalga *Chlorella protothecoides* by heterotrophic growth in fermenters, J. Biotechnol. 126 (2006) 499–507.
[154] Y. Lu, Y. Zhai, M. Liu, Q. Wu, Biodiesel production from algal oil using cassava (*Manihot esculenta* Crantz) as feedstock, J. Appl. Phycol. 22 (2010) 573–578.
[155] M.M. EL-Sheekh, M.Y. Bedaiwy, M.E. Osman, M.M. Ismail, Mixotrophic and heterotrophic growth of some microalgae using extract of fungal-treated wheat bran, Int. J. Rec. Org. Waste Agric. 1 (2012) 121–129.
[156] D. Pleissner, W.C. Lam, Z. Sun, C.S.K. Lin, Food waste as nutrient source in heterotrophic microalgae cultivation, Bioresour. Technol. 137 (2013) 139–146.
[157] A. Beopoulos, J. Cescut, R. Haddouche, J.L. Uribelarrea, C. Molina-Jouve, J.M. Nicaud, *Yarrowia lipolytica* as a model for bio-oil production, Prog. Lipid Res. 48 (2009) 375–387.
[158] L. Viikari, M. Alapuranen, T. Puranen, J. Vehmaanperä, M. Siika-Aho, Thermostable enzymes in lignocellulose hydrolysis, Adv. Biochem. Eng. Biotechnol. 108 (2007) 121–145.
[159] A.M. O'Reilly, J.A. Scott, Defined coimmobilization of mixed microorganism cultures, Enzyme Microb. Technol. 17 (1995) 636–646.
[160] M.A. Borowitzka, Commercial production of microalgae: ponds, tanks, tubes and fermenters, J. Biotechnol. 70 (1999) 313–321.
[161] G.W. Stratton, T.M. Smith, Interaction of organic solvents with the green alga *Chlorella pyrenoidosa*, Bull. Environ. Contam. Toxicol. 40 (1988) 736–742.
[162] A. El Jay, Effects of organic solvents and solvent-atrazine interactions on two algae, *Chlorella vulgaris* and *Selenastrum capricornutum*, Arch. Environ. Contam. Toxicol. 31 (1996) 84–90.
[163] I.S. Suh, S.B. Lee, A light distribution model for an internally radiating photobioreactor, Biotechnol. Bioeng. 82 (2003) 180–189.
[164] J.Y. Park, M.S. Park, Y.C. Lee, J.W. Yang, Advances in direct transesterification of algal oils from wet biomass, Bioresour. Technol. 184 (2015) 267–275.
[165] D.Y.C. Leung, X. Wu, M.K.H. Leung, A review on biodiesel production using catalyzed transesterification, Appl. Energy 87 (2010) 1083–1095.

Further reading

A. Yousuf, Biodiesel from lignocellulosic biomass—Prospects and challenges, Waste Manage. 32 (11) (2012) 2061–2067.

CHAPTER 5

Biobutanol from lignocellulosic biomass: bioprocess strategies

Jeyaprakash Dharmaraja[1], Sutha Shobana[2], Sundaram Arvindnarayan[3], Manokaran Vadivel[2], A.E. Atabani[4], Arivalagan Pugazhendhi[5] and Gopalakrishnan Kumar[6]
[1]Division of Chemistry, Faculty of Science and Humanities, Sree Sowdambika College of Engineering, Virudhunagar, India
[2]Department of Chemistry & Research Centre, Mohamed Sathak Engineering College, Ramanathapuram, India
[3]Department of Mechanical Engineering, Rajas Engineering College, Vadakkankulam, India
[4]Alternative Fuels Research Laboratory (AFRL), Energy Division, Department of Mechanical Engineering, Faculty of Engineering, Erciyes University, Kayseri, Turkey
[5]Innovative Green Product Synthesis and Renewable Environment Development Research Group, Faculty of Environment and Labour Safety, Ton Duc Thang University, Ho Chi Minh City, Vietnam
[6]Institute of Chemistry, Bioscience and Environmental Engineering, Faculty of Science and Technology, University of Stavanger, Stavanger, Norway

Contents

5.1 Introduction

Currently, the growth of human population is ineffective to the consumption of basic prerequisites, say food, water, energy, and other related products, for their developmental activities. In many countries the economic growth and development of industries and transportation sectors mainly depend on the long-term availability of petroleum-based fossil fuel resources [1]. On the contrary, owing to the rapid growth of human population, the exhaustion of nonrenewable fossil fuel (mainly derived from

Lignocellulosic Biomass to Liquid Biofuels
DOI: https://doi.org/10.1016/B978-0-12-815936-1.00005-8

petroleum and coal materials) utilization will lead to an energy crisis in the near future, that is, increasing the fossil fuel consumption would result in the rapid rise of fuel or oil prices, fast depletion of fossil fuel resources or reserves, and change the environmental condition, say global warming, which all consequently lead to the search for alternative strategies for energy production or generation from other sources [2–6]. Since the second half of the 20th century, many researchers as well as industrialists have been continuously searching for a substitute of petroleum-based fossil fuel sources with the development of novel, low-cost and eco-friendly, clean, and sustainable renewable energy sources that reduce the global GHG (greenhouse gas) emissions and also cause fulfillment of the global energy crisis. Many renewable clean energy group sectors, such as wind, solar, hydrothermal, and geothermal, have been mainly developed for the production of electricity but not for transportation fuels. The liquid biofuel or bio-oil generated from different kinds of solid biomass materials are regarded as promising alternative sustainable renewable energy sectors to fossil fuel sources [7–9]. Also, the biomass materials could assist as a source of foodstuffs for human and higher animal consumptions, raw materials for various developmental activities, and are utilized as energy sectors for heating, cooking, etc. [10]. Especially, the production of bioenergy from lignocellulosic biomass materials, such as barley straw, corn cobs, stover, straw and fiber, rice straw, green bean, giant reed leaves and stalk, wheat straw, bioenergy crops (switchgrass, *Miscanthus*), soybean straw, softwood stem, sunflower stalk, and sugarcane bagasse [11,12], are the most promising renewable resources, because they are cheap, nontoxic, widely dispersed, highly abundant, indigenous, autoregenerating (via photosynthesis), eco-friendly natural resources, which all can be used to replace or act as an alternative to fossil fuels and also reduce the GHGs during the combustion process [13–16]. Furthermore, lignocellulosic biomass materials comprise three major constituents, namely, cellulose (30%–50%), hemicellulose (15%–35%), and lignin (10%–20%); they may contain xylan, arabinan, galactan, glucuronic, acetic, ferulic acid, coumaric acid, etc., which are the potential substrates for the production of bio-oils, say ethanol and butanol [17–19].

Hence, many challenges are upgrading to generate alternative liquid transportation bio-oil or biofuel, which are more superior to bioethanol, but have similar fuel properties with those of petroleum-based gasoline. Among the various liquid biofuels, a four-carbon alcohol of biobutanol (C_4H_9OH), when the butanol or 1-butanol or butyl alcohol is generated from biomass materials (then often termed as biobutanol or second-generation biofuel), shows superior fuel properties than bioethanol [20–22]. In 1912–14 the first industrial production of biobutanol was explored via acetone, butanol, and ethanol (ABE) fermentation, using the molasses as well as cereal grains in the presence of *Clostridium acetobutylicum* (Weizmann's microorganism), but the large-scale production of biobutanol through microbial fermentation was reported by Louis Pasteur in 1861. Since the middle of the 20th century, some of the promising microbial strains, say *Clostridium beijerinckii, Clostridium saccharoperbutylacetonicum*, and *Clostridium saccharobutylicum*, have been identified and utilized for the production of biobutanol with maximum yield, although currently *C. acetobutylicum* has been widely utilized globally for the promising production of biobutanol.

In this chapter, we discuss the production of biobutanol from lignocellulosic biomass materials through (1) pretreatment and hydrolysis of different solid lignocellulosic biomass materials, (2) effects of degradation products on growth and biobutanol production by advanced ABE-fermenting microbial strains, and (3) strategies for improving the production of biobutanol.

5.2 Brief summary on lignocellulosic biomass material properties and their chemical compositions

Generally, all the lignocellulosic biomass materials (mainly from forest materials and residues, agricultural residues and grasses, starchy materials, bioenergy crops, and municipal wastes) are potential organic resources and are also utilized as an admirable rich source for the production of bio-oil or biofuel, which is due to their massive amount (~70%–80%) of fermentable sugar moieties [23]. Also, most of the commonly used cheap lignocellulosic biomass materials used as feedstocks are corn stover, sugarcane bagasse, wheat straw, rice straw, etc. for the production of bio-oil in developed as well as developing countries, say United States, South America, North America, Asia, and Europe [24,25]. The bioconversion of lignocellulosic biomass materials into bio-oils or valuable chemical compounds via pretreatment, enzymatic hydrolysis followed by fermentation processes are depicted in Fig. 5.1. The lignocellulosic biomass materials comprise three major forms of significant polymers, namely, cellulose, hemicellulose, and lignin, 30%–50%, 15%–35%, 10%–20%, respectively [7,14,18,19], with trace amounts of others such as pectin, protein,

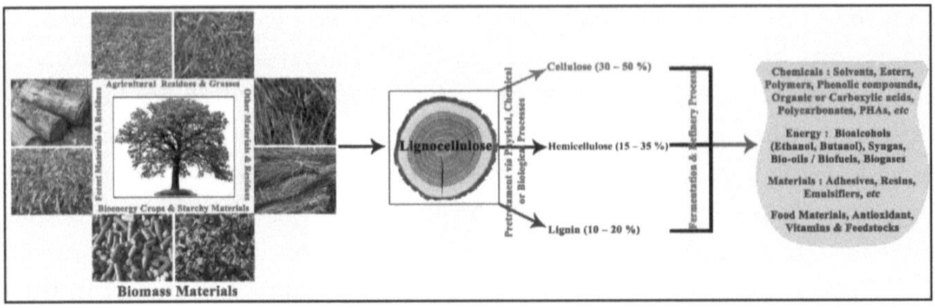

Figure 5.1 Schematic illustration of lignocellulosic biomass materials' biorefinery pathway.

extractives, and ash. Table 5.1 summarizes the percentage of cellulose, hemicellulose, and lignin content present in the various lignocellulosic biomass materials or feedstocks.

5.3 Pretreatment of lignocellulosic biomass feedstocks

The low-cost, highly abundant, renewable, eco-friendly, and carbon-rich lignocellulosic biomass sources are the most promising feedstocks for the production of biobutanol and other compounds. The conversion of biomass materials or feedstocks into fermentable energy-rich sugar moieties is carried out through pretreatment processes, that is, by opening of the biomass structure and releasing the sugar moieties from cellulose and hemicelluloses [75,76]. There are four types of pretreatment processes, which are well studied; (1) physical, (2) chemical, (3) physicochemical, and (4) biological or enzymatic pretreatment technologies (Fig. 5.2). The main aim of the preceding pretreatment technologies is to improve the fermentable sugar moieties and degrade other inhibitory compounds. The main advantages and disadvantages of various pretreatment processes are summarized in Table 5.2.

The physical pretreatment method is the key stage for lignocellulosic biomass processing; in this stage, the lignocellulosic biomass is reduced into a suitable powder form for better approachability of enzymes/microbes during hydrolysis process. Also, the size-reduced biomass materials help decrease the crystalline nature of celluloses, that is, to increase the biomass product yields. This pretreatment method includes milling or grinding, extrusion, microwave and ultrasonication processes; such methods are widely utilized for opening the structure of lignocellulosic

Table 5.1 Chemical compositions (in % dry weight) of various lignocellulosic biomass materials.

Lignocellulosic biomass materials	Cellulose (%)	Hemicellulose (%)	Lignin (%)	References
Alfalfa straw	31.00	10.00	10.00	[26]
Bagasse	54.87	16.52	23.33	[27–29]
Bamboo	45–50	18–20	23.00	[30,31]
	26–43	15–26	21–31	
Banana waste	13.20	14.80	14.00	[14,28–30,32]
Barley straw	45.00	38.00	19.00	[33]
Bermuda grass	32.40	24.80	20.33	[34]
Biomass sorghum	22.20	19.40	21.40	[35]
Corn cobs	45.00	35.00	15.00	[14,18,36–38]
Corn fiber	13.00	38.80	7.50	[39]
Corn stover	38.00	26.01	19.00	[29]
	34.50	27.70	16.50	[40]
Corn straw	35.00	28.00	17.00	[40]
Cotton stalk	31.00	11.00	30.00	[41]
Coffee pulp	35.00	46.30	18.80	[30]
Cotton seed hairs	80–95	5–20	0.00	[30,42]
Cucumber plant	17.00	17.00	3.00	[43–44]
Eucalyptus	38–45	12–13	25–37	[45]
Giant reed leaves	20.90	17.70	25.40	[46]
Giant reed stalk	33.00	18.50	24.50	[46]
Grasses	25–40	25–50	10–30	[7,29]
Green bean	17.00	16.00	8.00	[43,44]
Hardwood	40–55	24–40	18–25	[27–29]
Hybrid poplar	40.00	22.00	24.00	[47]
Jute fibers	45–53	18–21	21–26	[48]
Loblolly pine	35.00	16.80	29.00	[49]
Lodgepole pine	44.90	22.60	26.80	[50]
Malt residue	22.40	34.20	3.00	[51]
Miscanthus	38–48	19–30	12–25	[26,52–54]
Monterey pine	41.70	20.50	25.90	[47]
General MSW	33–49	9–16	10–14	[30]
Napier grass	45.70	33.70	20.60	[55,56]
Newspaper	40–55	25–40	18–30	[42,55]
Nut shells	25–30	–	30–40	[27–29]
Oat straw	33.00	23.00	21.00	[57–59]
Office paper	68.60	12.40	11.30	[60]
Oilseed rape	27.30	20.50	14.20	[61]
Olive mill waste	17.3–24.1	7.9–11.0	0.21–14.1	[62]
Pea vines	40.00	10.00	9.00	[59,63]

(Continued)

Table 5.1 (Continued)

Lignocellulosic biomass materials	Cellulose (%)	Hemicellulose (%)	Lignin (%)	References
Pepper plant	18.00	12.00	8.00	[43,64]
Pine	25–44	26–32	28–48	[54,59,63]
Poplar	45.00	21.00	24.00	[63,65]
Rapeseed stover	27.60	20.20	18.30	[66]
Rapeseed straw	37.00	25.00	17.00	[57,58]
Red maple/oak	39.00	33.00	23.00	[59,63]
Rice bran	15.50	31.10	11.50	[67]
Rice straw	32.10	24.00	18.00	[27–29]
Rye straw	38.00	36.90	17.60	[35,68]
Softwood	40–55	24–40	25–35	[7,14,18]
Softwood stem	40.00	30.00	30.00	[36,37]
Soybean straw	38.00	16.00	16.00	[59,63]
Sorghum straw	32–35	24–27	15–21	[69]
Sponge gourd fibers	66.59	17.44	15.46	[27]
Spruce	28–43	20–30	28–35	[54,59,63]
Sugarcane bagasse	42.00	25.00	20.00	[27,29,70,71]
	40–45	20–24	25–30	[72]
Sunflower stalk	31.00	15.60	29.20	[46]
Sweet sorghum	45.00	27.00	21.00	[27,28]
Switchgrass	36–45	28–30	12–26	[54,71,72]
Tomato plant	39.00	29.00	12.00	[73]
Tomato pomace	39.00	5.00	11.00	[59]
Waste paper from chemical pretreatment	60–70	10–20	5–10	[42]
Willow	49.30	14.10	20.00	[74]
Wheat straw	29–35	26–32	16–21	[27,28]

MSW, Municipal solid wastes.

biomasses [83–86]. The major chemical pretreatment technologies comprise acid hydrolysis (using mineral HCl, H_2SO_4), alkali treatment (using NaOH, KOH), ozonolysis, organosolvation process, and ionic liquids [87–89]. However, the utilization of various chemical pretreatments has some major drawbacks, which affect the entire bioconversion of lignocellulosic biomass feedstocks; hence, the combination of different processes for chemical and physical methods, in terms of physicochemical pretreatment processes, shows superior effects than individual pretreatment process. The major physicochemical pretreatment techniques comprise steam explosion (autohydrolysis with and without addition of chemicals), ammonia fiber explosion, supercritical CO_2 explosion, liquid hot water

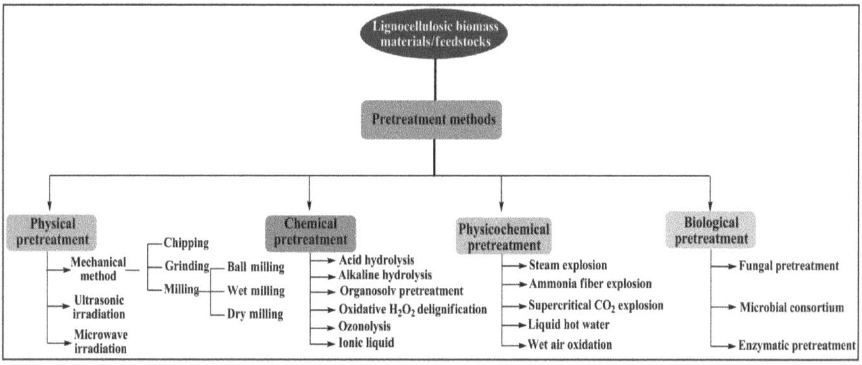

Figure 5.2 Schematic representation of various types of pretreatment process for lignocellulosic biomass feedstocks.

(hydrothermolysis, uncatalyzed solvolysis, aquasolv, aqueous fraction), and wet air oxidation [7,77,87,90].

In biological pretreatment the microorganisms (brown, white, and soft-rot fungi) are widely utilized for the treatment of lignocellulosic biomass feedstocks, which subsequently increase the enzymatic hydrolysis for enhancing the fermentable sugar moieties with high yield and effective delignification (removal of lignin from biomass feedstocks) process [91–93]. The main biological pretreatment process includes fungal, microbial consortium, and enzymatic techniques. Even though, such pretreatments possess several benefits as they are inexpensive, safer, belonging to low hydrolysis rate, of low energy input, eco-friendly, effective to delignification, of high yields without generating any inhibitory byproducts, easy to recover the microorganisms, etc.

5.4 Biobutanol as a valuable fuel and chemical source

Since 1916 the production of *n*-butanol, which has been considered an excellent solvent, as well as chemical feedstocks in many industries and all the isomers of *n*-butanol (except *tert*-butanol) are the most competitive fuel stabilizers for replacing petroleum-based gasoline, directly as well as superior to bioethanol [94–96]. Butanol is a colorless and highly flammable alcohol. In the 20th century, butanol has been widely utilized in the production of raw material (butadiene) of natural rubber in the rubber industry. Butanol has four different types of isomeric structures (Fig. 5.3), namely, (1) *n*-butanol or 1-butanol or butyl alcohol or primary butanol,

Table 5.2 Summary of the major advantages and disadvantages/limitations of various pretreatment processes.

Pretreatment method	Mechanism of action/ effects	Advantages	Disadvantages/ limitations	Remarks
I. Physical pretreatment process				
Mechanical pretreatment (chipping, grinding, and milling)	Reduces particle size	Reduces the final particle size and heat, mass transfer limitations, cellulose crystallinity, and reduces the cost of solid–liquid separation, to increase the external surface area and handling of materials easily	Cannot remove lignin and hemicelluloses, high energy consumption, that is, the power consumption is usually higher than the inherent biomass energy	Processing step for other pretreatment methods and large-scale operation is unfeasible
High energy electron radiation (g-rays, ultrasound, electron beam irradiation, pulsed electrical field, UV rays, microwave heating, or radiation)	Decrease of the DP, hydrolysis of hemicellulose, partial depolymerization of lignin, cleavage of β-1,4-glycosidic bonds	Simple operation, high energy efficiency, short time, improves accessibility and reactivity of cellulose, reduces cellulose polymerization degree	High cost, negative to enzymatic hydrolysis	
High-temperature pyrolysis		Decomposes cellulose rapidly	High energy consumption, low productivity	

II. Chemical pretreatment process				
Acid pretreatment (using dilute or conc. acids such as HCl, H_2SO_4)	To solubilize the lignocellulosic biomass materials, fractionation of hemicellulose and lignin	Effective hydrolysis of hemicellulose to xylose and lignin alteration, high sugar conversion with good yield, fast and does not need recycling of acid, alters the lignin structure	Highly toxic and formation of toxic compounds, expense of acids, high corrosive nature, hence, can easily corrode the reactor equipment, high cost, occurs at high temperature and pressure, formation of inhibitors, biomass material loss due to degradation	
Alkaline hydrolysis [using NaOH, KOH, Ca $(OH)_2$, lime]	To solubilize both lignin and hemicellulose easily	Effective as well as efficient removal of lignin and acetyl groups of hemicellulose, low inhibitor formation	More time consuming, destroys lignin, high cost of alkaline catalyst, considerable challenge to the neutralization process, generation of inhibitors	Unsuitable for biomass containing high lignin content
Organosolvent pretreatment (using organic solvents, say EtOH-	To break lignin–carbohydrate bonds	Hydrolysis to lignin and hemicellulose, easy to recover organic solvent with low	Highly expensive due to cleaning of solvent from reactor, certain effects on	Solvents are easily volatile in nature, and hence, digester leaks should be avoided

(Continued)

Table 5.2 (Continued)

Pretreatment method	Mechanism of action/ effects	Advantages	Disadvantages/ limitations	Remarks
H_2O, butanol-H_2O, ethylene-glycol)		boiling point by distillation and also recycle it, environment-friendly and also removes lignin effectively, more selective method with high yielding of fractions, produces low residual lignin substrates	environment and fermentation, handling of harsh organic solvents, formation of inhibitors	
Ozonolysis pretreatment (using O_3)	To solubilize the lignocellulosic biomass materials and also reduce the lignin content	Working under mild reaction or NTP condition, no formation of toxic compounds, increase the specific surface area, efficient lignin removal	Require large quantity of ozone, high cost of O_3 so this method is more expensive	
IL pretreatment	To reduce the cellulose crystallinity, removal of lignin, to compete with lignocellulosic components for H_2	Easy regeneration of cellulose and hemicellulose fibers, successful removal of lignin, increases the	High cost of ILs, process is more expensive	

	bonding, thereby disrupting its 3D network	surface area, environment-friendly, employed at high temperature range, easily recyclable and reusable, high immiscibility with a wide range of solvents		
III. Physicochemical pretreatment process				
Steam explosion	Treatment with high-pressure saturated steam by holding at high temperature (160°C–260°C) and moderate pressure (0.69–4.83 MPa) for a very short period of several seconds to few minutes followed by a rapid decomposition	Separation of individual cellulose fibers, alteration of lignin and hemicellulose removal, lignin transformation, hemicelluloses' solubilization, low cost	High generation of inhibitor compounds, high toxic nature, degradation of xylan portion, process occurs only at high temperature and pressure, incomplete disruption of the lignin–carbohydrate matrix, generation of compounds inhibitory to microorganisms	
AFEX	Lignin removal, exposure to liquid NH_3 at moderate	Solubilization of hemicellulose, cellulose	Ammonia handling and recovery, much less effective for	

(*Continued*)

Table 5.2 (Continued)

Pretreatment method	Mechanism of action/effects	Advantages	Disadvantages/limitations	Remarks
	temperature and pressure (80°C–150°C; 200–400 psi) for a very short period (5–30 min) followed by instantaneous pressure release	decrystallization, lignin depolymerization, high efficiency and selectivity for reaction with lignin, increased surface area of cellulose, absence of inhibition substances formed, ability to reduce, recover, and recycle the ammonia, needs moderate temperature, pH, and short time, high selectivity for reaction with lignin, total sugar yield is increased	softwood, high cost, not efficient for raw high lignin content material, environmental issues	
LHW	Removal of soluble lignin and hemicellulose hydrolysis, use of hot water to moderate	No additional chemicals, and minimum risk of inhibitory product formation, residual	Abundant water usage and energy input, incomplete destruction of lignin–carbohydrate	Large-scale operation is unfeasible

	temperatures instead of steam	lignin puts a negative effect on the subsequent enzymatic hydrolysis, requires low energy input, low cost of the solvent	matrix, destruction of a portion of the xylan in hemicellulose, formation of inhibitors at higher temperatures	
Supercritical CO_2 explosion	Injection of high pressure CO_2 followed by explosive decompression	Disruption of cellulosic structure and increase in accessible surface area	Not much effect on lignin and hemicellulose	
Wet air oxidation	Uses air in combination with water at elevated temperature (125°C–320°C) and pressure (0.5–2 MPa), leading to oxidative reaction	Opening of crystalline cellulose, solubilization of hemicellulose and lignin, limited formation of inhibitors, the oxidizing agents (air and O_2) are inexpensive and readily available in nature	Highly expensive because of equipment cost, high temperature and pressure require expensive materials for the reactor	
IV. Biological pretreatment process				
Fungi, microbial consortium, and bacteria species (using white-rot fungi, brown-rot fungi, and		Low energy is required, no chemicals are needed, operation at mild environmental	Very slow process, low rate of hydrolysis, large space is required	

(*Continued*)

Table 5.2 (Continued)

Pretreatment method	Mechanism of action/ effects	Advantages	Disadvantages/ limitations	Remarks
soft-rot fungi, *Sphingomonas paucimobilis*, *Bacillus circulans*)		conditions, degrades lignin and hemicellulose, eco-friendly process, no release of toxic compounds, no generation of fermentation inhibitors during the process, lignin degradation, low energy requirement		

AFEX, Ammonia fiber explosion; *DP*, degrees of polymerization; *IL*, ionic liquid; *LHW*, liquid hot water; *NTP*, normal temperature and pressure.
Source: Adapted from V.B. Agbor, N. Cicek, R. Sparling, A. Berlin, D.B. Levin, Biomass pretreatment: fundamentals toward application, Biotechnol. Adv. 29(6) (2011) 675–685; G. Cao, Y. Sheng, L. Zhang, J. Song, H. Cong, J. Zhang, Biobutanol production from lignocellulosic biomass: prospective and challenges, J. Bioremed. Biodegrad. 7(4) (2016) 363; C. Hongyan, L. Jinbao, C. Chang, C. DaMing, X. Yuan, L. Ping, et al., A review on the pretreatment of lignocellulose for high-value chemicals, Fuel Process. Technol. 160 (2017) 196–206; N.M. Huzir, M.M.A. Aziz, S. Ismail, B. Abdullah, N.A.N. Mahmood, N. Umor, et al., Agro-industrial waste to biobutanol production: eco-friendly biofuels for next generation. Renew. Sustain. Energy Rev. 94 (2018) 476–485; A. Morone, R.A. Pandey, Lignocellulosic biobutanol production: gridlocks and potential remedies. Renew. Sustain. Energy Rev. 37 (2014) 21–35; P. Bajpai, Pretreatment of Lignocellulosic Biomass for Biofuel Production, Springer, 2016 [77–82].

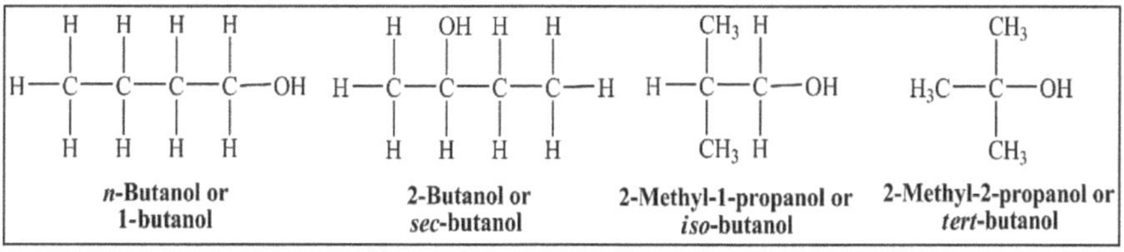

Figure 5.3 Schematic representation of different isomeric structures of butanol.

(2) 2-butanol or *sec*-butanol, (3) 2-methyl-1-propanol or *iso*-butanol, and (4) 2-methyl-2-propanol or *tert*-butanol [97], and all the isomers possess the same chemical formula but differ in terms of their manufacturing approach, energy content, and combustion characters.

Compared to ethanol, biobutanol shows superior biofuel properties as well as is considered as a next-generation biofuel as it (1) is more easily miscible with gasoline, thus being easy to directly blend (even at 100%) with gasoline at a higher concentration; (2) directly utilized in convention without any or very little modification in automobile engines; (3) easily stored and safely distributed due to its less or nonhygroscopic (thus does not pick up water) nature; (4) readily mixed with gasoline in any proportion; (5) safer to handle due to low Reid vapor pressures (7.5 times lower than ethanol) and flash point; (6) has a lower volatility than other conventional fuels [13.9, 13.5, and 6 times lesser than MeOH (methanol), gasoline, and EtOH (ethanol), respectively]; hence, it can be utilized safely in summer or winter conditions; (7) able to reduce the NO_x and approximately 20% of GHG emissions and soot creations from fuel evaporation or combustion in warm conditions; and (8) has higher energy content (~30% more energy than ethanol), polarity, combustion value, air-to-fuel ratio, and octane rating due to the presence of twice the number of C atoms (as compared to ethanol) in their molecular structure, that is, biobutanol is directly utilized in internal combustion engines. In addition, butanol and its blends are easily transported in existing fuel pipelines, tanks, and related infrastructures due to its soluble nature as well as very less corrosiveness, compared to ethanol, etc. [95,98,99]. Moreover, biobutanol can be utilized as a high value-added chemical feedstock in numerous industrial uses, such as paints, cosmetics (eye makeup, nail care products, lipsticks, shaving products), adhesives, inks, food flavors, solvents, surface coating (enamel and lacquer) materials, fragrances, and swelling agent in garments, and in pharmacological fields [94,100,101].

Even though there are some massive advantages associated with the utilization of liquid bio-alcohol as a fuel for the replacement of fossil fuels or petroleum-based gasoline or diesel, some problems are there while the productions of biobutanol, say low production yield and high substrate rate. Hence, many researchers and industrialists have continuously searched and identified low-cost substrates of renewable lignocellulosic biomass materials, such as maize stover, agricultural waste, rice straw, barley straw, and switchgrass, which are the feedstocks, utilized for the production of biobutanol [102–105]. At present the production of biobutanol from lignocellulosic biomass materials is extensively considered as an alternative feedstock, because the low-cost biomass materials present in high abundance are broadly distributed in nature [12,13,16]. The first generation–derived biobutanol from food and agricultural resources may cause some problems, that is, negative impact on food security, increased food prices, and net energy losses [106]. Globally, the production of biobutanol from renewable lignocellulosic biomass carbon feedstocks through the clostridial ABE fermentation process by solventogenic *Clostridium* species (namely, *Clostridium cellulolyticum* and *Clostridium cellulovorans*) since the ABE fermentation was considered as one of the ancient processes for large-scale industrial fermentation methods. Generally, the production of bioethanol from low-cost lignocellulosic biomass is more economical, but the same feedstocks for the production of higher bio-alcohols, say biobutanol, may cause some challenges due to their recalcitrance to degradation as well as release of fermentable sugars for further fermentation processes [12]. In recent years, some researchers have been investigating the production of ABE-fermented biobutanol from cellulosic and noncellulosic feedstocks. Globally, there are four leading biobutanol producers, namely, Butamax, Gevo, US Technology Corporation, and Green Biologics, in large-scale level production for the replacement of fossil fuel crises [107]. Furthermore, the production cost (including the oil price) for biobutanol from fossil fuel or petrochemical source is ~1.35 \$/L, whereas the cost of biobutanol production from bioenergetic crops through ABE fermentation process is 0.317 \$/L, but the production of bioethanol is 0.338 \$/L [108], that is, the utilization of lignocellulosic biomass materials derived from agricultural crops and residues, forestry plants and residues, plant biomass, etc. may reduce the production cost, renewability, and also promising materials for the production of biobutanol.

5.5 Production of biobutanol through microbial or acetone, butanol, and ethanol fermentation process

The production of biobutanol from the microbial fermentation process is more superior to the usual chemical routes [12,109,110], because this process is a more eco- as well as environment-friendly approach, and the utilized biological materials or feedstocks could reduce the GHG emissions. Hence, the microbial or ABE fermentation process is the promising tool for the production of eco-friendly biobutanol in many biofuel plants using the lignocellulosic feedstocks (corn cobs, corn stover, etc.) due to their high abundance, renewability, carbon richness, inexpensive nature, etc. [111]. The first and oldest industrial large-scale production of biobutanol via ABE fermentation using *Clostridium* strains (*C. acetobutylicum*) [102,112] and the biofuel plant was first developed in the United Kingdom in 1912. Prof. Chaim Weizmann (called the father of industrial fermentation for developing ABE fermentation process) developed the large-scale production of biobutanol from *C. acetobutylicum* in industrial sectors. During the ABE fermentation process, three major classes of products are formed, namely, (1) solvents (ABE), (2) organic acids (acetic, lactic, and butyric acids), and (3) gases (CO_2 and H_2 with 60:40 mole ratio) [12,113]. The main metabolic pathways for the production of biofuels through ABE fermentation of *Clostridia* are shown in Fig. 5.4.

5.6 Concluding remarks and future outlook

In the middle of the 20th century, many researchers and industrialists reported the low productivity of ABE products or biobutanol production from different lignocellulosic biomass feedstocks via ABE fermentation process. Currently, the utilization of enhanced pretreatment and genetically engineered processes has modified ABE fermentation process. Advanced recovery tools for the separation as well as purification of biobutanol with high yield will lead to an improved yield and production of ABE products or biobutanol. This could make the produced biobutanol a promising fuel resource for the replacement of petroleum-based fuels. In the current age the problems as well as challenges in the production of ABE products or biobutanol are feedstock availability (say inexpensive, renewable, high abundance, high fermentable sugar moieties), economical and environment-friendly advanced pretreatment and hydrolysis processes, eco-friendly or economic feasibility of ABE fermentation, and recovery of

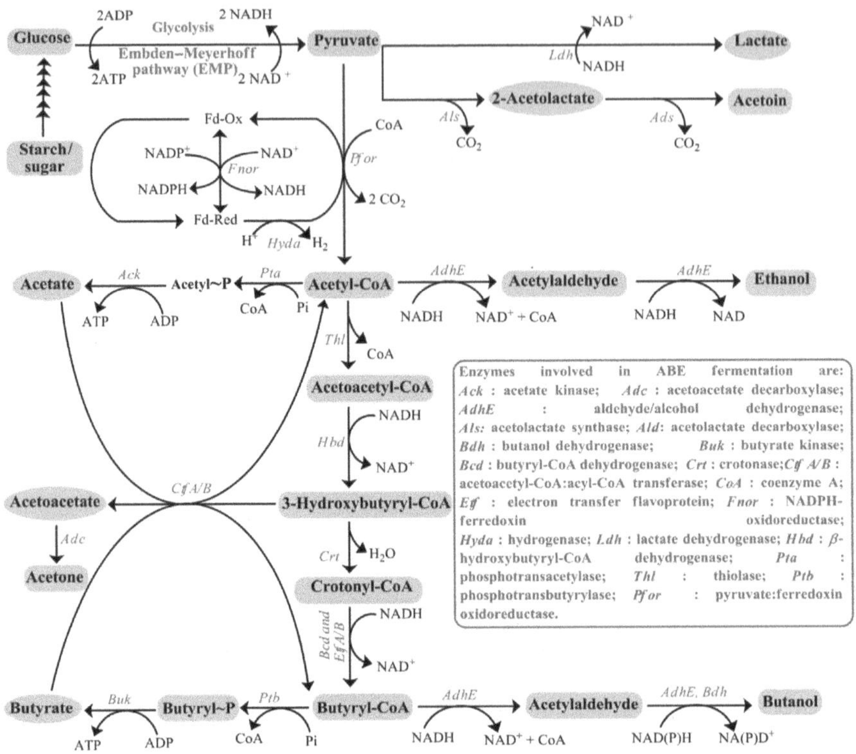

Figure 5.4 Schematic representation of microbial pathways of acid and solvent formation in ABE fermentation in *Clostridia* species. *ABE*, Acetone, butanol, and ethanol.

biobutanol with high purity. The high abundance and possibility of using waste cellulosic materials from agricultural and forest residues as feedstocks aids a new pathway in the production of biobutanol at industrial scale. The genetic engineering methods for the microbial strain improvement with supplement of biobutanol-producing gene of Clostridia in high butanol-tolerant microorganisms lead to substantial biobutanol production, which is a great opportunity. Recently, some researchers are investigating the aerobic production of biobutanol using some genetically engineered microorganisms. However, the emerging scope of process growth could be explored in developing novel continuous system, feedstock pretreatment, and efficient integrated biobutanol recovery for improving the productivity of biobutanol fermentation. Therefore more research, pertaining to the drawbacks, is required for the manipulation of these techniques, and consequently, the industrial-scale production of biofuels could be achieved. However, looking into the coming future, the production of biobutanol from various waste lignocellulosic biomass feedstocks shows superior capacity as an alternative fuel, when compared to ethanol- and petroleum-based fuels, considering the technical aspects and market value. Anticipating the merits and increasing demand, the market for butanol is expected to expand globally in coming years. Moreover, biobutanol will be the milestone for attracting the attention of the government, commercial, and research organizations for further support in implementing the innovative fermentation and extraction technology.

References

[1] G.W. Festel, Biofuels – economic aspects, Chem. Eng. Technol. 31 (5) (2008) 715–720.

[2] C. Dellomonaco, F. Fava, R. Gonzalez, The path to next generation biofuels: successes and challenges in the era of synthetic biology, Microb. Cell Fact. 9 (3) (2010) 1–15.

[3] T. Hong, C. Koo, T. Kwak, H.S. Park, An economic and environmental assessment for selecting the optimum new renewable energy system for educational facility, Renew. Sustain. Energy Rev. 29 (2014) 286–300.

[4] M. Höök, X. Tang, Depletion of fossil fuels and anthropogenic climate change—a review, Energy Policy 52 (C) (2013) 797–809.

[5] C. Kordulis, K. Bourikas, M. Gousi, E. Kordouli, A.J. Lycourghiotis, Development of nickel based catalysts for the transformation of natural triglycerides and related compounds into green diesel: a critical review, Appl. Catal. B Environ. 181 (2016) 156–196.

[6] E. Aransiola, T. Ojumu, O. Oyekola, T. Madzimbamuto, D. Ikhu-Omoregbe, A review of current technology for biodiesel production: state of the art, Biomass Bioenergy 61 (2014) 276–297.

[7] P. Kumar, D.M. Barrett, M.J. Delwiche, P. Stroeve, Methods for pretreatment of lignocellulosic biomass for efficient hydrolysis and biofuel production, Ind. Eng. Chem. Res. 48 (8) (2009) 3713–3729.

[8] A.E. Atabani, A.S. Silitonga, I.A. Badruddin, T. Mahlia, H. Masjuki, S. Mekhilef, A comprehensive review on biodiesel as an alternative energy resource and its characteristics, Renew. Sustain. Energy Rev. 16 (4) (2012) 2070–2093.

[9] A. Mishra, S. Ghosh, Bioethanol production from various lignocellulosic feedstocks by a novel "fractional hydrolysis" technique with different inorganic acids and co–culture fermentation, Fuel 236 (2019) 544–553.

[10] P.S. Nigam, A. Singh, Production of liquid biofuels from renewable resources, Prog. Energy Combust. Sci. 37 (1) (2011) 52–68.

[11] P. Wang, Y.M. Chen, Y. Wang, Y.Y. Lee, W. Zong, S. Taylor, et al., Towards comprehensive lignocellulosic biomass utilization for bioenergy production: efficient biobutanol production from acetic acid pretreated switchgrass with *Clostridium saccharoperbutylacetonicum* N1–4, Appl. Energy 236 (2019) 551–559.

[12] C. Xue, X.-Q. Zhao, C.-G. Liu, L.-J. Chen, F.-W. Bai, Prospective and development of butanol as an advanced biofuel, Biotechnol. Adv. 31 (8) (2013) 1575–1584.

[13] K. Huang, W. Won, K.J. Barnett, Z.J. Brentzel, D.M. Alonso, G.W. Huber, et al., Improving economics of lignocellulosic biofuels: An integrated strategy for coproducing 1,5-pentanediol and ethanol, Appl. Energy 213 (2018) 585–594.

[14] Z. Anwar, M. Gulfraz, M. Irshad, Agro-industrial lignocellulosic biomass a key to unlock the future bio-energy: a brief review, J. Radiat. Res. Appl. Sci. 7 (2) (2014) 163–173.
[15] M. Kumar, Y. Goyal, A. Sarkar, K. Gayen, Comparative economic assessment of ABE fermentation based on cellulosic and noncellulosic feedstocks, Appl. Energy 93 (2012) 193–204.
[16] S. Liu, A synergetic pretreatment technology for woody biomass conversion, Appl. Energy 144 (2015) 114–128.
[17] F.G. Calvo-Flores, J.A. Dobado, Lignin as renewable raw material, ChemSusChem 3 (11) (2010) 1227–1235.
[18] M. Kim, D.F. Day, Composition of sugar cane, energy cane, and sweet sorghum suitable for ethanol production at Louisiana sugar mills, J. Ind. Microbiol. Biotechnol. 38 (7) (2011) 803–807.
[19] J. Liu, F. Jin, M. Fan, L. Zhu, C. Tang, R. Chang, et al., Production of high–pure hydrogen by an integrated catalytic process: comparison of different lignocellulosic biomasses and three major components, Fuel 226 (2018) 322–330.
[20] L. Tao, E.C. Tan, R. McCormick, M. Zhang, A. Aden, X. He, et al., Techno-economic analysis and life-cycle assessment of cellulosic isobutanol and comparison with cellulosic ethanol and n-butanol, Biofuels Biofuel Bioprod. Biorefin. 8 (1) (2014) 30–48.
[21] P. Dürre, Fermentative butanol production: bulk chemical and biofuel, Ann. N.Y. Acad. Sci. 1125 (1) (2008) 353–362.
[22] P. Dürre, Biobutanol: an attractive biofuel, Biotechnol. J. 9 (12) (2007) 1525–1534.
[23] E.A. Bayer, R. Lamed, M.E. Himmel, The potential of cellulases and cellulosomes for cellulosic waste management, Curr. Opin. Biotechnol. 18 (3) (2007) 237–245.
[24] K.-K. Cheng, B.-Y. Cai, J.-A. Zhang, H.-Z. Ling, Y.-J. Zhou, J.-P. Ge, et al., Sugarcane bagasse hemicellulose hydrolysate for ethanol production by acid recovery process, Biochem. Eng. J. 38 (1) (2008) 105–109.
[25] S. Kim, B.E. Dale, Global potential bioethanol production from wasted crops and crop residues, Biomass Bioenergy 26 (4) (2004) 361–375.
[26] C. Sawatdeenarunat, K. Surendra, D. Takara, H. Oechsner, S.K. Khanal, Anaerobic digestion of lignocellulosic biomass: challenges and opportunities, Bioresour. Technol. 178 (2015) 178–186.
[27] M. Bertero, G. de la Puente, U. Sedran, Fuels from bio-oils: Bio-oil production from different residual sources, characterization and thermal conditioning, Fuel 95 (2012) 263–271.
[28] H. Muhammad Nasir Iqbal, M. Asgher, Characterization and decolorization applicability of xerogel matrix immobilized manganese peroxidase produced from *Trametes versicolor* IBL–04, Protein Pept. Lett. 20 (5) (2013) 591–600.
[29] S. Prasad, A. Singh, H.C. Joshi, Ethanol as an alternative fuel from agricultural, industrial and urban residues, Resour. Conserv. Recycl. 50 (1) (2007) 1–39.
[30] H. Zabed, J. Sahu, A. Boyce, G. Faruq, Fuel ethanol production from lignocellulosic biomass: an overview on feedstocks and technological approaches, Renew. Sustain. Energy Rev. 66 (2016) 751–774.
[31] E.F. Alves, S.K. Bose, R.C. Francis, J.L. Colodette, M. Iakovlev, A. Van Heiningen, Carbohydrate composition of eucalyptus, bagasse and bamboo by a combination of methods, Carbohydr. Polym. 82 (4) (2010) 1097–1101.
[32] G. Monsalve, F. John, P. Medina, C. Ruiz, A. Adriana, Ethanol production of banana shell and cassava starch, Dyna. Rev. Fac. Nac. Minas 73 (150) (2006) 21–27.
[33] R.M. Rowell, J. Rowell, Paper and Composites From Agro-Based Resources., CRC Press, 1996.

[34] Y. Sun, J.J. Cheng, Dilute acid pretreatment of rye straw and bermudagrass for ethanol production, Bioresour. Technol. 96 (14) (2005) 1599–1606.
[35] F. Monlau, A. Barakat, E. Trably, C. Dumas, J.-P. Steyer, H. Carrère, Lignocellulosic materials into biohydrogen and biomethane: impact of structural features and pretreatment, Crit. Rev. Environ. Sci. Technol. 43 (3) (2013) 260–322.
[36] R. Chandra, H. Takeuchi, T. Hasegawa, Hydrothermal pretreatment of rice straw biomass: a potential and promising method for enhanced methane production, Appl. Energy 94 (2012) 129–140.
[37] R. Chandra, H. Takeuchi, T. Hasegawa, Methane production from lignocellulosic agricultural crop wastes: a review in context to second generation of biofuel production, Renew. Sustain. Energy Rev. 16 (3) (2012) 1462–1476.
[38] M. Pointner, P. Kuttner, T. Obrlik, A. Jager, H. Kahr, Composition of corncobs as a substrate for fermentation of biofuels, Agron. Res. 12 (2) (2014) 391–396.
[39] H. Noureddini, J. Byun, T.-J. Yu, Stagewise dilute-acid pretreatment and enzyme hydrolysis of distillers' grains and corn fiber, Appl. Biochem. Biotechnol. 159 (2) (2009) 553–567.
[40] S. Kumar, U. Kothari, L. Kong, Y. Lee, R.B. Gupta, Hydrothermal pretreatment of switchgrass and corn stover for production of ethanol and carbon microspheres, Biomass Bioenergy 35 (2) (2011) 956–968.
[41] M. Rubio, J.F. Tortosa, J. Quesada, D. Gómez, Fractionation of lignocellulosics. Solubilization of corn stalk hemicelluloses by autohydrolysis in aqueous medium, Biomass Bioenergy 15 (6) (1998) 483–491.
[42] R. Howard, E. Abotsi, E.J. Van Rensburg, S. Howard, Lignocellulose biotechnology: issues of bioconversion and enzyme production, Afr. J. Biotechnol. 2 (12) (2003) 602–619.
[43] A.J. Callejón-Ferre, J. Carreño-Sánchez, F.J. Suárez-Medina, J. Pérez-Alonso, B. Velázquez-Martí, Prediction models for higher heating value based on the structural analysis of the biomass of plant remains from the greenhouses of Almería (Spain), Fuel 116 (2014) 377–387.
[44] A.J. Callejón-Ferre, B. Velázquez-Martí, J.A. López-Martínez, F. Manzano-Agugliaro, Greenhouse crop residues: energy potential and models for the prediction of their higher heating value, Renew. Sustain. Energy Rev. 15 (2) (2011) 948–955.
[45] O.P. Karthikeyan, C. Visvanathan, Bio-energy recovery from high-solid organic substrates by dry anaerobic bio-conversion processes: a review, Rev. Environ. Sci. Bio/Technol. 12 (3) (2013) 257–284.
[46] F. Monlau, A. Barakat, J.P. Steyer, H. Carrere, Comparison of seven types of thermo-chemical pretreatments on the structural features and anaerobic digestion of sunflower stalks, Bioresour. Technol. 120 (2012) 241–247.
[47] S.T. Merino, J. Cherry, Progress and challenges in enzyme development for biomass utilization, Biofuels, Springer, 2007, pp. 95–120.
[48] M. Mosihuzzaman, A. Quddus, N. Nahar, O. Theander, Comparative study of carbohydrates in the two major species of jute (*Corchorus capsularis* and *Corchorus olitorius*), J. Sci. Food Agric. 48 (3) (1989) 305–310.
[49] D. Rana, V. Rana, B.K. Ahring, Producing high sugar concentrations from loblolly pine using wet explosion pretreatment, Bioresour. Technol. 121 (2012) 61–67.
[50] X. Pan, D. Xie, R.W. Yu, J.N. Saddler, The bioconversion of mountain pine beetle-killed lodgepole pine to fuel ethanol using the organosolv process, Biotechnol. Bioeng. 101 (1) (2008) 39–48.
[51] A. Linan-Montes, S.M. De La Parra-Arciniega, M.T. Garza-González, R.B. García-Reyes, E. Soto-Regalado, F.J. Cerino-Córdova, Characterization and thermal analysis of agave bagasse and malt spent grain, J. Therm. Anal. Calorim. 115 (1) (2014) 751–758.

[52] W. De Jong, A. Pirone, M.A. Wojtowicz, Pyrolysis of *Miscanthus giganteus* and wood pellets: TG-FTIR analysis and reaction kinetics, Fuel 82 (9) (2003) 1139–1147.
[53] E. Hodgson, D. Nowakowski, I. Shield, A. Riche, A. Bridgwater, J. Clifton-Brown, et al., Variation in *Miscanthus* chemical composition and implications for conversion by pyrolysis and thermo-chemical bio-refining for fuels and chemicals, Bioresour. Technol. 102 (3) (2011) 3411–3418.
[54] S. Melin, Considerations for Grading Agricultural Residue, Ontario Federation of Agriculture, 2013.
[55] A. Janejadkarn, O. Chavalparit, Biogas production from Napier Grass (Pak Chong 1) (*Pennisetum purpureum* × *Pennisetum americanum*), Adv. Mater. Res. Trans. Tech. Publ. 856 (2014) 327–332.
[56] K.O. Reddy, C.U. Maheswari, M. Shukla, A.V. Rajulu, Chemical composition and structural characterization of Napier grass fibers, Mater. Lett. 67 (1) (2012) 35–38.
[57] A. Lehtomäki, S. Huttunen, T.M. Lehtinen, J.A. Rintala, Anaerobic digestion of grass silage in batch leach bed processes for methane production, Bioresour. Technol. 99 (8) (2008) 3267–3278.
[58] A. Lehtomäki, T.A. Viinikainen, J.A. Rintala, Screening boreal energy crops and crop residues for methane biofuel production, Biomass Bioenergy 32 (6) (2008) 541–550.
[59] T. Richard, N. Trautmann, M. Krasny, S. Fredenburg, C. Stuart, Cornell Composting Science and Engineering, 2013.
[60] A. Wiselogel, S. Tyson, D. Johnson, Biomass feedstock resources and composition, Handbook on Bioethanol., Routledge, 2018, pp. 105–118.
[61] A. Petersson, M.H. Thomsen, H. Hauggaard-Nielsen, A.-B. Thomsen, Potential bioethanol and biogas production using lignocellulosic biomass from winter rye, oilseed rape and faba bean, Biomass Bioenergy 31 (11–12) (2007) 812–819.
[62] A.G. Vlyssides, M. Loizidou, K. Gimouhopoulos, A. Zorpas, Olive oil processing wastes production and their characteristics in relation to olive oil extraction methods, Fresen. Environ. Bull. 7 (5) (1998) 308–313.
[63] R. Saidur, E.A. Abdelaziz, A. Demirbas, M.S. Hossain, S. Mekhilef, A review on biomass as a fuel for boilers, Renew. Sustain. Energy Rev. 15 (5) (2011) 2262–2289.
[64] A.A.T. Masiá, B.J.P. Buhre, R.P. Gupta, T.J. Wall, Characterising ash of biomass and waste, Fuel Process. Technol. 88 (11–12) (2007) 1071–1081.
[65] P. Sannigrahi, A.J. Ragauskas, G.A. Tuskan, Poplar as a feedstock for biofuels: a review of compositional characteristics, Biofuels Bioprod. Bioref. 4 (2) (2010) 209–226.
[66] H. Li, N.–J. Kim, M. Jiang, J.W. Kang, H.N. Chang, Simultaneous saccharification and fermentation of lignocellulosic residues pretreated with phosphoric acid-acetone for bioethanol production, Bioresour. Technol. 100 (13) (2009) 3245–3251.
[67] S. Sunphorka, W. Chavasiri, Y. Oshima, S. Ngamprasertsith, Protein and sugar extraction from rice bran and de-oiled rice bran using subcritical water in a semi-continuous reactor: optimization by response surface methodology, Int. J. Food Eng. 8 (3) (2012) 1–22.
[68] A.-S. Nizami, N.E. Korres, J.D. Murphy, Review of the integrated process for the production of grass biomethane, Environ. Sci. Technol. 43 (22) (2009) 8496–8508.
[69] A. Herrera, S. Téllez–Luis, J. Ramırez, M. Vázquez, Production of xylose from sorghum straw using hydrochloric acid, J. Cereal Sci. 37 (3) (2003) 267–274.
[70] D. Divya, L.R. Gopinath, P.M. Christy, A review on current aspects and diverse prospects for enhancing biogas production in sustainable means, Renew. Sustain. Energy Rev. 42 (2015) 690–699.

[71] S. Reshamwala, B.T. Shawky, B.E. Dale, Ethanol production from enzymatic hydrolysates of AFEX-treated coastal bermudagrass and switchgrass, Appl. Biochem. Biotechnol. 51 (1) (1995) 43.
[72] Zeshan, O.P. Karthikeyan, C. Visvanathan, Effect of C/N ratio and ammonia-N accumulation in a pilot-scale thermophilic dry anaerobic digester, Bioresour. Technol. 113 (2012) 294–302.
[73] R. Font, J. Moltó, A. Gálvez, M.D. Rey, Kinetic study of the pyrolysis and combustion of tomato plant, J. Anal. Appl. Pyrol. 85 (1–2) (2009) 268–275.
[74] T. Bridgeman, J. Jones, I. Shield, P.T. Williams, Torrefaction of reed canary grass, wheat straw and willow to enhance solid fuel qualities and combustion properties, Fuel 87 (6) (2008) 844–856.
[75] C. Birgen, P. Dürre, H.A. Preisig, A. Wentzel, Butanol production from lignocellulosic biomass: revisiting fermentation performance indicators with exploratory data analysis, Biotechnol. Biofuels 12 (1) (2019) 167.
[76] F. Carvalheiro, L.C. Duarte, F.M. Gírio, Hemicellulose biorefineries: a review on biomass pretreatments, J. Sci. Ind. Res. 67 (2008) 849–864.
[77] V.B. Agbor, N. Cicek, R. Sparling, A. Berlin, D.B. Levin, Biomass pretreatment: fundamentals toward application, Biotechnol. Adv. 29 (6) (2011) 675–685.
[78] G. Cao, Y. Sheng, L. Zhang, J. Song, H. Cong, J. Zhang, Biobutanol production from lignocellulosic biomass: prospective and challenges, J. Bioremed. Biodegrad. 7 (4) (2016) 1–6.
[79] C. Hongyan, L. Jinbao, C. Chang, C. DaMing, X. Yuan, L. Ping, et al., A review on the pretreatment of lignocellulose for high-value chemicals, Fuel Process. Technol. 160 (2017) 196–206.
[80] N.M. Huzir, M.M.A. Aziz, S. Ismail, B. Abdullah, N.A.N. Mahmood, N. Umor, et al., Agro-industrial waste to biobutanol production: eco-friendly biofuels for next generation, Renew. Sustain. Energy Rev. 94 (2018) 476–485.
[81] A. Morone, R.A. Pandey, Lignocellulosic biobutanol production: gridlocks and potential remedies, Renew. Sustain. Energy Rev. 37 (2014) 21–35.
[82] P. Bajpai, Pretreatment of Lignocellulosic Biomass for Biofuel Production, Springer, 2016.
[83] Y. Zhang, T. Hou, B. Li, C. Liu, X. Mu, H. Wang, Acetone-butanol-ethanol production from corn stover pretreated by alkaline twin-screw extrusion pretreatment, Bioprocess Biosyst. Eng. 37 (5) (2014) 913–921.
[84] X. Kong, H. Xu, H. Wu, C. Wang, A. He, J. Ma, et al., Biobutanol production from sugarcane bagasse hydrolysate generated with the assistance of gamma-valerolactone, Process. Biochem. 51 (10) (2016) 1538–1543.
[85] N. Qureshi, B.C. Saha, M.A. Cotta, Butanol production from wheat straw hydrolysate using *Clostridium beijerinckii*, Bioprocess Biosyst. Eng. 30 (6) (2007) 419–427.
[86] H. Su, G. Liu, M. He, F. Tan, A biorefining process: sequential, combinational lignocellulose pretreatment procedure for improving biobutanol production from sugarcane bagasse, Bioresour. Technol. 187 (2015) 149–160.
[87] D. Kumari, R. Singh, Pretreatment of lignocellulosic wastes for biofuel production: a critical review, Renew. Sustain. Energy Rev. 90 (2018) 877–891.
[88] K. Nwosu–Obieogu, L.I. Chiemenem, K.F. Adekunle, Utilization of agricultural waste for bioethanol production—a review, Int. J. Curr. Res. Rev. 8 (19) (2016) 1–5.
[89] W. Yuan, Z. Gong, G. Wang, W. Zhou, Y. Liu, X. Wang, et al., Alkaline organosolv pretreatment of corn stover for enhancing the enzymatic digestibility, Bioresour. Technol. 265 (2018) 464–470.
[90] N. Akhtar, K. Gupta, D. Goyal, A. Goyal, Recent advances in pretreatment technologies for efficient hydrolysis of lignocellulosic biomass, Environ. Prog. Sustain. 35 (2) (2016) 489–511.

[91] M. Balat, Production of bioethanol from lignocellulosic materials via the biochemical pathway: a review, Energy Convers. Manage. 52 (2) (2011) 858–875.
[92] H.K. Sharma, C. Xu, W. Qin, Biological pretreatment of lignocellulosic biomass for biofuels and bioproducts: an overview, Waste Biomass Valori. 10 (2) (2019) 235–251.
[93] R. Sindhu, P. Binod, A. Pandey, Biological pretreatment of lignocellulosic biomass—an overview, Bioresour. Technol. 199 (2016) 76–82.
[94] J. Zhang, S. Wang, Y. Wang, Biobutanol production from renewable resources: recent advances, Advances in Bioenergy., Elsevier, 2016, pp. 1–68.
[95] C. Jin, M. Yao, H. Liu, F.L. Chia-fon, J. Ji, Progress in the production and application of *n*-butanol as a biofuel, Renew. Sustain. Energy Rev. 15 (8) (2011) 4080–4106.
[96] A. Ranjan, S. Khanna, V.S. Moholkar, Feasibility of rice straw as alternate substrate for biobutanol production, Appl. Energy 103 (2013) 32–38.
[97] R. Grana, A. Frassoldati, T. Faravelli, U. Niemann, E. Ranzi, R. Seiser, et al., An experimental and kinetic modeling study of combustion of isomers of butanol, Combust. Flame 157 (11) (2010) 2137–2154.
[98] L.J. Visioli, H. Enzweiler, R.C. Kuhn, M. Schwaab, M.A. Mazutti, Recent advances on biobutanol production, Sustain. Chem. Process. 2 (1) (2014) 15.
[99] S.Y. Lee, J.H. Park, S.H. Jang, L.K. Nielsen, J. Kim, K.S. Jung, Fermentative butanol production by Clostridia, Biotechnol. Bioeng. 101 (2) (2008) 209–228.
[100] Y. Wang, H. Janssen, H.P. Blaschek, Fermentative biobutanol production: an old topic with remarkable recent advances, Bioprocessing of Renewable Resources to Commodity Bioproducts, Wiley Blackwell, 2014, pp. 227–260.
[101] M. Kirschner, *n*-Butanol. Chemical Market Reporter January 30–February 5, ABI/INFORM Global, 2006, p. 42.
[102] C.-L. Cheng, P.-Y. Che, B.-Y. Chen, W.-J. Lee, L.-J. Chien, J.–S. Chang, High yield bio-butanol production by solvent-producing bacterial microflora, Bioresour. Technol. 113 (2012) 58–64.
[103] C.-L. Cheng, P.-Y. Che, B.-Y. Chen, W.-J. Lee, C.-Y. Lin, J.-S. Chang, Biobutanol production from agricultural waste by an acclimated mixed bacterial microflora, Appl. Energy 100 (2012) 3–9.
[104] N. Qureshi, S. Liu, T. Ezeji, Cellulosic butanol production from agricultural biomass and residues: recent advances in technology, Advanced Biofuels and Bioproducts., Springer, 2013, pp. 247–265.
[105] A. Ranjan, V.S. Moholkar, Biobutanol: science, engineering, and economics, Int. J. Energy Res. 36 (3) (2012) 277–323.
[106] M.A. Martin, First generation biofuels compete, N. Biotechnol. 27 (5) (2010) 596–608.
[107] Y. Wang, S.-H. Ho, H.-W. Yen, D. Nagarajan, N.-Q. Ren, S. Li, et al., Current advances on fermentative biobutanol production using third generation feedstock, Biotechnol. Adv. 35 (8) (2017) 1049–1059.
[108] P.H. Pfromm, V. Amanor–Boadu, R. Nelson, P. Vadlani, R. Madl, Bio-butanol vs. bio-ethanol: a technical and economic assessment for corn and switchgrass fermented by yeast or *Clostridium acetobutylicum*, Biomass Bioenergy 34 (4) (2010) 515–524.
[109] C. Xue, J. Zhao, L. Chen, S.-T. Yang, F. Bai, Recent advances and state-of-the-art strategies in strain and process engineering for biobutanol production by *Clostridium acetobutylicum*, Biotechnol. Adv. 35 (2) (2017) 310–322.
[110] M. Kumar, K. Gayen, Developments in biobutanol production: new insights, Appl. Energy 88 (6) (2011) 1999–2012.

[111] C. Sun, S. Zhang, F. Xin, S. Shanmugam, Y.-R. Wu, Genomic comparison of *Clostridium* species with the potential of utilizing red algal biomass for biobutanol production, Biotechnol. Biofuels 11 (1) (2018) 42.
[112] H.G. Moon, Y.-S. Jang, C. Cho, J. Lee, R. Binkley, S.Y. Lee, One hundred years of clostridial butanol fermentation, FEMS Microbiol. Lett. 363 (3) (2016) 1–15.
[113] M.C. Mansur, M.K. O'Donnell, M.S. Rehmann, M. Zohaib. ABE Fermentation of Sugar in Brazil, 2010.

CHAPTER 6

Syngas fermentation to bioethanol

Minhaj Uddin Monir[1,2], Abu Yousuf[3] and Azrina Abd Aziz[1]
[1]Faculty of Engineering Technology, Universiti Malaysia Pahang, Gambang, Malaysia
[2]Department of Petroleum and Mining Engineering, Jessore University of Science and Technology, Jessore, Bangladesh
[3]Department of Chemical Engineering and Polymer Science, Shahjalal University of Science and Technology, Sylhet, Bangladesh

Contents

Lignocellulosic Biomass to Liquid Biofuels
DOI: https://doi.org/10.1016/B978-0-12-815936-1.00006-X

6.1 Introduction

Syngas fermentation is one of the most favorable biochemical conversion techniques for the production of biofuels [1–4]. In this process, syngas is used as a substrate for microorganisms [2], which is produced through a thermochemical process from biomasses [5–8]. Commonly, carbon-containing lignocellulosic biomass (forest residue, coconut shell, empty fruit bunch of palm oil, municipal solid waste, etc.) is converted into gases, such as carbon dioxide (CO_2), carbon monoxide (CO), and methane (CH_4) [5,6,9–11], and it is further converted into biofuels by utilizing carbon-fixing microbes [1,2,12].

Biomass-derived syngas fermentation from gasification of carbonaceous feedstocks is the most promising conversion technologies of biomass to liquid biofuels. Bioethanol along with acetate, butanol, butyrate, formaldehyde, peptone, and methane (produced from chemical catalytic and biosynthetic processes) is converted to clean and sustainable transportation fuel produced from the lignocellulosic biomasses, such as forest or agricultural biomass [4,5]. Syngas comprises various mixture of CO, CO_2, H_2, and CH_4, which can be produced through gasification of lignocellulosic biomass [6,13]. The composition of syngas varies with the type of biomass used as the feedstock. Different types of gasifiers, such as downdraft, fluidized-bed, and fixed-bed, are used to produce syngas, and it goes through several cleaning stages before entering to the fermenter. Up to this time, it is an on-going research at laboratory scale and novel concepts are integrating to develop the commercial scale.

The acetogenic bacteria reduce H_2-CO_2/CO to acetate in their metabolic pathway [14], which is the primary stage of syngas fermentation to bioethanol. Commonly, the conversion of this substrate (syngas) results into organic acid (acetic acid and butyric acid) and alcohols, such as ethanol, hexanol, and butanol [15]. These products are usually used as the generation of electricity, transportation fuels, and commodity chemicals [16]. Therefore syngas fermentation has a broad interest both in the scientific and industrial fields as an alternative technology to produce renewable bioenergy over the last decade.

6.1.1 Microbiology of syngas fermentation

There are various types of microorganisms that are involved in syngas fermentation. They have the capabilities of utilizing CO, H_2, and CO_2 as metabolic building blocks, both in the case of aerobic or anaerobic species. The anaerobic acetogenic, hydrogenogenic, and methanogenic microorganisms are able to produce different chemicals from biomass-derived syngas. Drzyzga et al. [17] reported that in most of the cases, expensive chemical catalyst is used for syngas fermentation for the conversion of syngas (C1 components) into various types of multicarbon compounds. Therefore research is still undergoing for syngas fermentation for the enhancement of microorganism's productivity and their metabolisms process.

Microorganisms are unable to degrade the biomass entirely through the direct fermentation of biomass, which contained lignin. The gasification is the technique where lignin-based biomass converted into syngas by using fixed-bed (downdraft and updraft gasifier), fluidized-bed, and entrained bed gasifier [9,18–21]. Therefore cogasification technique is employed for enhancing the quality of syngas by using biomass, natural coal, and by-product charcoal [9,22–24]. Produced syngas contains some impurities, such as tar (higher hydrocarbon) and small particles [25–27], which hinder the metabolic functions of microorganisms.

Currently, syngas fermentation is concerning to enhance the yield by optimizing the bioreactor operational parameters [17]. Mesophilic and thermophilic type microbes are used for this process based on the fermentation conditions. The common microorganisms, employed for syngas fermentation, are *Clostridium autoethanogenum*, *Clostridium carboxidivorans*, *Acetobacterium woodii*, *Clostridium ragsdalei*, *Butyribacterium methylotropphicum*, *Clostridium ljungdahlii*, etc. [4]. Heiskanen et al. [28] also reported that the bioethanol production from syngas is more efficient by using biological catalysts (*A. woodii* and *C. carboxidivorans*) than chemical catalysts (copper, cobalt, or iron). The detailed list of common microorganisms that are involved in syngas fermentation with their optimum parameters of temperature, pH, and yield is shown in Table 6.1.

6.1.2 Microbial culture medium

The microbial culture media is the necessary part for syngas fermentation. The fermentation medium is prepared by the mixture of nutrients, substrates, and microorganisms. The ratio and type of nutrient are depending on the specific type of microorganisms that are responsible for the production of ethanol or other products. Reinforced clostridial medium (RCM) is usually used as a medium for syngas fermentation. The composition of the medium is yeast extract (3.0 g/L), lab-lemco powder (10.0 g/L),

Table 6.1 Production of liquid fuels by microbial syngas fermentation with various carbon sources and operational parameters.

Microorganisms	Syngas composition (%)	Fermentation mode	T_{opt} (°C)	pH_{opt}	Products	Reference
Mesophilic bacteria						
Clostridium autoethanogenum	$CO:CO_2:H_2:N_2$ (30:10:20:40)	Fed-batch	30	6	Acetic acid, ethanol	[29]
C. autoethanogenum	$CO:H_2:CO_2$	Batch	37	5.8	Ethanol	[30]
C. autoethanogenum	$CO:H_2:CO_2:N_2$ (36.2: 23.0: 15.4: 11.3)	Batch	37	6	Bioethanol	[31]
Acetobacterium woodii	$CO:H_2:CO_2$	Batch	30	6.8	Acetate	[32]
A. woodii	CO_2	Batch	30	7.2	Acetate	[33]
Butyribacterium methylotropphicum	CO (100%)	Batch	37	6	Ethanol, acetate, butyrate, butanol	[32]
Clostridium carboxidivorans	$CO:H_2:CO_2$	Batch	38	6.2	Acetate, ethanol, butyrate, butanol	[32]
C. carboxidivorans	$CO:CO_2:H_2:N_2$ (30:10:20:40)	Continuous	33	5	Acids and alcohols	[34]
Clostridium ragsdalei	$CO:H_2:CO_2$ (40:30:30)	Fed-batch	37	5.5	Ethanol	[35]
C. ragsdalei	$CO:CO_2:H_2:N_2$ (38:28.5:28.5: 5)	Semicontinuous	37	5.8–4.6	Ethanol	[3]
Clostridium ljungdahlii	$CO:CO_2:H_2$	Batch	37	6.8	Acetic acid, ethanol	[36]
C. ragsdalei	$CO:H_2:CO_2$ (40:30:30)	Batch	37	5.7–6.4	Acetate, ethanol	[35]
Eubacterium limosum	$CO:H_2:CO_2$	Batch	38–39	7.0–7.2	Acetate	[32]
Mesophilic bacterium P7	$CO:CO_2:N_2$ (25:15:60)	Batch	37	5.7–5.8	Acetate, ethanol, butyrate, butanol	[37]
Thermophilic bacteria						
Clostridium thermocellum	$N_2:CO_2:H_2$ (85:10:5)	Batch	55	7.0	Ethanol	[38]
Moorella thermoacetica	$CO:H_2:CO_2$	Batch	55	6.8–7.8	Acetate	[39]
Moorella thermoautotrophica	$CO:H_2:CO_2$	Batch	58	6.1	Acetate	[40]

peptone (10.0 g/L), soluble starch (1.0 g/L), glucose (5.0 g/L), cysteine hydrochloride (0.5 g/L), sodium chloride (5.0 g/L), sodium acetate (3.0 g/L), and agar (0.5 g/L) mixed with deionized water at a ratio of 38 g/L [41]. The medium is needed to be sterilized by autoclaving at 121°C for 20 min. In addition, microorganisms are to be reactivated by transferring 2 mL of the stock culture into 20 mL of RCM prior to cultivation. Subsequently, cultural serum bottle is cleaned with nitrogen (N_2) gas for 2 min to create anaerobic condition and inoculate at optimized temperature.

The medium for syngas fermentation contains (g/L) mineral solution (10 mL), trace element solution (10 mL), vitamin solution (10 mL), peptone (2 g), yeast extract (1 g), L-cysteine-HCl (0.5 g), morpholinoethanesulfonic acid (5 g), xylose (5 g), and resazurin (1 mL; 0.1%), which is added as a redox potential indicator [41]. The pH of the medium has been controlled as per fermentation process condition. The additional nutrient is also needed for the experimental medium preparation.

- The mineral solution added to the medium (per liter): NaCl (80 g), NH_4Cl (100 g), KCl (10 g), KH_2PO_4 (10 g), $MgSO_4 \cdot 7H_2O$ (20 g), and $CaCl_2 \cdot 2H_2O$ (4 g).
- The trace element solution added to the medium (per liter): nitrilotriacetic acid (2 g), $MnCl_2 \cdot 4H_2O$ (1.3 g), $CoCl_2 \cdot 6H_2O$ (0.2 g), $ZnSO_4 \cdot 7H_2O$ (0.2 g), $FeCl_3 \cdot 6H_2O$ (0.4 g), $CuCl_2 \cdot 2H_2O$ (0.02 g), $NiCl_2 \cdot 6H_2O$ (0.02 g), $Na_2MoO_4 \cdot 2H_2O$ (0.02 g), and $Na_2WO_4 \cdot 2H_2O$ (0.025 g).

The prepared fermentation medium is required to be autoclaved at 121°C for 20 min and cooled down to room temperature under the biosafety hood. Subsequently, the syngas is passed through the broth medium through the specific types of bioreactor for the production of acetic acid or bioethanol.

6.1.3 Microbial cultivation system

Following are the three main types for microbial growth cultivation system:

1. batch cultivation,
2. fed-batch cultivation, and
3. continuous cultivation.

1. *Batch cultivation:* Batch culture is the closed system where there is no interaction between system and surrounding during the experiments. In this technique, fermentation broth medium is prepared initially and then

the cultured organism is inoculated to the medium. During the process the reactor is aerated, though no further additions of medium are made. Once a production cycle is terminated, the spent medium is removed to add fresh medium to the cultivation vessel. The cultivation medium is prepared and sterilized before fermentation run. In this cultivation, microorganisms are inoculated into the bioreactor before the process started. Since there is no fresh media added after the experimental run, the concentration of nutrition is decreased continuously. The volume of culture usually remains constant. Batch fermentation gives characteristics growth curve with lag, exponential, stationary, and death phases. Finally, the microbial cells grow and produce the yield.

Advantages: The conversions of substrate occur completely which are the main advantages of this system. It has properly sterilized and low risk of infection from microbial strain.

Disadvantages: The disadvantages of this system are as follows: the labor cost is high. Every batch needed to be sterilized, growth, and cleaning the system.

2. *Fed-batch cultivation:* The fed-batch (or semiclosed) system is a culture where substrates are inoculated to the bioreactor after some interval. In this system, nutrient media is prepared, and organisms are inoculated to the broth medium and then incubated. In the course of incubation, nutrients are fed at given intervals. As a result, the volume of the culture is continuously increased. This technique is applicable to various fermentation processes when some nutrients, though essential for biomass growth, may inhibit the microbial growth if their concentration is too high. In this case, lower initial concentrations of these nutrients can be adopted, adding them continuously or discontinuously during the fermentation. The parameters, such as temperature, pH, substrate inoculation interval time, are needed to be investigated. The growth phase of the microorganisms is monitored enormously.

Advantages: The toxic and concentrated microorganisms are suitable for this fermentation system.

Disadvantages: More attention should be necessary when toxic microorganism is used. Experiment handling is not an easy task. Sometimes microorganisms are expensive.

3. *Continuous cultivation:* A continuous (or open) system is a culture allowing the continuous production of products. This system is feasible for syngas fermentation, especially for industrial purposes. Fresh sterile medium is fed continuously to the vessel and spent fermentation medium

is continuously removed. As a result, the volume always remains constant within the fermenter. The bacterial growth occurs mainly during the log phase. Substrates are gone through the bioreactor continuously. The continuous fermenters allow a steady-state microbial growth if the input flow rate is kept constant. Under steady-state conditions the microbial cell density remains constant.

Advantages: This system works all the time. As a result, labor cost always remains low and perfect utilization of the bioreactor. The productivity is high and maintained constant product quality.

Disadvantages: The substrate must be inoculated continuously for the continuous production of product. Sometimes, the selected microorganisms are contaminated with the nonproducing microbial strain.

6.2 Fermenter for syngas fermentation

A fermenter is basically a device, in which the microorganisms are cultured for the production of desired products. This system is usually designed to give the right environment for optimal microbial cell growth and their metabolic activity of the organism. There are various types of fermenter that are usually used for syngas fermentation [4].

6.2.1 Continuous stirred-tank reactor

The continuous stirred-tank reactor is one of the most common fermenters used for the estimation of key unit operation variables (Fig. 6.1). In this system, one or more reactants (inlet syngas, nutrients) are introduced into a reactor equipped with an agitator fixed with stirrer bars and the ultimate products are removed continuously. The agitator rotates the stirrer bar to ensure the perfect mixing of inlet syngas, nutrients, and the fermentation broth uniformly throughout the whole fermentation. As a result, the composition of the product is uniform.

6.2.2 Bubble column reactor

The bubble column reactor (BCR) comprises a vessel containing a liquid or a liquid–solid suspension at the bottom, which has the capabilities for the distribution of gas. The schematic diagram of the BCR is represented in Fig. 6.2. Chemical and biochemical industries usually use this type of fermenter, where the reactions are involving as oxidation, fermentation chlorination, Fischer–Tropsch synthesis polymerization, hydrogenation,

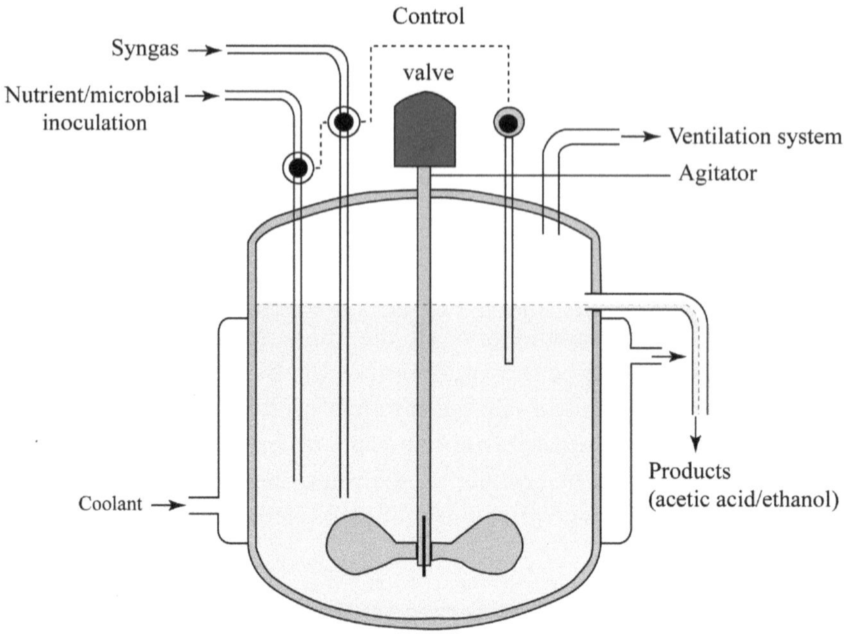

Figure 6.1 Schematic diagram of continuous stirred-tank reactor.

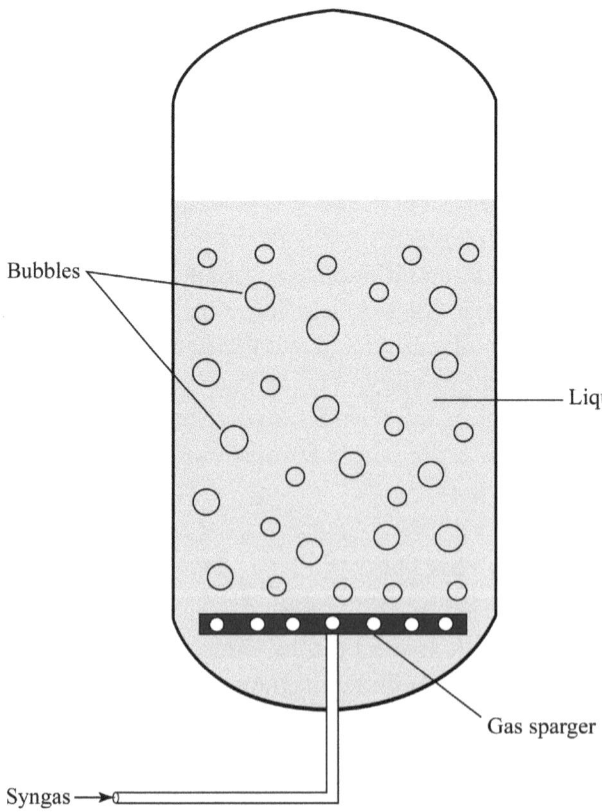

Figure 6.2 Schematic diagram of bubble column reactor.

wastewater treatment, etc. This fermenter is used because of its excellent heat and mass-transfer characteristics, less operational cost, and durability of any type of solid material [42]. In addition, this type of fermenter has the facilities for the addition or removal of any catalyst from the column. Nevertheless, there are some limitations or challenges that are associated with this fermenter. The syngas is fed to the column through the gas sparger at the bottom of the column. The gas in the column may form separate bubbles that rise and spread out due to their buoyancy. It also induces mixing of the continuous liquid phase, as reported by Holland [42].

6.2.3 Monolithic biofilm reactor

Monolithic biofilm reactor (MBR) is a type of fermenter integrating a cordierite monolithic packing material within a bubble column. It offers a perfect way to sustain the high cell density systems due to its high mass-transfer capacity. The monolithic column is fixed inside the plexiglass column by two symmetrical block rings located at the top and bottom of the monolithic column. The heat loss is accurately minimized by covering the monolithic column with an insulation sheet. The medium is also perfectly circulated between the column and the vessel during syngas fermentation. The disadvantage of the fermenter is clogging due to biofilm formation during the biological process. The schematic diagram of the MBR is shown in Fig. 6.3. In addition, a BCR is established as a control to gauge the mass transfer and the fermentation of syngas. The parameters of the column and the operational conditions were identical to MBR reported by Shen et al. [43].

6.2.4 Trickle-bed reactor

The trickle-bed reactor is one type of chemical reactor which involved the downward movement of a liquid and the downward or upward movement of syngas over a catalytical packed bed (Fig. 6.4). This type of fermenter is intrinsically nonstop in operation. Due to the low operating cost and nominal plug flow of both gas and liquid phases, a good control occurs over the process. As a result, it maintains good product quality. It has the facilities of random packed catalytical bed. Therefore it has the ability to inherently a single production plant or single catalytical plant. On the contrary, it has no heat transfer facilities within the bed. Thus the designs for trickle bed and batch plants are based on various flow rates, design specification, and cost extrapolations.

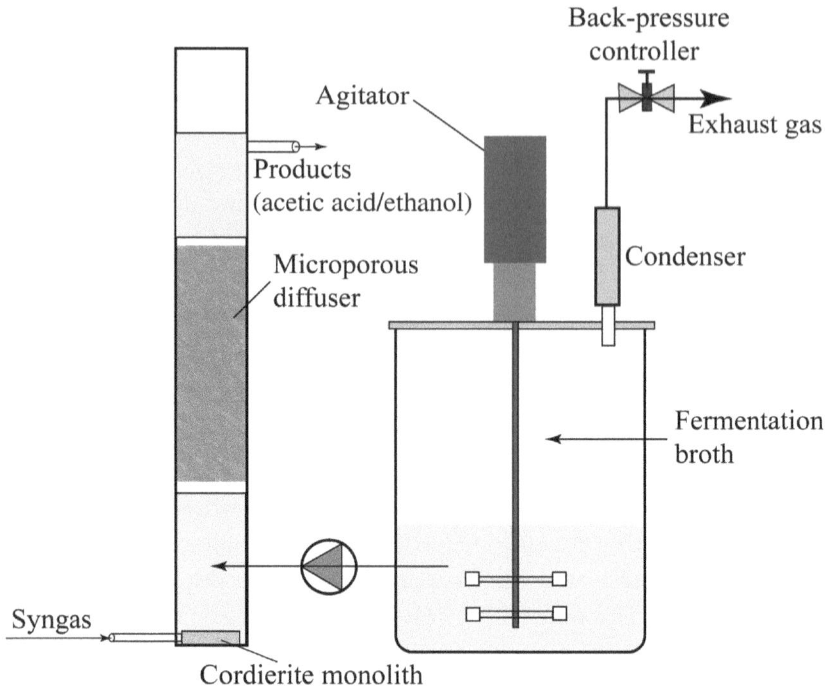

Figure 6.3 Schematics diagram of monolithic biofilm reactor.

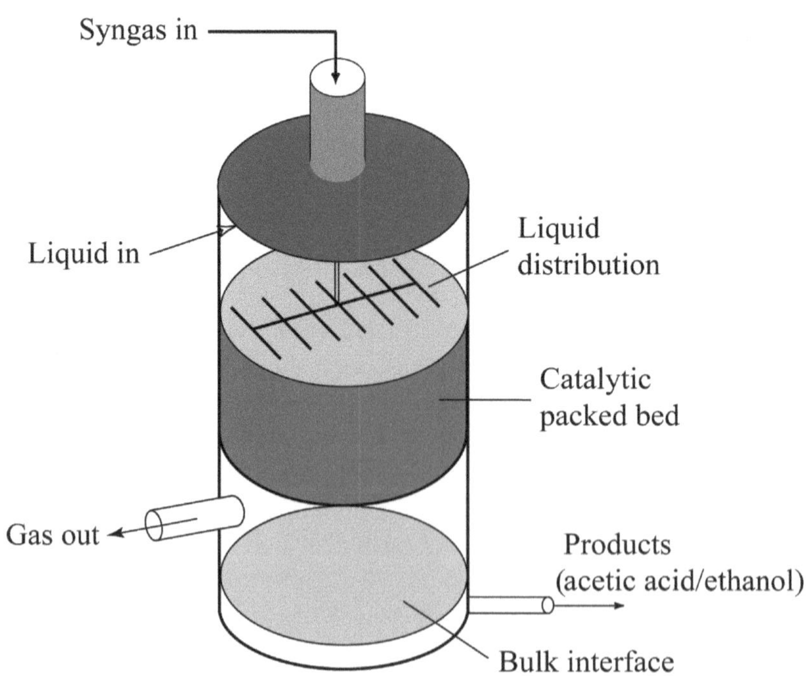

Figure 6.4 Schematic diagram of a trickle-bed reactor.

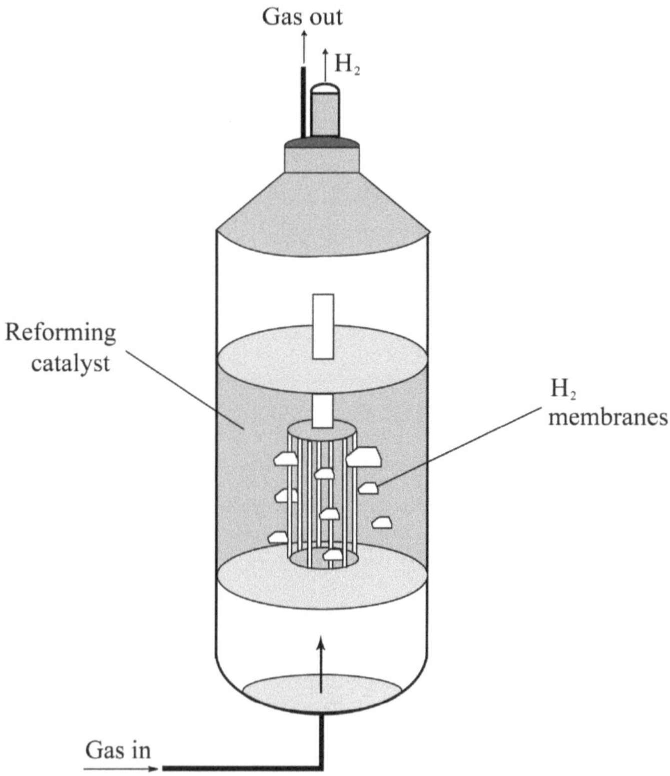

Figure 6.5 Schematic diagram of membrane-based system reactor.

6.2.5 Membrane-based system reactor

The schematic design of membrane-based system reactor is shown in Fig. 6.5. It is compared to the fuel processor, which generates H_2 by the dilution of CO_2 and similar types of gases. This type of fermenter allows the fuel cell to operate at high-voltage electricity and fuel consumption factors, which reduces the degradation rates due to the CO harming. This type of bioreactor has major advantages to achieving a maximum yield and reaction rate. It has the maximum tolerance to toxic compounds that exist in the syngas.

6.3 Microbial pathway for acetic acid and ethanol production

The syngas fermentation for the production of acetic acid and bioethanol, CO, H_2, and CO_2, followed a series of elementary chemical reactions

Table 6.2 Reactions involved for acetic acid and ethanol production.

Reactions	Products	Reference
$4CO + 2H_2O \rightarrow CH_3COOH + 2CO_2$ $3CO + H_2 + H_2O \rightarrow CH_3COOH + CO_2$ $2CO + 2H_2 \rightarrow CH_3COOH$ $CO + 3H_2 + CO_2 \rightarrow CH_3COOH + H_2O$ $H_2 + 2CO_2 \rightarrow CH_3COOH + 2H_2O$	Acetic acid	[4,44,45]
$6CO + 3H_2O \rightarrow CH_3CH_2OH + 4CO_2$ $5CO + H_2 + 2H_2O \rightarrow CH_3CH_2OH + 3CO_2$ $4CO + 2H_2 + H_2O \rightarrow CH_3CH_2OH + 2CO_2$ $3CO + 3H_2 \rightarrow CH_3CH_2OH + CO_2$ $2CO + 4H_2 \rightarrow CH_3CH_2OH + H_2O$ $CO + 5H_2 + CO_2 \rightarrow CH_3CH_2OH + 2H_2O$ $6H_2 + 2CO_2 \rightarrow CH_3CH_2OH + 3H_2O$	Ethanol	[4,44]
$CO + CO_2 + 6H^+ + 6e^- \rightarrow CH_3COOH + H_2O$ $CO + CO_2 + 10H^+ + 10e^- \rightarrow CH_3CH_2OH + 2H_2O$	Acetic acid, ethanol	[44]

(Table 6.2, Fig. 6.6). Each reaction involved some microbial metabolism process within the microbial cell that occurred in the cytoplasm or surface of the cell membrane. The cells act individually, but integral action of all cells sets condition in the fermentation bulk liquid. The reactions occur under certain optimum condition (pH and temperature) for various microorganisms. The inorganic substrates, CO, H_2, and CO_2, are converted to acetyl-CoA and then organic products of acetic acid and ethanol (Fig. 6.6). Moreover, acetyl-CoA is converted to complex organic cell components, carbohydrates, proteins, and lipids. Most of the gas molecules provide the energy for cell function and finally resulting in the acetic acid and ethanol.

6.4 Syngas impurities

Syngas is the main product produced from the gasification and consists of hydrogen (H_2), carbon monoxide (CO), carbon dioxide (CO_2), and methane (CH_4) [5,6,9,10,47]. However, produced syngas contained some by-products that can be expected along with the syngas. In general, these by-products are ash (solid product), tars (liquid products) and ethane, hydrogen sulfide, benzene, acetylene, ammonia, hydrogen cyanide, sulfur dioxide, nitrous oxide, nitrogen, methane, ethylene, etc. (gaseous products)

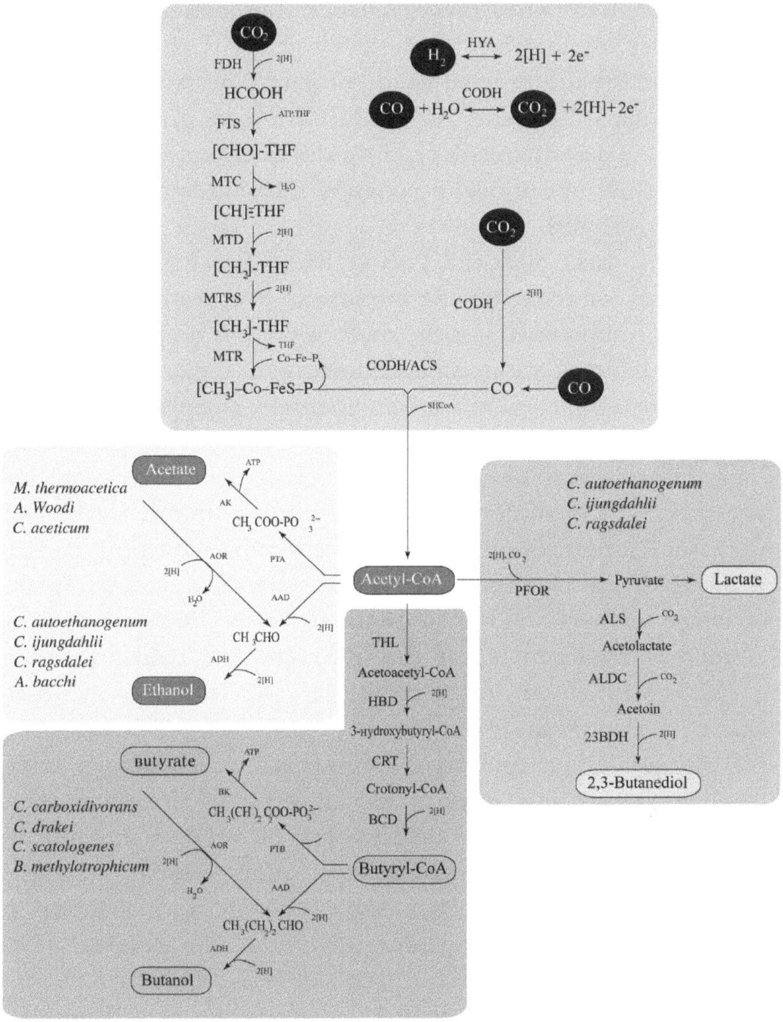

Figure 6.6 Wood–Ljungdahl pathway of acetogens and their metabolic end products [46].

[6,21,23,24,47,48]. The most common unwanted syngas impurities are as follows:

Dust/particles: Dusts are fine grained particles. It could be organic or inorganic. It is primarily affected by the quality of the biomass-based syngas. During the thermal conversion of biomass the diameter of the particles decreases by increasing gasification temperature, though the reduction of particle-size is significantly affected by elutriation, in particular when pellets and chips in entrained-flow gasifiers and fluidized-bed by fragmentation. Moreover, temperature expressively alters its chemical properties and finally increases the residual solid content. The level of undesirable components of tars, dust, and ash produced during gasification depends on some factors. These factors depend on the gasifying feedstocks, type of reactors, and various operational parameters, such as temperature, pressure, and gasifying agents [49].

Tar: Tar is a dark, oily, and viscous flammable liquid distilled from biomass, charcoal, or coal. It is the mixture of hydrocarbons, resins, alcohols, and other compounds. It is the most important by-product that produced during biomass gasification. Sometimes, these components are mixed with syngas which clog the transportation engines, when it is used directly. This by-product also used for road-making and for coating and preserving the timbers. The quality of the tar is affected by temperature, residence time, and equivalent ratios that promote the thermal cracking. Among these, temperature plays a vital role for the reduction of tar content in producer gas by promoting thermal cracking. Usually, with the increase in the temperature, the concentration of tar decreases. Equivalent ratio is another significant effect that promotes the oxidation reactions during char volatilization. As a result, tar concentration increases by increasing equivalent ratios [48].

NH_3: Ammonia is a compound of nitrogen and hydrogen with the formula NH_3. During biomass gasification it is formed along with the production of syngas. This type of impurity occurs because of the presence of nitrogen in the feedstocks (biomass or coal). It also depends on the concentration of nitrogen in various feedstocks. Moreover, operating parameters impact the yield of nitrogenous impurities. The existence of steam during gasification promotes the formation of NH_3. These syngas impurities are generally produced due to the decomposition of proteins or heterocyclic aromatic structures. Moreover, nitrogen-containing compounds in syngas may deactivate catalytic activity, and ultimately it may cause of air pollution. Therefore the quantity of sulfur-based impurities in biomass producer gas is lower in comparison to that of coal-based producer gas.

H_2S, COS: These are the poisonous and flammable syngas impurities that affect the quality of syngas. During gasification, primary gases are involved in different reactions with H_2S and other sulfur impurities. As a result, these gases are influencing the yield.

Mercury: This is another type of toxic impurities that exists the producer syngas. The producer gas contains several heavy metals in trace quantities, such as Hg, As, Se, Cu, Pd, Cd, and Zn. It is more difficult to eliminate oxidized Hg in the form of HgS.

HCl: Hydrochloric acid is a corrosive, strong mineral acid that sometimes forms during gasification with the syngas. Hydrogen halides concentration in syngas is affected by temperature. As a result, removal of this impurity is very important to diminish corrosion and slagging of the equipment surfaces (filters, turbine blades, and heat exchanger).

6.5 Syngas purification

The syngas quality is very much depending on the presence of these impurities in the produced syngas. The syngas cleaning is required for chemical-production process, while combustion on high-ranked coal-fired power stations almost requires no cleaning. It is expected that the syngas for biological fermentation process toward the ethanol production will require some cleaning but not very stringent. The syngas-cleaning processes are as follows:

Wet scrubbing: This is one of the most effective techniques for removing particles from syngas. A wet scrubber acts by introducing the syngas with a scrubbing liquid (i.e., water). Purified syngas is separated, and particles are collected with the scrubbing liquid.

Catalytic tar removal: The tar concentration is removed by catalytic cracking of biomass over the char-supported Ni catalyst in a lab-scale fixed-bed reactor. Recent studies have investigated the effect of catalytic cracking temperature, Ni loading, and residence time of gas on product distribution and gas composition, as reported by Hu et al. [50]. They also found the optimum conditions for catalytic cracking a catalytic cracking temperature of 800°C, 6 wt.% Ni loading, and a gas residence time of 0.5 s. Baidya et al. [51] also studied on the high-performance Ni–Fe–Mg catalyst for tar removal in producer gas.

Thermal tar removal: Tar is eliminated by thermal decomposition combined with physical adsorption, using a reformer as first step and a fixed-bed absorber as second step. The required temperature for thermal tar decomposition is about 800°C. The operational temperature has a significant influence on tar decomposition. The gasifying tar was efficiently decomposed by improving the efficiency of tar reduction. To this scope, either steam or air was introduced into the reactor as a reforming agent. Tar decomposition leads to the reduction of tar from the syngas that is required to prevent damage to downstream equipment.

Tar removal by oil washing: This is one of the techniques for the removal of tar from syngas [50]. Unyaphan et al. [52] reported the improvement of tar-removal performance of oil scrubber by producing microbubbles.

Chemical absorption: This technique is applied for the absorption process where atoms, molecules, or ions enter some bulk phase-liquid or solid materials. Pallozzi et al. [53] stated on the combined gas conditioning and cleaning for the reduction of tars during biomass gasification.

6.6 Factors affecting syngas fermentation

6.6.1 Effect of organic source

The substrate containing organic source is biomass-based syngas. The carbonic acid and carbonate formation are depending on the concentrations of carbon dioxide and high acetate, which may potentially inhibit microbiological actions in the fermentation media. Biomass-based syngas can constrain hydrogen production and adapt product distribution of ethanol and acetate. It has also potential to retain microbial cells in an inactive stage during the bioethanol production.

6.6.2 pH level of the medium

The pH level is one of the most important factors, which is affected on syngas fermentation. It depends on the various types of medium, nutrient, or substrate. As a result, the productivity of ethanol or acetic acid from biomass-based syngas depends on the pH level. It also affects the substrate metabolism and other critical factors, such as pH, membrane, and proton motive force. The biological actions can be affected by small changes of pH. Obviously, large changes of pH can lead to the death of microbes or at least can inhibit the generation of the desired products. The acetogenic bacteria are widely used in syngas fermentation, where the ethanol production level is high at lower pH levels. The optimum pH value for the generation of acetate is from 5 to 7 (growth level), whereas for ethanol it is 4–4.5 (nongrowth level) by using specific types of bacteria.

6.6.3 Temperature of the medium

This parameter is also one of the key factors that affect the production of yield. The specific type of microorganisms is survived on specific temperature. The optimum temperature for mesophilic microorganisms is needed to maintain within the ranges of 37°C–40°C, whereas for thermophilic bacteria it is 55°C–80°C [4].

6.6.4 Gas flow rate

The microbial growth rate and its metabolic activities are affected by the substrates used in syngas fermentation. Consequently, the partial pressures of syngas components should be carefully controlled and kept at optimum values.

6.6.5 Mass transfer

The vital factor for syngas fermentation is limiting the mass-transfer rate from gas to liquid. The efficiency of mass-transfer rate among different fermenter can be compared by the evaluation of gas–liquid volumetric mass-transfer coefficient. This parameter also gives information about the hydrodynamic condition of the bioreactor. Thus the design of the bioreactor is very much significant for syngas fermentation.

6.6.6 Trace metals

The trace metal components contained in syngas and acted as an impurity. It plays a significant role in enhancing the microbial growth during the syngas fermentation. It has the ability to adopt an iron concentration of 10 times higher in comparison to the standard medium used for the fermentation of *C. carboxidivorans*. As a result, the ethanol production is doubled, and the production of acetic acid and butyric acid is reduced.

6.6.7 Reducing agent

The bioethanol production during syngas fermentation may affect the level of oxygen concentration, which is required to be optimized. As a result, the growth rate of the microorganisms can be maximized. In addition, higher levels of oxygen concentration can decrease the growth rate of anaerobic bacteria.

6.7 Roles of nanoparticles on syngas fermentation

Nowadays, more attention has been taken for the enhancement of ethanol yield produced from the syngas fermentation. In this regard, different types of nanoparticles are usually used for enhancing the production rate. Zhu et al. [54] reported that the increase of H_2 yield is due to enhanced

CO−water mass transfer through the addition of nanoparticles. Recently, methyl-functionalized silica and methyl-functionalized cobalt ferrite−silica ($CoFe_2O_4@SiO_2-CH_3$) types of nanoparticles are used to increase the syngas−water mass-transfer rate successfully. Each type of nanoparticle is efficient for syngas fermentation, and effective production of ethanol and acetic acid that enhanced during syngas fermentation using *C. ljungdahlii* [55].

6.8 Integrated biorefinery

A biorefinery is the integration of the conversion process, which is associated with equipment or specific reactor for the production of various types of biofuels, chemicals, and other bio-based products from lignocellulosic feedstocks [56−58]. Biorefineries apply the biomass-based feedstocks without producing environmental pollutant and toxic wastes. Among these, bioethanol is the most valuable product from biorefinery and used as the transportation fuels with the mixture of gasoline with various ratios. The energy content of ethanol is about 2/3 in comparison to gasoline. As a result, ethanol mixed with gasoline up to 10% is usually used in the gasoline engines, where there is no modification needed to the existing design [59]. Syngas fermentation also produces some by-products (acetic and butyric acids, and butanol) which are also valuable [60]. For the enhancing of yield of bioethanol/biobutanol, various types of nanoparticles are used for syngas fermentation. It is happening due to the significantly higher surface area-to-mass ratio of amine group, which absorbs acetic acid from the fermentation medium. As a result, absorbed acid recovered readily as a by-product [32,60]. The nanomaterials can be the ability to regenerate by controlling the pH level of the fermentation broth. The lignin-based biomass is always considered as the low-value feedstocks, which are difficult to convert entirely in biochemical-based ethanol production plant. Therefore, the integrated biorefinery concept to be applicable for the mostly available lignocellulosic biomasses to have maximum yield of the product and to make the process more efficient [61].

6.9 Conclusion

The fermentation of biomass-based syngas is a promising technology for the production of bioethanol. This process offers significant advantages as it allows the conversion of whole biomass, including lignin and it avoids complex pretreatment stages. In addition, it allows the use of the existing fermentation technology and is targeted to ethanol well-established market. Different studies suggested the use of mesophilic and thermophilic microorganisms to optimize syngas fermentation in industrial scale. In addition, the syngas production by gasification and cogasification offers flexible yield, according to the market demand. Ethanol is the main product of this process; other products, such as butanol, 2,3-butanediol, acetate, lactate, butyrate, and other biofuels, can be obtained by changing the microorganisms used and the operational conditions. In order to develop a full commercialization of this market, some major limitations have to be overcome, such as low yield, formation of syngas impurities, reduced mass transfer from gas to liquid, inefficient microbial consumption, and product recovery. By overcoming all challenges the fermentation of biomass-based syngas can become a profitable process to produce ethanol and other valuable products that are useful to fulfill the future energy demand.

References

[1] Y. Shen, R.C. Brown, Z. Wen, Syngas fermentation by *Clostridium carboxidivorans* P7 in a horizontal rotating packed bed biofilm reactor with enhanced ethanol production, Appl. Energy 187 (2017) 585–594.

[2] J.R. Phillips, R.L. Huhnke, H.K. Atiyeh, Syngas fermentation: a microbial conversion process of gaseous substrates to various products, Fermentation 3 (2) (2017) 28.

[3] M. Devarapalli, H.K. Atiyeh, J.R. Phillips, R.S. Lewis, R.L. Huhnke, Ethanol production during semi-continuous syngas fermentation in a trickle bed reactor using *Clostridium ragsdalei*, Bioresour. Technol. 209 (2016) 56–65.

[4] B. Acharya, P. Roy, A. Dutta, Review of syngas fermentation processes for bioethanol, Biofuels 5 (5) (2014) 551–564.

[5] M.U. Monir, A. Abd Aziz, R.A. Kristanti, A. Yousuf, Syngas production from co-gasification of forest residue and charcoal in a pilot scale downdraft reactor, Waste Biomass Valorization (2018).

[6] M.U. Monir, A. Abd Aziz, R.A. Kristanti, A. Yousuf, Gasification of lignocellulosic biomass to produce syngas in a 50 kW downdraft reactor, Biomass Bioenergy 119 (2018) 335–345.

[7] K. Asimakopoulos, H.N. Gavala, I.V. Skiadas, Reactor systems for syngas fermentation processes: a review, Chem. Eng. J. 348 (2018) 732–744.

[8] V.S. Sikarwar, M. Zhao, P.S. Fennell, N. Shah, E.J. Anthony, Progress in biofuel production from gasification, Prog. Energy Combust. Sci. 61 (2017) 189–248.

[9] M.U. Monir, A. Abd Aziz, R.A. Kristanti, A. Yousuf, Co-gasification of empty fruit bunch in a downdraft reactor: a pilot scale approach, Bioresour. Technol. Rep. 1 (2018) 39–49.

[10] K. Kundu, A. Chatterjee, T. Bhattacharyya, M. Roy, A. Kaur, Thermochemical conversion of biomass to bioenergy: a review, in: A.P. Singh, R.A. Agarwal, A.K. Agarwal, A. Dhar, M.K. Shukla (Eds.), Prospects of Alternative Transportation Fuels, Springer, Singapore, 2018, pp. 235–268.

[11] H.M. Zakir Hossain, Q. Hasna Hossain, M.M. Uddin Monir, M.T. Ahmed, Municipal solid waste (MSW) as a source of renewable energy in Bangladesh: revisited, Renew. Sustain. Energy Rev. 39 (2014) 35–41.
[12] N. Shen, K. Dai, X.-Y. Xia, R.J. Zeng, F. Zhang, Conversion of syngas (CO and H_2) to biochemicals by mixed culture fermentation in mesophilic and thermophilic hollow-fiber membrane biofilm reactors, J. Cleaner Prod. 202 (2018) 536–542.
[13] J.G. Speight, Chapter 6—Production of syngas, synfuel, bio-oils, and biogas from coal, biomass, and opportunity fuels A2, in: J. Oakey (Ed.), Fuel Flexible Energy Generation, Woodhead Publishing, Boston, MA, 2016, pp. 145–174.
[14] V. Müller, Energy conservation in acetogenic bacteria, Appl. Environ. Microbiol. 69 (11) (2003) 6345–6353.
[15] D. Kennes, H.N. Abubackar, M. Diaz, M.C. Veiga, C. Kennes, Bioethanol production from biomass: carbohydrate vs syngas fermentation, J. Chem. Technol. Biotechnol. 91 (2) (2016) 304–317.
[16] P.J. Woolcock, R.C. Brown, A review of cleaning technologies for biomass-derived syngas, Biomass Bioenergy 52 (2013) 54–84.
[17] O. Drzyzga, O. Revelles, G. Durante-Rodriguez, E. Diaz, J.L. Garcia, A. Prieto, New challenges for syngas fermentation: towards production of biopolymers, J. Chem. Technol. Biotechnol. 90 (10) (2015) 1735–1751.
[18] G. Xu, T. Murakami, T. Suda, Y. Matsuzaw, H. Tani, Two-stage dual fluidized bed gasification: its conception and application to biomass, Fuel Process. Technol. 90 (1) (2009) 137–144.
[19] J. Zhou, Q. Chen, H. Zhao, X. Cao, Q. Mei, Z. Luo, et al., Biomass–oxygen gasification in a high-temperature entrained-flow gasifier, Biotechnol. Adv. 27 (5) (2009) 606–611.
[20] K. Patil, P. Bhoi, R. Huhnke, D. Bellmer, Biomass downdraft gasifier with internal cyclonic combustion chamber: design, construction, and experimental results, Bioresour. Technol. 102 (10) (2011) 6286–6290.
[21] A. Susastriawan, H. Saptoadi, Small-scale downdraft gasifiers for biomass gasification: a review, Renew. Sustain. Energy Rev. 76 (2017) 989–1003.
[22] M.U. Monir, A. Yousuf, A.A. Aziz, S.M. Atnaw, Enhancing co-gasification of coconut shell by reusing char, Indian J. Sci. Technol. 10 (6) (2017) 1–4.
[23] E. Beagle, Y. Wang, D. Bell, E. Belmont, Co-gasification of pine and oak biochar with sub-bituminous coal in carbon dioxide, Bioresour. Technol. 251 (2018) 31–39.
[24] Y. Zhang, Y. Zheng, Co-gasification of coal and biomass in a fixed bed reactor with separate and mixed bed configurations, Fuel 183 (2016) 132–138.
[25] M. Kuba, H. Hofbauer, Experimental parametric study on product gas and tar composition in dual fluid bed gasification of woody biomass, Biomass Bioenergy 115 (2018) 35–44.
[26] J. Moon, W. Jo, S. Jeong, B. Bang, Y. Choi, J. Hwang, et al., Gas cleaning with molten tin for hydrogen sulfide and tar in producer gas generated from biomass gasification, Energy 130 (2017) 318–326.
[27] A.S. Al-Rahbi, P.T. Williams, Hydrogen-rich syngas production and tar removal from biomass gasification using sacrificial tyre pyrolysis char, Appl. Energy 190 (2017) 501–509.
[28] H. Heiskanen, I. Virkajärvi, L. Viikari, The effect of syngas composition on the growth and product formation of *Butyribacterium methylotrophicum*, Enzyme Microb. Technol. 41 (3) (2007) 362–367.
[29] B. Lagoa-Costa, H.N. Abubackar, M. Fernández-Romasanta, C. Kennes, M.C. Veiga, Integrated bioconversion of syngas into bioethanol and biopolymers, Bioresour. Technol. 239 (2017) 244–249.

[30] F. Liew, A.M. Henstra, M. Köpke, K. Winzer, S.D. Simpson, N.P. Minton, Metabolic engineering of *Clostridium autoethanogenum* for selective alcohol production, Metab. Eng. 40 (2017) 104–114.

[31] H. Xu, C. Liang, Z. Yuan, J. Xu, Q. Hua, Y. Guo, A study of CO/syngas bioconversion by *Clostridium autoethanogenum* with a flexible gas-cultivation system, Enzyme Microb. Technol. 101 (2017) 24–29.

[32] P.C. Munasinghe, S.K. Khanal, Biomass-derived syngas fermentation into biofuels, Biofuels 101 (13) (2011) 79–98.

[33] S. Hoffmeister, M. Gerdom, F.R. Bengelsdorf, S. Linder, S. Flüchter, H. Öztürk, et al., Acetone production with metabolically engineered strains of *Acetobacterium woodii*, Metab. Eng. 36 (2016) 37–47.

[34] H.N. Abubackar, M.C. Veiga, C. Kennes, Production of acids and alcohols from syngas in a two-stage continuous fermentation process, Bioresour. Technol. 253 (2018) 227–234.

[35] X. Sun, H.K. Atiyeh, A. Kumar, H. Zhang, Enhanced ethanol production by *Clostridium ragsdalei* from syngas by incorporating biochar in the fermentation medium, Bioresour. Technol. 247 (2018) 291–301.

[36] Y.-K. Kim, S.E. Park, H. Lee, J.Y. Yun, Enhancement of bioethanol production in syngas fermentation with *Clostridium ljungdahlii* using nanoparticles, Bioresour. Technol. 159 (2014) 446–450.

[37] R.S. Lewis, R.S. Tanner, R.L. Huhnke, Indirect or Direct Fermentation of Biomass to Fuel Alcohol, Google Patents, 2007.

[38] L. Tian, B. Papanek, D.G. Olson, T. Rydzak, E.K. Holwerda, T. Zheng, et al., Simultaneous achievement of high ethanol yield and titer in *Clostridium thermocellum*, Biotechnol. Biofuels 9 (1) (2016) 116.

[39] B. Molitor, E. Marcellin, L.T. Angenent, Overcoming the energetic limitations of syngas fermentation, Curr. Opin. Chem. Biol. 41 (2017) 84–92.

[40] S. Sakai, Y. Nakashimada, H. Yoshimoto, S. Watanabe, H. Okada, N. Nishio, Ethanol production from H_2 and CO_2 by a newly isolated thermophilic bacterium, *Moorella* sp. HUC22-1, Biotechnol. Lett. 26 (20) (2004) 1607–1612.

[41] R.M. Atlas, Handbook of Microbiological Media, CRC Press, 2010.

[42] D.J. Holland, Chapter 18—Applications of tomography in bubble column and trickle bed reactors A2, in: M. Wang (Ed.), Industrial Tomography, Woodhead Publishing, 2015, pp. 477–507.

[43] Y. Shen, R. Brown, Z. Wen, Enhancing mass transfer and ethanol production in syngas fermentation of *Clostridium carboxidivorans* P7 through a monolithic biofilm reactor, Appl. Energy 136 (2014) 68–76.

[44] O. Pardo-Planas, H.K. Atiyeh, J.R. Phillips, C.P. Aichele, S. Mohammad, Process simulation of ethanol production from biomass gasification and syngas fermentation, Bioresour. Technol. 245 (2017) 925–932.

[45] D.K. Kundiyana, R.L. Huhnke, M.R. Wilkins, Syngas fermentation in a 100-L pilot scale fermentor: design and process considerations, J. Biosci. Bioeng. 109 (5) (2010) 492–498.

[46] J. Chen, M.A. Henson, In silico metabolic engineering of *Clostridium ljungdahlii* for synthesis gas fermentation, Metab. Eng. 38 (2016) 389–400.

[47] A. Ramos, E. Monteiro, V. Silva, A. Rouboa, Co-gasification and recent developments on waste-to-energy conversion: a review, Renew. Sustain. Energy Rev. 81 (2018) 380–398.

[48] S. Adhikari, N. Abdoulmoumine, H. Nam, O. Oyedeji, Chapter 16—Biomass gasification producer gas cleanup, Bioenergy Systems for the Future, Woodhead Publishing, 2017, pp. 541–557.

[49] R. Rauch, J. Hrbek, H. Hofbauer, Biomass gasification for synthesis gas production and applications of the syngas, Wiley Interdiscip. Rev. Energy Environ. 3 (4) (2014) 343–362.

[50] M. Hu, M. Laghari, B. Cui, B. Xiao, B. Zhang, D. Guo, Catalytic cracking of biomass tar over char supported nickel catalyst, Energy 145 (2018) 228–237.

[51] T. Baidya, R.J. Cattolica, R. Seiser, High performance Ni-Fe-Mg catalyst for tar removal in producer gas, Appl. Catal. A: Gen. 558 (2018) 131–139.

[52] S. Unyaphan, T. Tarnpradab, F. Takahashi, K. Yoshikawa, Improvement of tar removal performance of oil scrubber by producing syngas microbubbles, Appl. Energy 205 (2017) 802–812.

[53] V. Pallozzi, A. Di Carlo, E. Bocci, M. Carlini, Combined gas conditioning and cleaning for reduction of tars in biomass gasification, Biomass Bioenergy 109 (2018) 85–90.

[54] H. Zhu, B.H. Shanks, D.W. Choi, T.J. Heindel, Effect of functionalized MCM41 nanoparticles on syngas fermentation, Biomass Bioenergy 34 (11) (2010) 1624–1627.

[55] Y.-K. Kim, H. Lee, Use of magnetic nanoparticles to enhance bioethanol production in syngas fermentation, Bioresour. Technol. 204 (2016) 139–144.

[56] M. Riazi, D. Chiaramonti, Biofuels Production and Processing Technology, CRC Press, 2017.

[57] N. Dahmen, E. Henrich, T. Henrich, Synthesis gas biorefinery, Advances in Biochemical Engineering/Biotechnology, Springer Berlin Heidelberg, Berlin, Heidelberg, 2017, pp. 1–29.

[58] A. Yousuf, M.R. Khan, A. Islam, M.U. Monir, Z. Ab Wahid, D. Pirozzi, Application of electroporation technique in biofuel processing, in: MATEC Web of Conferences, EDP Sciences, 2017, p. 01085.

[59] P.C. Munasinghe, S.K. Khanal, Biomass-derived syngas fermentation into biofuels: opportunities and challenges, Bioresour. Technol. 101 (13) (2010) 5013–5022.

[60] S. De Tissera, M. Köpke, S.D. Simpson, C. Humphreys, N.P. Minton, P. Dürre, Syngas biorefinery and syngas utilization, Adv. Biochem. Eng. Biotechnol. 166 (2017) 247–280.

[61] N. Gaurav, S. Sivasankari, G.S. Kiran, A. Ninawe, J. Selvin, Utilization of bioresources for sustainable biofuels: A Review, Renew. Sustain. Energy Rev. 73 (2017) 205–214.

CHAPTER 7

Fischer–Tropsch synthesis of syngas to liquid hydrocarbons

Sanjeet Mehariya[1,2], Angela Iovine[1,2], Patrizia Casella[1], Dino Musmarra[2], Alberto Figoli[3], Tiziana Marino[3], Neeta Sharma[4] and Antonio Molino[1]
[1]ENEA, National Agency for New Technologies, Energy and Sustainable Economic Development, Department of Sustainability, Portici, Italy
[2]Department of Engineering, University of Campania "Luigi Vanvitelli", Aversa, Italy
[3]CNR, Institute on Membrane Technology, National Research Council, University of Calabria, Rende, Italy
[4]ENEA, National Agency for New Technologies, Energy and Sustainable Economic Development, Department of Sustainability, ENEA Research Centre Trisaia, Rotondella, Italy

Contents

7.1 Introduction

In a growing economy, the prime issue of concern is the rapid depletion of fossil fuels due to urbanization and industrialization, which gradually increases the prices of crude oil [1–5]. Transportation represents the second largest sector for the global energy consumption. The predicted fraction of global transportation fuel will be increased from 54% in 2008 to 60% by 2035, which corresponds to 82% of the overall surge in global liquid fuel consumption [6]. These directives are forcing to develop

Lignocellulosic Biomass to Liquid Biofuels
DOI: https://doi.org/10.1016/B978-0-12-815936-1.00007-1

advanced and effective technologies for the production of liquid transportation fuels. In recent year, Fischer–Tropsch (FT) process has received the scientific attention for the production of sulfur less diesel fuel to offset the fossil-fuel demand. FT synthesis represents one of the most promising and sustainable solutions for the production of ultraclean fuel at economically feasible cost [6–21]. The FT fuels can be used in combustion engines, which exhibit lower greenhouse gas emission levels compared to petroleum-based fuels, because FT fuels are sulfur free and contain a very low concentration of aromatics and nitrogen [6,7,22–24].

FT synthesis was developed by Franz Fischer and Hans Tropsch in 1925 at Kaiser Wilhelm Institute, Germany. FT is a heterogeneous catalytic process that converts syngas into a variety of products, such as alcohols, aldehydes, olefins, paraffins, and especially liquid transportation fuels. The syngas could be derived from different sources, such as biomass, coal, coal-bed gas, natural gas or shale gas, through steam reforming, partial or auto thermal oxidation, or gasification process [1,6,25,26]. The large-scale FT plants do not have major technical issues and biomass gasification coupled with FT process assures the production of green liquid fuels. The biomass is gasified to produce syngas that is further transformed into green liquid transportation fuels through FT synthesis [6,14,15,17]. Therefore metal catalysts are necessary to carry out FT catalytic reactions at the temperatures of 150°C–300°C and selective pressures, in the range of 5–60 bar, produce water as major by-product of the process [6,22–24,27–31]. Moreover, cobalt (Co) and iron (Fe) are more vital catalytic components for industrial-scale synthesis of FT process. The Fe catalysts could operate under a wider range of temperatures and H_2/CO ratios with low CH_4 selectivity compared to Co catalysts. Particularly, Fe catalysts show a higher activity for the water–gas shift (WGS) reaction under high temperature, which is more helpful for the conversion of syngas with lower H_2/CO ratios resulting from biomass or coal gasification [1,6,10,28], although the syngas derived from biomass gasification could be H_2 deficient, which will be demanding the Co catalysts for FT synthesis. The Co catalysts-based FT process shows greater productivity than Fe-based catalysts [6,32]. Moreover, certain chemical promoters, such as basic and transition metals (K, Cu, and Mn), as well as structure promoters (Al, Si, and Zn) are added into the Fe catalysts to improve FT synthesis performances [2,6,13,24,25,30,33–47]. The overall schematic representation of the FT synthesis technology is shown in Fig. 7.1. However, the industrialization of FT process was started by using a Ruhrchemie atmospheric

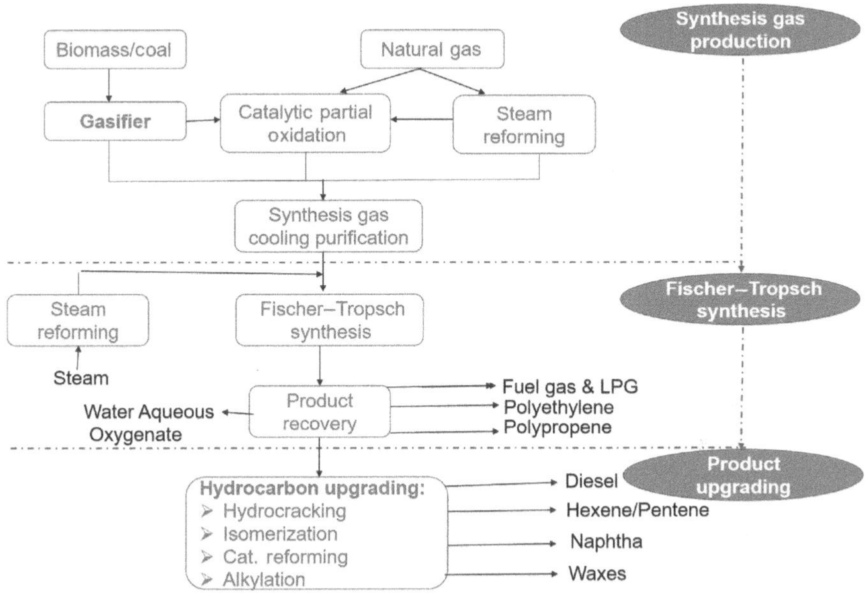

Figure 7.1 Schematic line-up of the integrated FT synthesis process [2]. *FT*, Fischer–Tropsch.

fixed-bed reactor in 1936 in Germany with a capacity of ~274–330 MT/day, and gasoline was the major product of the process and constituted 72% of the total production. Besides, the first coal-based production plant was constructed in Secunda, South Africa in 1955 by South African Coal and Oil Company due to the huge availability of coal in this country. Shell in Malaysia was the second earliest commercial plant that was producing liquid hydrocarbons using FT synthesis process, while circulating fluidized bed (CFB) reactors were used for world's largest FT application known as "Synthol reactors" [6,7,11,12,48].

This chapter discusses the technical details linking to FT synthesis along with its recent developments and presents the current status on syngas to liquid hydrocarbons. Recently, the interest towards FT synthesis has increased as a result of eco-friendly concerns and the higher fossil fuels consumption. This chapter mainly aimed to provide a detailed summary of (1) FT synthesis of syngas to liquid hydrocarbons, (2) role of different catalysts in FT synthesis, (3) kinetic modeling and process simulation for FT synthesis, (4) role of carbon nanofibers (CNFs)/carbon felt (CF) composites, and (5) plug the knowledge gaps and provide new research directions to improve the FT synthesis.

7.2 Fischer–Tropsch chemistry and reaction

The FT chemistry is often regarded as the vital technological input for converting syngas into liquid fuels in the range of C_1-C_{100} [6,7,49]. Also the FT reactions are considered as highly exothermic ($\Delta_r H \approx -200$ kJ/mol) surface polymerization reaction. The reactants, gas mixture, absorb and dissociate at the surface of the catalyst and react with chain initiator [2,6]. The FT synthesis commonly contains the following reactions:

$$(2n+1)H_2 + nCO \rightarrow C_nH_{2n+2} + nH_2O \tag{7.1}$$

$$2nH_2 + nCO \rightarrow C_nH_{2n} + nH_2O \tag{7.2}$$

Toward the synthesis of alkanes and alkenes, these are the end molecules for FT process [Eq. (7.3)], where water is the main oxygenated product.

$$2nH_2 + nCO \rightarrow C_nH_{2n+2}O + (n-1)H_2O \tag{7.3}$$

Moreover, the WGS reaction, shown in Eq. (7.4), also occurs over most FT catalysts and is a reversible reaction, where CO_2 is fundamentally produced with respect of CO by WGS reaction. Also, iron catalysts show a key role in FT synthesis, which are used as a catalyst for WGS reaction during FT synthesis.

$$CO + H_2O \rightarrow CO_2 + H_2 \tag{7.4}$$

In general, the chemical reactions of the FT synthesis are controlled during the conversion of CO and H_2. Moreover, the usage ratio significantly depends on the extent of the other reactions and is influenced by the WGS reaction. The WGS reaction arises simultaneously during FT synthesis by reducing the usage ratio over the iron catalysts. This reaction permits to use the H_2/CO ratio less than 2.1 of syngas. Co catalysts show low activity for the WGS reaction; therefore the scope of WGS reaction is negligible.

Since the ideal case the syngas generated from biomass gasification and subsequently undergoes FT synthesis, the reaction is as follows:

$$2C + \frac{1}{2}O_2 + H_2O \rightarrow -CH_2- + CO_2 \tag{7.5}$$

The yields of FT synthesis in an extensive variety of hydrocarbon products, such as conventional crude oil, could not refer to a pure product formation. FT products formation depends on the used catalysts and the reaction conditions. Therefore the FT synthesis step directly influences the product qualitatively and quantitatively. The FT synthesis is kinetically and mechanically controlled reaction. The kinetic and mechanical factors affect the products formation during the FT synthesis [25,50–52].

7.3 Roles of catalyst during Fischer–Tropsch synthesis

The heterogeneous catalysts play a major role in FT process, which should have the optimum hydrogenation activity to catalyze the hydrogenation of CO to higher hydrocarbons. The selectivity of FT synthesis primarily depends on transition metals used during the reaction. There are some metals of VIII transition group, which show optimum hydrogenation activity for FT synthesis, which are cobalt (Co), iron (Fe), nickel (Ni), and ruthenium (Ru) [25,27,47,53–56]. However, other elements, such as rhodium (Rh), iridium (Ir), palladium (Pd), and platinum (Pt), are also used in FT synthesis, which have higher selectivity compared to Ru, Ni, Co, and Fe but not considered as industrial catalysts due to high cost. Among all catalysts, Fe is the most commonly applied catalyst, due to low cost and easily availability in respect to other catalysts. However, ruthenium shows maximum catalytic reaction during FT synthesis, but it is an extremely expensive catalyst. Nickel has very high activity for hydrogenation but higher CH_4 selectivity and is more costly compared to Co or Fe [27,49,55,57]. Therefore Co and Fe are the favorable catalysts and possible options for FT synthesis that could be used for the operation of FT plants. Overall, Co is considered to be more active than Fe, and the economics of operation plant during FT synthesis with Co catalysts are significantly higher than on Fe catalysts. Several salient features of Co and Fe catalysts for FT synthesis are summarized in Table 7.1. Moreover, the physico-chemical properties of the catalyst, specifically the rate of diffusion, influence the overall FT synthesis and diffusion rate [6,7,54,58,59]. As observed in the Thiele–Wheeler graph, shown in Fig. 7.2, the catalyst efficiency drops less than one as the Thiele modulus, φ, and increases above unity. It is reflected that intraparticle diffusion plays a major role for FT catalyst particle with diameters relatively greater than 500 μm. Therefore creating intraparticle diffusion is crucial for selecting catalyst particle size and shape for a fixed-bed FT process [6,60].

Table 7.1 Comparison of some salient features of cobalt and iron Fischer–Tropsch catalysts [6].

Operative parameter	Cobalt catalyst	Iron catalyst
Feed gas	Lower H_2:CO ratio in the range of 2.0–2.3, due to minor WGS reaction	Wide range of H_2:CO ratio from 0.5 to 2.5, because of high WGS reaction
Operating pressure	5–15 bar and higher pressure decrease the CH_4 selectivity	10–60 bar and higher pressure could increase the selectivity of C_+ hydrocarbons
Operative temperature (°C)	190–240 Used only in LTFT reactors High temperature increases CH_4 selectivity and causes catalyst deactivation	200–350 Operates both in HTFT and LTFT reactors
Activity	Highly active at higher CO conversions, that is, lower space velocity	Activity improved compared to cobalt at higher space velocity
Product spectrum	Major products are *n*-paraffins with minor accumulation of α-olefins Higher paraffin/olefin ratio	Major products are *n*-paraffins with significant accumulation of α-olefins Lower paraffin/olefin ratio
Operating plants	Shell Middle Distillate Synthesis, Oryx-GTL facility—Sasol	Sasol Slurry process (LTFT), Sasol-SAS (HTFT), Mossgass facility
Promoters	Noble metals (Ru, Rh, Pt, and Pd); oxide promoters (ZrO_2, La2O_3, and CeO_2)	Alkali metals (Li, Na, K, Rb, and Ca)
Attrition resistance	Highly resistant	Not very resistant
Cost and life	More expensive and longer life time	Less expensive and lower life time

HTFT, High-temperature Fischer–Tropsch; *LTFT*, low-temperature Fischer–Tropsch; *WGS*, water–gas shift.

7.3.1 Role of iron catalyst

High reactivity of iron carbides enhances the FT synthesis rate, and it is easily accessible and relatively low-cost catalyst [32,61–63]. The main role of iron catalysts is to stimulate the WGS reaction, which is favored for lower ratio of H_2/CO feed. The WGS reaction provides external hydrogen for FT synthesis, which is necessary for syngas production from

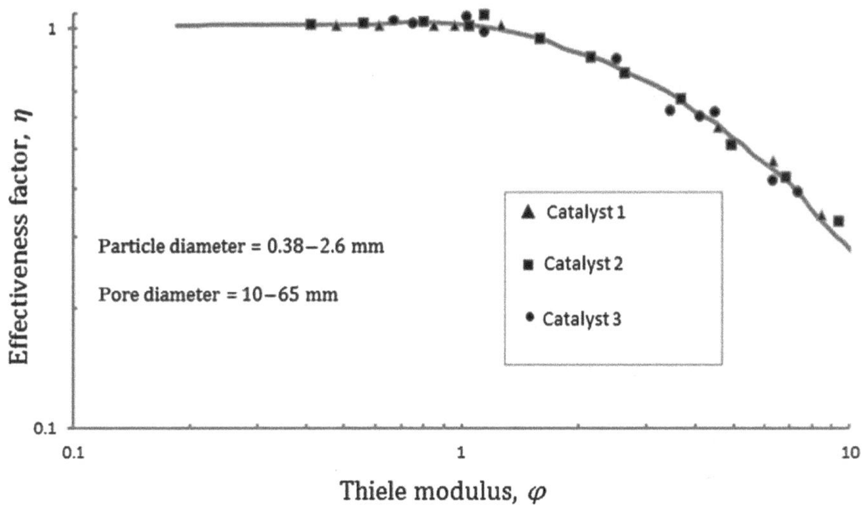

Figure 7.2 Catalyst effectiveness φ as a function of the Thiele modulus for various cobalt and iron catalysts [6,60].

coal gasification. FT process with iron catalysts could operate under both low-temperature range (220°C–270°C) and high-temperature range (300°C–350°C) [6,8,46,55,64–66]. The iron catalysts could be synthesized by several methods, such as precipitation, continuous precipitation, continuous coprecipitation, and spray-drying method, which required an aqueous solution. Moreover, the alkali metal ions as promoters are essential to attain high activity and stability of iron catalysts. The alkali metals act as electronic supporters to affect the electronic properties of iron, enhancing the chemisorption of CO. The role of several alkali metal ions on the catalytic performance of a precipitated iron catalyst was investigated by several authors [67–72], and the role of different promoters are summarized in Table 7.2. The addition of Cu seemed to promote iron reduction, while Mg improved FT synthesis and Mn could be applied to control the selectivity towards the desired hydrocarbons [2,6,36,39,58,68,69].

Chun et al. [58] synthesized the iron catalyst using the aqueous solution of Na_2CO_3, which added to solution containing both $Fe(NO_3)_3$ and $Cu(NO_3)_2$ in the desired ratio at 80°C ± 1°C until the pH reached to 8.0 ± 0.1. The precipitate was filtered and added with the mixture of K_2CO_3 solution, and colloidal suspension of SiO_2 was dried at 200°C (inlet) and 95°C (outlet) temperature. The calcination was carried out

Table 7.2 Effects of alkali promoters on catalytic behaviors of Fe-based Fischer–Tropsch (FT) catalysts.

Promoter	Possible functions	References
		[67,70,71]
Mn	Decreases deactivation by forming minor Fe carbide species by the formation of mixed oxides with iron Enhances CO conversions that increase the selectivity of C_2-C_4 Increases dispersion and increases selectivity to lower alkenes	[70,72]
Mg	Improves the FT synthesis process, while reduces WGS reaction Enhances the production of olefins Changes the selectivity for production of gasoline	[55,67,73,74]
Cu or Ru	Enhances the reduction and carburization of iron Increases the selectivity to heavier hydrocarbons and the fraction of olefins in lighter hydrocarbons, possibly by enhancing the basicity together with alkali metal ions Promotes CO hydrogenation and WGS activity Promotes the Fe-catalyzed FT reaction by a combination of enhanced structural integrity and catalytic properties	[50,75]
K^+ and alkali metal ions	Enhance the syngas conversion to light olefins and improved olefins selectivity Improve the activity of Fe in FT synthesis and enhance WGS reaction Enhance the olefin/paraffin ratio in C_2-C_4 hydrocarbons Increase the selectivity to C_{5+} hydrocarbons and decreased CH_4 production Increase surface basicity that enhances olefin production Reduce rapid catalyst deactivation	

WGS, Water–gas shift.

with dried sample at 400°C for 8 h, which was pressed into pellets and crushed and sieved to obtain 300–600 μm particles for FT synthesis in a fixed-bed reactor. Luo and Davis [73] prepared Fe catalysts using an aqueous solution of ferric nitrate obtained from Fe $(NO_3)_3 \cdot 9H_2O$ and tetraethyl or thosilicate and mixed to attain a preferred Fe:Si ratio. The mixture of iron nitrate and tetraethyl orthosilicate was added to a continuously stirred tank reactor (CSTR) precipitation vessel together with a stream of 30% ammonium hydroxide to achieve pH of 9.0. The mixture from the CSTR was filtered using a vacuum drum filter and washed twice with deionized water, and then it was dried for overnight at 110°C. Also, the Fe:Si catalyst powder was impregnated in an appropriate amount of aqueous K_2CO_3 and $Cu(NO_3)_2$ solution to produce the desired atomic ratio of Fe:Si:K:Cu (100:4.6:1.44:2.0). The final catalyst was dried for overnight at 110°C and calcined at 350°C in an air flow for 4 h [73]. However, the choice and level of promoters are important for catalyst preparation with low selectivity to CH_4 and high selectivity to heavy hydrocarbon production with preferred olefin and oxygenate content in the products. Also, higher level of promoter could have adverse effect on catalyst activity; an optimum level is required [74–76]. Several promoters, such as copper, manganese, silica, and potassium, are added for increased activity and improved stability of iron catalyst. Copper improves the reduction of iron oxide phase from hematite (Fe_2O_3) to magnetite (Fe_3O_4) then further to iron metal or iron carbide [55]. Silica is a structural promoter used to stabilize the surface area and also has the chemical effect on Fe catalyst properties [46,62,77]. The initial activity of FT synthesis is reduced with increased manganese due to weak carburization. Also, it is able to limit the reoxidation of Fe carbides to Fe_3O_4 and augment additional carburization of the catalyst, which increase light olefin selectivity. Potassium is reflected to promote CO adsorption and reduce H_2 adsorption, which could enhance chain growth [75,78]. It increases both the FT activity and the olefin yield via inhibition of secondary hydrogenation reactions and decreases the CH_4 selectivity. Potassium could also enhance the catalytic activity for FT synthesis and the rate of WGS reactions [6,7,25,62–64,71].

The iron catalysts are used for two pathways selectivity toward the desired products. One pathway supports the production of lower molecular weight olefinic hydrocarbons during FT synthesis, and it is known as Sasol–Synthol process. High reaction temperature (~340°C) promotes Sasol–Synthol process, and the average molecular weight of products is comparatively lower because liquid phase does not occur under reaction conditions. The small catalyst particles (~100 μm diameter) are exciting in the reactor and carbon deposition on the catalyst surface does not interrupt operation process. The second pathway of Fe catalyst supports the highest catalyst activity where major hydrocarbon is in the liquid phase under low reaction temperature (~230°C). The low-temperature reaction supports paraffin production and could be further processed for obtaining pure products. Thus iron catalysts are very flexible and could produce a wide variety of petroleum products, such as olefin and paraffins [6,54,55].

7.3.2 Role of Co catalyst

FT synthesis using the Co catalysts shows high activity up to 60%−70% conversion in a single process, along with significantly high stability and selectivity of linear hydrocarbons, and it has insignificant activity on CO conversion. FT synthesis over Co catalysts increases the CH_4 formation at higher temperatures, which limits its application at lower temperatures respect to the Fe catalysts [54,79−81]. The C_2-C_4 and CH_4 selectivity are higher over Co catalysts at 250°C or above temperatures, while the CH_4 selectivity is less than 15% at 200°C and below. However, pressure plays a crucial role in FT synthesis for products selectivity [6]. The deactivation and regeneration of Co catalysts is challenging, which involves sequential treatment of hydrogen and steam or solvent wash [82,83], and the deactivation is a major challenge due to high costs of Co-based catalysts [84]. Oxidation of Co forms H_2O and CO_2 and leads to carbon deposition on the catalysts surface, which is the major source of deactivation of the catalyst though long-term effect on Co-based catalysts [85]. The metal-supported Co-synthesized catalysts could improve the activity and the extensive life of the catalysts, together with the products selectivity. However, the performance and structure of Co catalysts depend on the type of catalytic support material. The key function of the catalyst support is to diffuse Co and produce stable Co metal particles in the catalysts after reduction [25,86].

Several studies have demonstrated that the addition of noble metals and transition metal oxides as distinctive promoters, such as Pt, Ru, and Re, can increase the reduction of Co and known as active stage of FT process. Also, the properties of different promoters are summarized in Table 7.3. Tsubaki et al. [91] found that the addition of Co/SiO_2 could

Table 7.3 Role of different promoters on catalytic behaviors of Co-based Fischer–Tropsch (FT) catalysts.

Promoters	Prospective effect	References
Earth oxides (CeO_2 and La_2O_3)	The appropriate amount of CeO_2 increases the active sites of the catalysts due to the increase of reducibility and dispersion, which increases the FT synthesis activity. Inhibit H_2O reoxidation as well as further the aggregation of cobalt metal particles during the reaction. Also, the CO conversion increased with a proper amount and preparation method. The selectivity of C_{5+} hydrocarbons is increased. Also, the ratio of olefin/paraffin ratio increased after modifying the Co support interactions with La_2O_3.	[41,87–90]
Noble metals (Pt, Pd, Re, and Ru)	The Pt resulted in a significant increase in the FT reaction rate but decrease the selectivity of C_{5+} hydrocarbon. Also, Pt and Pd showed higher CH_4 selectivity and hardly exerted any effect on the degree of cobalt reduction. Pt and Pd increased the degree of cobalt reduction but decreased the TOF. Re enhanced the activity for FT synthesis, which increases the selectivity of C_{5+} and decreases CH_4 selectivity. Also, Re raised the dispersion of Co catalysts. Ru increased the catalytic activity and TOF, but the CH_4 selectivity was unchanged.	[53,91–94]
ZrO_2	Increases the Co metal active sites and reducibility, leading to the increase of CO conversion and C_{5+} selectivity. The increasing content of Zr decreased the CH_4 selectivity. Also it decreases the ratio of olefin to paraffin in the products. Increasing the selectivity of C_{5+} due to the increased Co site density and the lowered hydrogenation activity.	[87,95,96]
Mn	Increases the selectivity to C_{5+} and decreases CH_4 production due to decreasing the hydrogenation reaction, which increases the ratio of olefin to paraffin. Possibly decreases the CO conversion activity if the Mn content is high.	[97–99]

TOF, Turnover frequency.

increase hydrogenation reaction of CO. The addition of low-quantity ruthenium could increase Co reduction and enhance hydrogen turnover frequency (TOF). Furthermore, inclusion of platinum and palladium improves Co reduction, while it reduces the TOF [91]. In general, the addition of platinum as promoter remarkably enhances the FT synthesis process and methane production but reduces the C_{5+} production [92]. Also, rhenium is frequently used as promoter for Co catalysts, which enhances the diffusion of Co supported on TiO_2 by avoiding the cluster of CoO_x [53]. Storsæter et al. [93] reported that rhenium doping on CO catalysts could promote selectivity of C_{5+} hydrocarbons and increase the production rate.

However, reaction rate of Co catalysts could be increased by the addition of certain metal oxides during FT synthesis [25,47,53]. The addition of ZrO_2 metal oxide as promoter increased the C_{5+} selectivity and consumption of CO during FT synthesis with Co/SiO_2 catalysts [95]. Jongsomjit et al. [96] observed that the addition of ZrO_2 on a 20 wt.% Co/Al_2O_3 saturated γ-Al_2O_3 and enhanced the reducibility of cobalt by avoiding the origination of aluminate on catalysts surface. They also reported that a higher amount of zirconium increased the olefin:paraffin ratio in C_2-C_{17} and reduced the hydrogen production ability of cobalt catalysts, which was promoted by ZrO_2 [87]. It was also reported in literature that the addition of manganese could enhance the reaction rate of FT synthesis with higher CO conversion [97]. The paraffin-to-olefin ratio in hydrocarbons could be significantly improved by supplement with a large amount of manganese in $Mn-Co/TiO_2$. The inclusion of MnO_x decreased the CH_4 selectivity and increased the C_{5+} selectivity, which could be attained by the selection of the optimum preparation method [36,97,98].

Furthermore, the rare earth oxides could be useful as promoters for cobalt-based catalysts during FT synthesis. The supplementation of La_2O_3 to 20 wt.% Co/SiO_2 increased the Co time yield and decreased the CH_4 and C_2-C_4 hydrocarbons selectivity. The increasing ratio of La/Co ratio enhanced the olefin to paraffin ratio in C_2-C_5 hydrocarbons [88,89]. The presence of La_2O_3 could not modify the inherent properties of the cobalt species but probably increased the concentration of active sites. The La_2O_3-promoted Co/Al_2O_3 catalysts improved the selectivity toward hydrocarbons, while the formation of olefins was increased with the increase in La content. However, a very high amount of La_2O_3 loading could increase the CH_4 selectivity by decreasing the C_{5+} selectivity.

The addition of La_2O_3 decreased the reducibility of Co species to enhance their dispersion [87]. The addition of CeO_2 into cobalt catalysts could alter the hydrocarbon productivity [100]. However, the selectivity toward CH_4 and C_2-C_4 hydrocarbons increased significantly in the presence of CeO_2 modifier [100]. The role of promoters, which were commonly used in cobalt catalysts for FT synthesis, is reported in Table 7.3 [75,94,99,100].

7.4 Kinetic modeling of Fischer–Tropsch synthesis

The kinetic modeling is a major challenge to describe the FT synthesis due to the complexity of the reaction mechanism and the combination of a number of steps, which results in empirical power-law expressions for the kinetics. Several kinetic modeling studies were conducted using single and bimetallic catalysts, including Co, Fe, Ni, and Ru. The selectivity of FT products and its kinetics constitute two pillars in the reaction mechanism. Among various studies investigated on product distribution and reaction kinetics of the FT synthesis, continuous laboratory reactors were used. Major studies were carried in atmospheric gas–solid packed bed reactors. Packed beds at high pressures were applied, but integral reactors were not unsuitable for kinetic studies of the FT synthesis [17,101–104]. Continuous recycle reactors are also suitable for gas–solid kinetic measurements, such as spinning basket reactor and Berty reactor. Slurry phase FT synthesis was carried out in slurry-phase reactors (SPRs), and continuous process was carried out with constant gas phase, while batch process was carried out in liquid phase [35,105–107].

FT synthesis kinetic modeling is a prerequisite for industrialization the process, for process design, operation, and simulation [35,107]. The kinetics of cobalt-based FT catalysts has been the subject of many researches in the last decades. The kinetics of the FT synthesis has been studied extensively to describe the reaction rate using a power law rate equation. However, other rate equations are also applied, such as Langmuir Hinshelwood Hougen Watson (LHHW) and Eley Rideal, based on a reaction mechanism for the hydrocarbon formation [35,107]. Mostly, formation of the monomer accrues during the rate-determining step and conversion of syngas commonly varies due to monomer structure [104,108–110]. The atomic mass balances of C, H, and O are essential to see on reliability and stability FT synthesis. The kinetic and selectivity models should be developed on the basis of an extensive experimental data set, and periodic standards are required to check the variations. Also process parameters such as velocity, temperature, CO, and H_2 reactor pressure require investigations to develop kinetic models.

A kinetic model was developed by Kellner and Bell [111] and Takoudis [112] for the hydrocarbons production without assumptions on a rate-determining step. However, several assumptions were introduced to solve the resulting set of equations. In general, the development of kinetic model should be based on rate-determining steps with particular mechanistic scheme during FT synthesis.

The FT synthesis could be expressed as the combination of the FT reaction and the WGS reaction [Eqs. (7.6) and (7.7)].

$$CO + \left(1 + \frac{m}{2n}\right) H_2 \rightarrow \frac{1}{n} C_n H_m + H_2O \ (R_{FT}) \tag{7.6}$$

$$CO + H_2O \leftrightarrow CO_2 + H_2 \ (R_{WGS}) \tag{7.7}$$

Kinetic equations and model could be developed on the overall syngas consumption [68,108,112–116], which could be individualistic of the WGS equilibrium for FT products. The FT reaction rate allows the variation in syngas consumption rate by reaction stoichiometry, which varies depending on the catalyst and is presented in Table 7.4. Several kinetic studies were carried out using fixed-bed reactors at high syngas conversion. However, some reactor types such as plug flow reactor (PFR) cannot be used for studying the intrinsic kinetics of the FT synthesis due to the CO and H_2 partial pressure, which vary along the axis of the reactor [104]. Some other difficulties could arise, because of heat and mass-transfer rate, secondary reactions, and self-product reticence. In order to achieve a compromise between

Table 7.4 Kinetic equations rate for syngas consumption rate, which proposed in Table 7.5 [35].

Kinetic equations	Eq. no.	References
kP_{H_2}	(7.8)	[64,68]
$kP_{H_2}P_{CO}/P_{CO} + aP_{H_2O}$	(7.9)	[68,105,113]
$kP_{H_2}^2 P_{CO}/P_{CO}P_{H_2} + aP_{H_2O}$	(7.10)	[105,108,114]
$kP_{H_2}P_{CO}/P_{CO} + aP_{CO_2}$	(7.11)	[68,106,109,114]
$kP_{CO}^{1/2}p_{H_2}^{1/2}/(1a+P_{CO}^{1/2}+bP_{H_2}^{1/2})^2$	(7.12)	[110]
$kP_{CO}p_{H_2}^{1/2}/(1a+P_{CO}+bP_{H_2}^{1/2})^2$	(7.13)	[104]
$kP_{CO}P_{H_2}/(1+bP_{CO})^2$	(7.14)	[115,116]

economic feasibility and efficiency in process, it is of primary importance to accurately formulate the integral kinetic models of the appropriate catalyst based on elementary steps, keeping in mind the influence of transport of reactants on reaction kinetics. Therefore reaction kinetics is essential for optimizing the catalyst design based on the knowledge of the rate-determining steps and kinetic trends. In general, the composition of catalyst as well as influence of transport of reactants on reaction condition can control the numerical values of the intrinsic kinetic rate parameters. Still the equations rate is not equivalent; thus *k* could be considered as combinations of different adsorption and kinetic constants. Several studies showed that the FT synthesis activity could be determined on the preparation method, metal loading on the catalyst, and catalyst support [35,107].

7.4.1 Kinetics of Fe- and Co-based catalysts

The first kinetic equations for Fe-based catalysts were proposed by Huff and Satterfield [108] and Zimmerman and Bukur [68], which are summarized in Table 7.5. In general, the FT reaction rate increases with the H_2 partial pressure and decreases with the H_2O partial pressure. It was observed that addition of 27 mol.% water vapor to syngas decreased the catalyst activity, and after the addition of 42 mol.% water vapor, the catalyst did not regain its initial activity. Addition of water vapor decreased CH_4 selectivity and increased oxygenate selectivity and the rate of the WGS [117]. The mechanistic kinetic rate expression of Fe catalysts is dependent on the construction of monomer and different models for the formation of the monomer species are assumed in literature: (1) carbide mechanism, where CO is adsorbed and dissociates on the metal surface and the intermediate carbon species undergo to hydrogenation, forming methylene species [111], (2) combined carbide mechanism, where methylene groups are the result of water elimination from the enol [118]. However, it could be summarized that the development of FT kinetic equations still requires additional research during the FT synthesis on iron catalysts.

Few kinetic studies on Co catalysts were carried out and were summarized in Table 7.5, and kinetic equations of Co- and Fe-based catalysts are different. Commonly, the LHHW kinetic expression depends on a rate-determining step, which involves uppermost layer reaction of double molecule, while H_2O inhibition rate on Co catalysts is not observed because the WGS reaction hardly shows effect on Co catalysts and

Table 7.5 Kinetic studies for the Fischer–Tropsch synthesis on Co and Fe catalysts [35].

Catalyst	Properties	Reactor	Operative conditions			Kinetic expression from Table 7.4	References
			Temperature (°C)	Pressure (bar)	H_2/CO feed molar ratio		
Fe/K	Fused	Fixed bed	225–265	100–180	1.2–7.2	(7.8)	[64]
Fe/K/Cu	Precipitated	Slurry	235–265	150–300	0.6–1.0	(7.8)	[68]
Fe/K/Cu	Precipitated	Slurry	250	150–300	0.6–1.0	(7.9)	[68]
Fe/K	Fused	Fixed bed	250–315	200	2.0	(7.9)	[113]
Iron	Fused	Slurry	232–263	40–150	0.5–1.8	(7.10)	[108]
Fe	Precipitated	Slurry	220–260	100	0.5–0.6	(7.11)	[106]
Fe	Precipitated	Slurry	220–280	50–120	0.5–3.5	(7.9)	[109]
Fe	Fused	Slurry	210–270	50–550	0.5–3.5	(7.11)	[109]
Co/Kieselguhr	–	Berty	190	20–150*	0.5–8.5	(7.12) and (7.13)	[104,110]
Co/MgO/ SiO_2	–	Slurry	220–240	150–350	1.5–3.5	(7.14)	[116]
Co/Zr/SiO_2	–	Slurry	220–280	210	0.5–2.0	(7.10)	[95]
Fe/Cu/K	Precipitated	Gradient less	230–264	100–260	1.1–2.4	(7.10)	[105]
Fe/Cu/K	Precipitated	Gradient less	230–264	100–260	1.1–2.4	(7.9)	[105]
Fe	Precipitated	Slurry	220–260	–	0.5–0.8	(7.11)	[114]
Fe	Precipitated	Slurry	220–260	–	0.8–2.0	(7.9)	[114]
Fe	Precipitated	Fixed bed	250–350	60–210	3.0–6.0	–	[101]

formation of CO_2 is inhibited [35,107,119]. The kinetic modeling of FT synthesis over an industrial Fe–Cu–K catalyst was based on the carbide polymerization mechanism [120]. The performance of the catalytic FT catalyst over Ru/Co/Zr/SiO_2 described the kinetic model with the estimation of kinetic parameters to predict the effect of the operating conditions on hydrocarbon distribution and CO consumption [79]. The kinetic study of Co-based catalysts is inclusive as comparison to Fe catalysts due to the absence of the WGS reaction and less different catalytic sites. Also, LHHW kinetic equations on Co-based catalysts are a reliable mechanistic reaction for the CO conversion, and this is an appropriate approach for modeling of multipart reaction steps, such as FT process [35,104,110,121].

7.4.2 Kinetics of water–gas shift reaction

The WGS reactions have major effect on the partial pressure of FT synthesis reactants to understanding its kinetics reaction. Several studies on WGS kinetics were reported, and the proposed rate expressions are summarized in Table 7.6. Co catalysts were not active in contrast to Fe-based catalysts [35,68,123–127]. The WGS reaction could change the FT synthesis rate due to adsorption and desorption reactions, which allows the separation of H_2O, H_2, and CO_2. The individual WGS kinetics was widely reported and suggested the appearance of formiate species [35]. The formation of formiate could be due to reaction between hydroxyl groups or H_2O and CO in any phase. The WGS reaction was studied on several supported and unsupported iron oxide and zinc oxide catalyst. Some authors recommended the WGS reaction over unproven magnetite process through straight oxidation reaction, while Fe catalysts operate via oxidation of formiate because of the partial oxidation of Fe cations [128]. It could be concluded that the WGS kinetics under FT synthesis still requires additional research. Also, the knowledge about the dynamic positions for FT synthesis with WGS reaction could provide more information for the kinetic modeling of FT synthesis process. Several studies on kinetics focused on empirical rate equations, which are known for CO conversion; a popular one utilized a simple first order in CO [102,122]. In general, two WGS mechanisms are applied for kinetic reaction, which convert CO through formiate reaction [128].

In the first model developed by Bohlbro [34], which was established on straight CO conversion process, CO reacted with adsorbed oxygen from gas phase, the reaction formed by water decomposition with H_2

Table 7.6 Kinetic rate equation expressions of water–gas shift for Fischer–Tropsch synthesis [62].

Rate of equation expressions	E_a (kJ/mol)	References
$R_{WGS} = k_{WGS} P_{CO}$	124	[68,122]
$R_{WGS} = k_{WGS}(P_{CO}P_{H_2O} - P_{CO2}P_{H_2}/k_{WGS}/P_{CO} + a \cdot P_{H_2O} + b \cdot P_{CO_2} + c \cdot P_{H_2})$	–	[34]
$R_{WGS} = k_{WGS}(P_{CO}P_{H_2O} - P_{CO2}P_{H_2}/K_{WGS})$	–	[68]
$R_{WGS} = k_{WGS}(P_{CO}P_{H_2O} - P_{CO2}P_{H_2}/K_{WGS}/P_{CO} + a \cdot P_{H_2O} + b \cdot P_{CO_2})$	125	[68,105]
$R_{WGS} = k_{WGS}(P_{CO}P_{H_2O} - P_{CO2}P_{H_2}/K_{WGS}/P_{H_2O}P_{CO} + a \cdot P_{H_2O})$	88	[68,105]
$R_{WGS} = k_{WGS}(P_{CO}P_{H_2O} - P_{CO_2}P_{H_2}^{1/2}/K_{WGS}/(1 + a \cdot P_{H_2O}/P_{H_2}^{1/2})^2)$	27.7	[101]
$R_{WGS} - k_{WGS}(P_{CO}P_{H_2O} - P_{CO2}P_{H_2}/K_{WGS}/(1 + a \cdot P_{CO} + b \cdot P_{H_2O})^2)$	–	[107]
$R_{WGS} - k_{WGS}(P_{CO}P_{H_2O}/P_{H_2}^{1/2} - P_{CO_2}/P_{H_2}^{1/2}/K_{WGS}/(1 + a \cdot P_{CO} + b \cdot P_{H_2O})^2)$	–	[107]
$R_{WGS} - k_{WGS}(P_{CO}P_{H_2O} - P_{CO_2}P_{H_2}/K_{WGS}/(1 + a \cdot P_{H_2O} + b \cdot P_{H_2O}/P_{H_2}^{1/2})^2)$	–	[61]

discharge. During the straight conversion process, it has been considered that the catalyst surface is oxidized by water to produce H_2 and then subsequently reduced by CO, forming CO_2 [32,61]. Several experimental outcomes with different supported iron catalysts showed limited change in Fe cations; Rethwisch and Dumesic [128] discharged the direct oxidation process that was a favorable formiate with WGS reaction [35,107].

7.5 Process simulation for Fischer–Tropsch synthesis

The FT synthesis process is extremely reliant on several parameters as temperature, partial pressure, feed flow rate, molar ratio of CO and H_2O, and property of catalyst. To obtain a better product yield, researchers often conduct the experiments studying the effect of many parameters, especially for the development of new catalysts and could not afford the time and cost to carry out all the parameter combinations. Therefore computer modeling is necessary to explore the several parameters to understand the effects of new processing conditions, catalyst kinetics, and different kinetic mechanisms. There are four types of FT reactors available in commercial applications: (1) circulating fluidized bed reactor, (2) standard fluidized bed reactor, (3) fixed packed bed reactor (FPBR), and (4) SPR [79]. Furthermore, FBR is recognized as nonfavorable for the production of liquid transportation fuels, because liquid phase products may cause catalyst agglomeration and a loss of fluidization [23,28,42,48,59]. Moreover, FT synthesis reactors are also described as low temperature (LTFT) or high temperature (HTFT) reactors. The main difference is that a liquid phase forms when operating in LTFT reactors, while HTFT reactors operate entirely in the gas phase. LTFT fixed bed and slurry-phase are suitable for producing liquid hydrocarbons products. However, FPBRs could deactivate the catalyst, which may lead to temperature increase. On the other hand, these reactors show several benefits, such as ease operation, extra product separation device not required, and easy scale-up process. Several research studies for design and modeling of SPR exist in literature, while limited investigations on deigning and modeling are available on the FPBR. Furthermore, heterogeneous and single-dimensional PFR was developed to optimize the operational parameters on large-scale reactors [12,42,43,113,126,129–131]. First two-dimensional plug flow, pseudo homogeneous model without intraparticle diffusion limitations was developed by Bub et al. [102]. Jess et al. [132] also described the use of a two-dimensional, pseudo homogeneous model to account for intraparticle dispersion restrictions, which further explored by Wang et al. [133]. Co catalysts-based heterogeneous and single-dimensional model for FPBR was proposed by de Swart et al. [134]. A comprehensive software package developed by Fernandes and Cartaxo [135] for FT process in a SPR was used to simulate the overall reactor conditions. This study integrated three phases (gas, slurry, and solid phase) and different catalysts and could be adopted different synergy of these mechanisms. However, it was demonstrated that these three mechanisms are not accurate to describe the overall process and the kinetic parameters in their software could not be modified. Furthermore, a numerical model is simulated to evaluate the FT process in tubular multitube reactor using Fe-based catalyst to investigate the impact of process parameters on hydrocarbon production [48]. Kwack et al. [79] used lsqcurvefit in MATLAB to simulate the product distribution during FT synthesis on Co catalyst using a revised mechanism. Also, Kshetrimayum et al. [136] proposed a detailed computational fluid dynamics simulated modeling for FT process in packed microchannel reactor, considering both single and multichannel reactor. Therefore it was suggested that the temperature effect on CO conversion and the selectivity of hydrocarbon products revealed necessity for maintaining a reaction channel temperature below 250°C, during FT process under low-temperature conditions. In general, the research on simulations was fragmented, and there was no software packages to perform complete simulations, which should include the feedstock and kinetic parameter specifications, modeling of diffusion limitation graphical description of syngas consumption, and hydrocarbon production rate [79,136,137].

7.6 Carbon nanofibers/Carbon felt reactors

Natural gas, coal, or biomass that is converted into syngas could be further transformed in liquid hydrocarbons via FT synthesis. It is extremely exothermic reaction and effective heat transfer is prerequisite for successful operation. The selectivity for C_{5+} hydrocarbons highly deepens on appropriate catalysts selection, temperature profile of reactor, and gas–liquid mass transfer rate, and reactors deign is key factor to achieve high selectivity. Mostly, the fixed-bed and slurry SPRs were used for the FT synthesis. To attain high catalyst efficiency and C_{5+} hydrocarbons selectivity the heat removal is a major challenge in these reactors. Moreover, large units are favored but applications such as offshore production could benefit from compact and modular conversion technology. Space and weight as well as demanding for greatly combined, solid, light-weight, and nontoxic approach represent the substantial limitations [18,28,53,138–140]. Monolithic/Microstructured reactors were considered to meet such requirements and comprise the advantages of FBR and SPR reactor. Monolithic reactors have numerous promising benefits in respect to traditional FT reactors, such as higher gas–liquid mass transfer rates, short diffusion distance, plug flow characteristics, low pressure drip, wax-catalyst separation not necessary, and precise temperature control by direct cooling of catalyst with the liquid medium and external heat removal [59,141,142]. However, the low thermal conductivity of the ceramic monoliths and small-scale production capacity are the major drawbacks of the monolithic reactor [57,84].

Over the past decade, CNFs and carbon nanotubes (CNTs) have been reflected as a potential support material for active metals and oxides catalysts. This specific interest is attributed to their wide properties, such as relatively high surface area, high purity, chemical resistance to acids and bases, and good electric and thermal conductivity [9,57,59,143]. The CNFs showed an enhanced catalytic activity and high C_{5+} selectivity compared to conventional support materials, such as alumina, silica, or activated carbon [22,127,144–152]. Yu et al. [153] reported that CNF-supported cobalt catalyst demonstrated high activity and high selectivity to C_{5+} hydrocarbons compared with conventional cobalt catalysts supported on alumina for FT synthesis. However, CNF-supported Co catalysts deactivate quickly. Their activity could be improved using hierarchically structured CNF/CF composites as compact FT reactors [154]. Hierarchically structured CNFs/CF composites in the FT synthesis showed several advantages, such as improved dispersion and reducibility of cobalt nanoparticles, which results in high activities and selectivity [155,156]. Although industrial applications of CNTs and CNFs in powder form have several difficulties, such as difficult to handling and transportation problem, quick pressure drop for gas phase, and accumulation with time on stream [144]. Although the health risk and safety concerns associated to the large-scale airborne presence of CNTs/CNFs, these aspects limit the applicability of such catalysts for industrialization. To overcome these limitations, several researchers observed the growth of CNTs/CNFs on structured supports, where macroscopic shape offers easy handling and transportation.

CNTs/CNFs-supported Fe catalysts are preferred compared to Co catalysts for FT synthesis, which produce light olefins [157]. The Fe nanoparticles on CNF with sulfur and sodium promoters could achieve the 60% selectivity of light olefins [66]. Recently, various studies have examined the influence for insertion of iron catalysts, with supporters and promoters onto CNTs [158]. The preparation of Fe/CNT by using Cu seemed to improve the FT synthesis and WGS reaction frequency, even though it did not influence the selectivity towards selected hydrocarbons. In contrast the preparation of Fe/CNT with K^+ could increase the selectivity of C_5-C_{11} up to ~52% and of C_{12+} up to 8% and decrease the selectivity of C_2-C_4 and CH_4 [158]. Also, the CNT-supported bimetallic Fe–Ru catalyst with the addition of K^+ increased the fraction of C_2H_4 up to ~70% in total C_2 hydrocarbons and significantly improved the formation of C_{5+} and the ratio of olefins to paraffin [159]. Fe/CNT was compared to Co/CNT, both prepared by the impregnation method and observed that Fe/CNT exhibited as effective catalyst for FT synthesis compared to Co/CNT catalysts. The amount of C_2-C_6 olefins in the total C_2-C_6 hydrocarbons was relatively high up to greater than 85%. Overall, it showed higher C_2-C_4 and C_{5+} selectivity as well as better olefin selectivity and lower CH_4 selectivity [74,160]. CNFs and CNTs supported on cobalt catalysts were studied by several researchers for FT process and constant activity was observed. The herringbone type CNF-supported Co catalyst increased the selectivity of C_{5+} by 86% during FT synthesis with Co/CNF catalyst [161]. The Mn-promoted Co/CNF catalysts could be produced by two-step impregnation method, and Mn could be directly associated with Co surface [156,161]. However, after the reduction, Mn remained on Co particles over CNFs, possibly due to feeble interactions between CNFs and Mn [156]. The selectivity of C_{5+} could be increased by the addition of Mn in small amount for the Co/CNF catalyst. Therefore the CNFs could allow to preparing FT catalysts with robust relationship in active metal and modifier without intrusion of support effects. In addition, these catalysts could have higher C_{5+} selectivity and higher CO conversion during FT process. The reactive activity of the Co catalysts loaded on CNFs/CNTs compared with traditional supports metals, such as metal oxide, showed stable catalytic activity as well as the higher C_{5+} selectivity. Furthermore, CNFs with various configurations, such as platelet and herringbone, were compared with those on Al_2O_3, and it was observed that Co particles were smaller over the Co/CNF platelets respect to Co/CNT herringbone, whereas the selectivity of C_{5+} was similar over both catalysts.

An extensive study carried out by Trépanier et al. [56] on CNTs-supported Co catalysts with different loadings of Co, Ru, and K by coimpregnation method for FT synthesis. During the addition of 15 wt.% Co/CNT catalyst, metal particles were homogeneously distributed on the inner surface CNTs and remained on the CNTs outer surface. After increasing the Co loading to 30 wt.% increased the cobalt cluster sizes and decreased the reduction temperature and dispersion. The increasing of Co loading from 15 to 30 wt.% increased the C_{5+} selectivity from 70% to 77% and CO conversion from 48% to 86%. Ru enhanced the reducibility of Co_3O_4 to CoO, which favored the dispersion and decreased the average cobalt cluster size. However, K was responsible in shifting the reduction temperatures to higher temperatures. Both promoters enhanced the selectivity of FT synthesis toward the higher molecular weight hydrocarbons, but the effect of Ru is less pronounced. K increased the olefin to paraffin ratio from 0.73 to 3.5 and the C_{5+} selectivity from 70% to 87% [56].

7.7 Conclusions and perspectives

The current environmental scenario needs to flourish the sustainable approaches for the generation of transportation fuels. Recently, FT synthesis has received renewed interests because of the global demand for a decreased dependence on petroleum for the production of fuels. FT process is a heterogeneous catalytic reaction, which transforms syngas into green transportation liquid fuels. The syngas could be produced by various feedstocks, that is, agricultural biomass, coal, and natural gas. The production of liquid fuels and activity of the FT synthesis depend on chemical and physical properties of catalysts and operational conditions, which affects the selectivity of the FT products distribution. The products selectivity could be regulated by kinetic modeling and simulation for FT synthesis. This chapter summarizes the role of catalysts, kinetic modeling, and simulation for FT synthesis particularly for the products selectivity and highlights recent developments of carbon nanofibers with an aim to control the product selectivity. The development of effective catalysts with high stability and activity is required to attain the desired product selectivity.

References

[1] T. Fu, Z. Li, Highly dispersed cobalt on N-doped carbon nanotubes with improved Fischer–Tropsch synthesis activity, Catal. Commun. 47 (2014) 54–57.

[2] M.A. Fahim, T.A. Alsahhaf, A. Elkilani, Clean fuels, Fundamentals of Petroleum Refining, Elsevier, Amsterdam, 2010, pp. 303–324.
[3] S. Mehariya, A.K. Patel, P.K. Obulisamy, E. Punniyakotti, J.W.C. Wong, Co-digestion of food waste and sewage sludge for methane production: current status and perspective, Bioresour. Technol. 265 (2018) 519–531.
[4] O. Parthiba Karthikeyan, E. Trably, S. Mehariya, N. Bernet, J.W.C. Wong, H. Carrere, Pretreatment of food waste for methane and hydrogen recovery: a review, Bioresour. Technol. 249 (2018) 1025–1039.
[5] G. Liu, B. Yan, G. Chen, Technical review on jet fuel production, Renew. Sustain. Energy Rev. 25 (2013) 59–70.
[6] S.S. Ail, S. Dasappa, Biomass to liquid transportation fuel via Fischer Tropsch synthesis – technology review and current scenario, Renew. Sustain. Energy Rev. 58 (2016) 267–286.
[7] A.S. Snehesh, H.S. Mukunda, S. Mahapatra, S. Dasappa, Fischer–Tropsch route for the conversion of biomass to liquid fuels—technical and economic analysis, Energy 130 (2017) 182–191.
[8] G.F. Botes, T.C. Bromfield, R.L.J. Coetzer, R. Crous, P. Gibson, A.C. Ferreira, Development of a chemical selective iron Fischer Tropsch catalyst, Catal. Today 275 (2016) 40–48.
[9] W. Chen, T. Lin, Y. Dai, Y. An, F. Yu, L. Zhong, et al., Recent advances in the investigation of nanoeffects of Fischer-Tropsch catalysts, Catal. Today 311 (2018) 8–22.
[10] W. Ma, G. Jacobs, D.E. Sparks, J.L.S. Klettlinger, C.H. Yen, B.H. Davis, Fischer–Tropsch synthesis and water gas shift kinetics for a precipitated iron catalyst, Catal. Today, 275, 2016, pp. 49–58.
[11] N. Moazami, M.L. Wyszynski, K. Rahbar, A. Tsolakis, Parametric study and multi-objective optimization of fixed-bed Fischer–Tropsch (FT) reactor: the improvement of FT synthesis product formation and synthetic conversion, Ind. Eng. Chem. Res. 56 (2017) 9446–9466.
[12] A.A. Mirzaei, B. Shirzadi, H. Atashi, M. Mansouri, Modeling and operating conditions optimization of Fischer–Tropsch synthesis in a fixed-bed reactor, J. Ind. Eng. Chem. 18 (2012) 1515–1521.
[13] W. Ma, G. Jacobs, T.K. Das, C.M. Masuku, J. Kang, V.R.R. Pendyala, et al., Fischer–Tropsch synthesis: kinetics and water effect on methane formation over 25% $Co/\gamma\text{-}Al_2O_3$ catalyst, Ind. Eng. Chem. Res. 53 (2014) 2157–2166.
[14] K. Kim, Y. Kim, C. Yang, J. Moon, B. Kim, J. Lee, et al., Long-term operation of biomass-to-liquid systems coupled to gasification and Fischer–Tropsch processes for biofuel production, Bioresour. Technol. 127 (2013) 391–399.
[15] M. Rafati, L. Wang, D.C. Dayton, K. Schimmel, V. Kabadi, A. Shahbazi, Techno-economic analysis of production of Fischer-Tropsch liquids via biomass gasification: the effects of Fischer-Tropsch catalysts and natural gas co-feeding, Energy Convers. Manage. 133 (2017) 153–166.
[16] Y. Xue, H. Ge, Z. Chen, Y. Zhai, J. Zhang, J. Sun, et al., Effect of strain on the performance of iron-based catalyst in Fischer-Tropsch synthesis, J. Catal. 358 (2018) 237–242.
[17] A. Sharma, V. Pareek, D. Zhang, Biomass pyrolysis—a review of modelling, process parameters and catalytic studies, Renew. Sustain. Energy Rev. 50 (2015) 1081–1096.
[18] H. Jahangiri, J. Bennett, P. Mahjoubi, K. Wilson, S. Gu, A review of advanced catalyst development for Fischer-Tropsch synthesis of hydrocarbons from biomass derived syn-gas, Catal. Sci. Technol. 4 (2014) 2210–2229.
[19] H. Mahmoudi, M. Mahmoudi, O. Doustdar, H. Jahangiri, A. Tsolakis, S. Gu, et al., A review of Fischer Tropsch synthesis process, mechanism, surface chemistry and catalyst formulation, Biofuels Eng. 2 (2017) 11–31.

[20] A. Chiodini, L. Bua, L. Carnelli, R. Zwart, B. Vreugdenhil, M. Vocciante, Enhancements in biomass-to-liquid processes: gasification aiming at high hydrogen/carbon monoxide ratios for direct Fischer-Tropsch synthesis applications, Biomass Bioenergy 106 (2017) 104–114.
[21] Y. Xu, D. Liu, X. Liu, Conversion of syngas toward aromatics over hybrid Fe-based Fischer-Tropsch catalysts and HZSM-5 zeolites, Appl. Catal., A: Gen. 552 (2018) 168–183.
[22] D. Nabaho, J.W. Niemantsverdriet, M. Claeys, E. van Steen, Hydrogen spillover in the Fischer–Tropsch synthesis: an analysis of platinum as a promoter for cobalt–alumina catalysts, Catal. Today 261 (2016) 17–27.
[23] B. Todic, L. Nowicki, N. Nikacevic, D.B. Bukur, Fischer–Tropsch synthesis product selectivity over an industrial iron-based catalyst: effect of process conditions, Catal. Today 261 (2016) 28–39.
[24] M.K. Gnanamani, G. Jacobs, U.M. Graham, M.C. Ribeiro, F.B. Noronha, W.D. Shafer, et al., Influence of carbide formation on oxygenates selectivity during Fischer-Tropsch synthesis over Ce-containing Co catalysts, Catal. Today 261 (2016) 40–47.
[25] Q. Zhang, J. Kang, Y. Wang, Development of novel catalysts for Fischer–Tropsch synthesis: tuning the product selectivity, ChemCatChem 2 (2010) 1030–1058.
[26] G.P. Huffman, Incorporation of catalytic dehydrogenation into Fischer–Tropsch synthesis of liquid fuels from coal to minimize carbon dioxide emissions, Fuel 90 (2011) 2671–2676.
[27] M.E. Dry, FT catalysts, in: A.P. Steynberg and M. Dry (Eds.), Fischer-Tropsch Technology, Elsevier, Amsterdam, 2004, pp. 533–600.
[28] T. Fu, Z. Li, Review of recent development in Co-based catalysts supported on carbon materials for Fischer–Tropsch synthesis, Chem. Eng. Sci. 135 (2015) 3–20.
[29] N. Moazami, H. Mahmoudi, K. Rahbar, P. Panahifar, A. Tsolakis, M.L. Wyszynski, Catalytic performance of cobalt–silica catalyst for Fischer–Tropsch synthesis: effects of reaction rates on efficiency of liquid synthesis, Chem. Eng. Sci. 134 (2015) 374–384.
[30] N. Moazami, M.L. Wyszynski, H. Mahmoudi, A. Tsolakis, Z. Zou, P. Panahifar, et al., Modelling of a fixed bed reactor for Fischer–Tropsch synthesis of simulated N_2-rich syngas over Co/SiO_2: hydrocarbon production, Fuel 154 (2015) 140–151.
[31] Q. Chen, G. Liu, S. Ding, M. Chanmiya Sheikh, D. Long, Y. Yoneyama, et al., Design of ultra-active iron-based Fischer-Tropsch synthesis catalysts over spherical mesoporous carbon with developed porosity, Chem. Eng. J. 334 (2018) 714–724.
[32] F.G. Botes, J.W. Niemantsverdriet, J. van de Loosdrecht, A comparison of cobalt and iron based slurry phase Fischer–Tropsch synthesis, Catal. Today 215 (2013) 112–120.
[33] J. Tu, M. Ding, T. Wang, L. Ma, Y. Xu, S. Kang, et al., Direct conversion of bio-syngas to gasoline fuels over a Fe_3O_4@C Fischer-Tropsch synthesis catalyst, in: J. Yan, F. Sun, S.K. Chou, U. Desideri, H. Li, P. Campana, R. Xiong (Eds.), Proceeding of the Eight International Conference on Applied Energy, ICAE2016, Beijing, China, October 8–11, 2016, Energy Procedia 105 (2017) 82–87.
[34] H. Bohlbro, An Investigation on the Kinetics of the Conversion of Carbon Monoxide with Water Vapour Over Iron Oxide Based Catalysts: A Contribution from The Haldor Topsøe Research Laboratory: Vedbæk, Gjellerup, 1969.
[35] G.P. van der Laan, A.A.C.M. Beenackers, Kinetics and selectivity of the Fischer–Tropsch synthesis: a literature review, Catal. Rev. 41 (1999) 255–318.
[36] V.R.R. Pendyala, G. Jacobs, M.K. Gnanamani, Y. Hu, A. MacLennan, B.H. Davis, Selectivity control of Cu promoted iron-based Fischer-Tropsch catalyst by tuning the oxidation state of Cu to mimic K, Appl. Catal., A: Gen. 495 (2015) 45–53.

[37] M. Lualdi, G. Di Carlo, S. Lögdberg, S. Järås, M. Boutonnet, V. La Parola, et al., Effect of Ti and Al addition via direct synthesis to SBA-15 as support for cobalt based Fischer-Tropsch catalysts, Appl. Catal., A: Gen. 443 (2012) 76–86.
[38] N. Osakoo, R. Henkel, S. Loiha, F. Roessner, J. Wittayakun, Effect of support morphology and Pd promoter on Co/SBA-15 for Fischer–Tropsch synthesis, Catal. Commun. 56 (2014) 168–173.
[39] A. Dinse, M. Aigner, M. Ulbrich, G.R. Johnson, A.T. Bell, Effects of Mn promotion on the activity and selectivity of Co/SiO_2 for Fischer–Tropsch synthesis, J. Catal. 288 (2012) 104–114.
[40] S. Bai, C. Huang, J. Lv, Z. Li, Comparison of induction behavior of Co/CNT and Co/SiO_2 catalysts for the Fischer–Tropsch synthesis, Catal. Commun. 22 (2012) 24–27.
[41] X. Zhang, H. Su, Y. Zhang, X. Gu, Effect of CeO_2 promotion on the catalytic performance of Co/ZrO_2 catalysts for Fischer–Tropsch synthesis, Fuel 184 (2016) 162–168.
[42] B. Todic, T. Bhatelia, G.F. Froment, W. Ma, G. Jacobs, B.H. Davis, et al., Kinetic model of Fischer–Tropsch synthesis in a slurry reactor on Co–Re/Al_2O_3 catalyst, Ind. Eng. Chem. Res. 52 (2013) 669–679.
[43] W. Qian, H. Zhang, W. Ying, D. Fang, The comprehensive kinetics of Fischer–Tropsch synthesis over a Co/AC catalyst on the basis of CO insertion mechanism, Chem. Eng. J. 228 (2013) 526–534.
[44] A. Mosayebi, A. Haghtalab, The comprehensive kinetic modeling of the Fischer–Tropsch synthesis over Co@Ru/γ-Al_2O_3 core–shell structure catalyst, Chem. Eng. J. 259 (2015) 191–204.
[45] Z. Wu, Z. Zou, J. Huang, F. Gao, Fe-doped NiO mesoporous nanosheets array for highly efficient overall water splitting, J. Catal. 358 (2018) 243–252.
[46] H. Wang, J. Wang, Y. Yuan, Q. Zhao, X. Teng, S. Huang, et al., Shape-selective FeMnK/Al_2O_3@silicalite-2 core-shell catalyst for Fischer-Tropsch synthesis to lower olefins, Catal. Today (2018) (*In Press*).
[47] Y. Bi, Z. Cai, D. Zhou, Y. Tian, Q. Zhang(m), Q. Zhang(f), et al., Understanding the incorporating effect of Co^{2+}/Co^{3+} in NiFe-layered double hydroxide for electrocatalytic oxygen evolution reaction, J. Catal. 358 (2018) 100–107.
[48] A. Moutsoglou, P.P. Sunkara, Fischer–Tropsch synthesis in a fixed bed reactor, Energy Fuels 25 (2011) 2242–2257.
[49] P.K. Swain, L.M. Das, S.N. Naik, Biomass to liquid: a prospective challenge to research and development in 21st century, Renew. Sustain. Energy Rev. 15 (2011) 4917–4933.
[50] C.H. Zhang, Y. Yang, B.T. Teng, T.Z. Li, H.Y. Zheng, H.W. Xiang, et al., Study of an iron-manganese Fischer–Tropsch synthesis catalyst promoted with copper, J. Catal. 237 (2006) 405–415.
[51] Q. Zhang, K. Cheng, J. Kang, W. Deng, Y. Wang, Fischer–Tropsch catalysts for the production of hydrocarbon fuels with high selectivity, ChemSusChem 7 (2014) 1251–1264.
[52] A.A. Muleja, Y. Yao, D. Glasser, D. Hildebrandt, A study of Fischer-Tropsch synthesis: product distribution of the light hydrocarbons, Appl. Catal., A: Gen. 517 (2016) 217–226.
[53] E. Iglesia, Design, synthesis, and use of cobalt-based Fischer-Tropsch synthesis catalysts, Appl. Catal., A: Gen. 161 (1997) 59–78.
[54] R. Luque, A.R. de la Osa, J.M. Campelo, A.A. Romero, J.L. Valverde, P. Sanchez, Design and development of catalysts for biomass-to-liquid-Fischer–Tropsch (BTL-FT) processes for biofuels production, Energy Environ. Sci. 5 (2012) 5186–5202.

[55] A. Comazzi, C. Pirola, M. Longhi, C.L.M. Bianchi, K.S. Suslick, Fe-based heterogeneous catalysts for the Fischer-Tropsch reaction: sonochemical synthesis and bench-scale experimental tests, Ultrason. Sonochem. 34 (2017) 774–780.
[56] M. Trépanier, A. Tavasoli, A.K. Dalai, N. Abatzoglou, Co, Ru and K loadings effects on the activity and selectivity of carbon nanotubes supported cobalt catalyst in Fischer–Tropsch synthesis, Appl. Catal., A: Gen. 353 (2009) 193–202.
[57] S. Zarubova, S. Rane, J. Yang, Y. Yu, Y. Zhu, D. Chen, et al., Fischer–Tropsch synthesis on hierarchically structured cobalt nanoparticle/carbon nanofiber/carbon felt composites, ChemSusChem 4 (2011) 935–942.
[58] D.H. Chun, J.C. Park, S.Y. Hong, J.T. Lim, C.S. Kim, H.T. Lee, et al., Highly selective iron-based Fischer–Tropsch catalysts activated by CO_2-containing syngas, J. Catal. 317 (2014) 135–143.
[59] E. Garcia-Bordeje, Y. Liu, D.S. Su, C. Pham-Huu, Hierarchically structured reactors containing nanocarbons for intensification of chemical reactions, J. Mater. Chem. A 5 (2017) 22408–22441.
[60] S.T. Sie, R. Krishna, Fundamentals and selection of advanced Fischer–Tropsch reactors, Appl. Catal., A: Gen. 186 (1999) 55–70.
[61] F.G. Botes, Water–gas-shift kinetics in the iron-based low-temperature Fischer–Tropsch synthesis, Appl. Catal., A Gen. 328 (2007) 237–242.
[62] D.B. Bukur, B. Todic, N. Elbashir, Role of water-gas-shift reaction in Fischer–Tropsch synthesis on iron catalysts: a review, Catal. Today 275 (2016) 66–75.
[63] M.G.A. Cruz, A.P.S. de Oliveira, F.A.N. Fernandes, F.F. de Sousa, A.C. Oliveira, J. M. Filho, et al., Fe-containing carbon obtained from ferrocene: influence of the preparation procedure on the catalytic performance in FTS reaction, Chem. Eng. J. 317 (2017) 143–156.
[64] M.E. Dry, T. Shingles, L.J. Boshoff, Rate of the Fischer-Tropsch reaction over iron catalysts, J. Catal. 25 (1972) 99–104.
[65] M.G.A. Cruz, F.A.N. Fernandes, A.C. Oliveira, J.M. Filho, A.C. Oliveira, A.F. Campos, et al., Effect of the calcination temperatures of the Fe-based catalysts supported on polystyrene mesoporous carbon for FTS synthesis, Catal. Today 282 (2017) 174–184.
[66] H.M.T. Galvis, J.H. Bitter, C.B. Khare, M. Ruitenbeek, A.I. Dugulan, K.P. de Jong, Supported iron nanoparticles as catalysts for sustainable production of lower olefins, Science 335 (2012) 835–838.
[67] W. Ngantsoue-Hoc, Y. Zhang, R.J. O'Brien, M. Luo, B.H. Davis, Fischer – Tropsch synthesis: activity and selectivity for Group I alkali promoted iron-based catalysts, Appl. Catal., A: Gen. 236 (2002) 77–89.
[68] W.H. Zimmerman, D.B. Bukur, Reaction kinetics over iron catalysts used for the Fischer-Tropsch synthesis, Can. J. Chem. Eng. 68 (1990) 292–301.
[69] T. Li, H. Wang, Y. Yang, H. Xiang, Y. Li, Effect of manganese on the catalytic performance of an iron-manganese bimetallic catalyst for light olefin synthesis, J. Energy Chem. 22 (2013) 624–632.
[70] N. Lohitharn, J.G. Goodwin, Effect of K promotion of Fe and FeMn Fischer–Tropsch synthesis catalysts: analysis at the site level using SSITKA, J. Catal. 260 (2008) 7–16.
[71] F. Liu, Q. Hao, H. Wang, Y. Yang, L. Bai, Y. Zhu, et al., Effect of potassium promoter on reaction performance of iron-based catalyst for Fischer-Tropsch synthesis in slurry reactor, Chin. J. Catal. 11 (2004) 7.
[72] N. Lohitharn, J.G. Goodwin, Impact of Cr, Mn and Zr addition on Fe Fischer–Tropsch synthesis catalysis: investigation at the active site level using SSITKA, J. Catal. 257 (2008) 142–151.

[73] M. Luo, B.H. Davis, Fischer–Tropsch synthesis: Group II alkali-earth metal promoted catalysts, Appl. Catal., A: Gen. 246 (2003) 171–181.
[74] R.M. Malek Abbaslou, J. Soltan, A.K. Dalai, Iron catalyst supported on carbon nanotubes for Fischer–Tropsch synthesis: effects of Mo promotion, Fuel 90 (2011) 1139–1144.
[75] S. Li, A. Li, S. Krishnamoorthy, E. Iglesia, Effects of Zn, Cu, and K promoters on the structure and on the reduction, carburization, and catalytic behavior of iron-based Fischer–Tropsch synthesis catalysts, Catal. Lett. 77 (2001) 197–205.
[76] A.P. Steynberg, Introduction to Fischer–Tropsch technology, in: A.P. Steynberg and M. Dry (Eds.), Fischer-Tropsch Technol., Elsevier, Amsterdam, 2004: pp. 1–63.
[77] A. Alayat, D.N. Mcllroy, A.G. McDonald, Effect of synthesis and activation methods on the catalytic properties of silica nanospring (NS)-supported iron catalyst for Fischer-Tropsch synthesis, Fuel Process. Technol. 169 (2018) 132–141.
[78] D.A. Wesner, F.P. Coenen, H.P. Bonzel, Influence of potassium on carbon monoxide hydrogenation over iron: a surface analytical study, Langmuir 1 (1985) 478–487.
[79] S.H. Kwack, M.J. Park, J.W. Bae, K.S. Ha, K.W. Jun, Development of a kinetic model of the Fischer-Tropsch synthesis reaction with a cobalt-based catalyst, React. Kinet. Mech. Catal. 104 (2011) 483–502.
[80] N. Fischer, E. van Steen, M. Claeys, Structure sensitivity of the Fischer–Tropsch activity and selectivity on alumina supported cobalt catalysts, J. Catal. 299 (2013) 67–80.
[81] Z. Wang, S. Skiles, F. Yang, Z. Yan, D.W. Goodman, Particle size effects in Fischer–Tropsch synthesis by cobalt, Catal. Today 181 (2012) 75–81.
[82] E. Iglesia, S.L. Soled, R.A. Fiato, G.H. Via, Bimetallic synergy in cobalt ruthenium Fischer-Tropsch synthesis catalysts, J. Catal. 143 (1993) 345–368.
[83] V.R.R. Pendyala, G. Jacobs, U.M. Graham, W.D. Shafer, M. Martinelli, L. Kong, et al., Fischer–Tropsch synthesis: influence of acid treatment and preparation method on carbon nanotube supported ruthenium catalysts, Ind. Eng. Chem. Res. 56 (2017) 6408–6418.
[84] J. Zhu, J. Yang, A.H. Lillebø, Y. Zhu, Y. Yu, A. Holmen, et al., Compact reactor for Fischer–Tropsch synthesis based on hierarchically structured Co catalysts: towards better stability, Catal. Today 215 (2013) 121–130.
[85] Y. Yao, X. Liu, D. Hildebrandt, D. Glasser, The effect of CO_2 on a cobalt-based catalyst for low temperature Fischer–Tropsch synthesis, Chem. Eng. J. 193–194 (2012) 318–327.
[86] A. Feller, M. Claeys, E. van Steen, Cobalt cluster effects in zirconium promoted Co/SiO_2 Fischer–Tropsch catalysts, J. Catal. 185 (1999) 120–130.
[87] H. Xiong, Y. Zhang, K. Liew, J. Li, Catalytic performance of zirconium-modified Co/Al_2O_3 for Fischer–Tropsch synthesis, J. Mol. Catal. A: Chem. 231 (2005) 145–151.
[88] G.J. Haddad, B. Chen, J.G. Goodwin, Characterization of La^{3+}-promoted Co/SiO_2 catalysts, J. Catal. 160 (1996) 43–51.
[89] G.J. Haddad, B. Chen, J.G. Goodwin, Effect of La^{3+} promotion of Co/SiO_2 on CO hydrogenation, J. Catal. 161 (1996) 274–281.
[90] S. Vada, B. Chen, J.G. Goodwin, Isotopic transient study of La promotion of Co/Al_2O_3 or CO hydrogenation, J. Catal. 153 (1995) 224–231.
[91] N. Tsubaki, S. Sun, K. Fujimoto, Different functions of the noble metals added to cobalt catalysts for Fischer–Tropsch synthesis, J. Catal. 199 (2001) 236–246.
[92] W. Chu, P.A. Chernavskii, L. Gengembre, G.A. Pankina, P. Fongarland, A.Y. Khodakov, Cobalt species in promoted cobalt alumina-supported Fischer–Tropsch catalysts, J. Catal. 252 (2007) 215–230.

[93] S. Storsæter, Ø. Borg, E.A. Blekkan, A. Holmen, Study of the effect of water on Fischer–Tropsch synthesis over supported cobalt catalysts, J. Catal. 231 (2005) 405–419.
[94] A. Scarabello, D. Dalle Nogare, P. Canu, R. Lanza, Partial oxidation of methane on Rh/ZrO_2 and $Rh/Ce-ZrO_2$ on monoliths: catalyst restructuring at reaction conditions, Appl. Catal., B: Environ. 174–175 (2015) 308–322.
[95] H.P. Withers Jr, K.F. Eliezer, J.W. Mitchell, H.P. Withers, K.F. Eliezer, J.W. Mitchell, Slurry-phase Fischer-Tropsch synthesis and kinetic studies over supported cobalt carbonyl derived catalysts, Ind. Eng. Chem. Res. 29 (1990) 1807–1814.
[96] B. Jongsomjit, J. Panpranot, J.G. Goodwin, Effect of zirconia-modified alumina on the properties of Co/γ-Al_2O_3 catalysts, J. Catal. 215 (2003) 66–77.
[97] F. Morales, F.M.F. de Groot, O.L.J. Gijzeman, A. Mens, O. Stephan, B.M. Weckhuysen, Mn promotion effects in Co/TiO_2 Fischer–Tropsch catalysts as investigated by XPS and STEM-EELS, J. Catal., 230, 2005, pp. 301–308.
[98] F. Morales, E. de Smit, F.M.F. de Groot, T. Visser, B.M. Weckhuysen, Effects of manganese oxide promoter on the CO and H_2 adsorption properties of titania-supported cobalt Fischer–Tropsch catalysts, J. Catal. 246 (2007) 91–99.
[99] T.E. Feltes, L. Espinosa-Alonso, E. de Smit, L. D'Souza, R.J. Meyer, B.M. Weckhuysen, et al., Selective adsorption of manganese onto cobalt for optimized $Mn/Co/TiO_2$ Fischer–Tropsch catalysts, J. Catal. 270 (2010) 95–102.
[100] B. Ernst, L. Hilaire, A. Kiennemann, Effects of highly dispersed ceria addition on reducibility, activity and hydrocarbon chain growth of a Co/SiO_2 Fischer–Tropsch catalyst, Catal. Today 50 (1999) 413–427.
[101] E.S. Lox, G.F. Froment, Kinetics of the Fischer-Tropsch reaction on a precipitated promoted iron catalyst. 2. Kinetic modeling, Ind. Eng. Chem. Res. 32 (1993) 71–82.
[102] G. Bub, M. Baerns, B. Büssemeier, C. Frohning, Prediction of the performance of catalytic fixed bed reactors for Fischer-Tropsch synthesis, Chem. Eng. Sci. 35 (1980) 348–355.
[103] D.B. Bukur, S.A. Patel, X. Lang, Fixed bed and slurry reactor studies of Fischer-Tropsch synthesis on precipitated iron catalyst, Appl. Catal. 61 (1990) 329–349.
[104] B.W. Wojciechowski, The kinetics of the Fischer-Tropsch synthesis, Catal. Rev. 30 (1988) 629–702.
[105] W.J. Shen, J.L. Zhou, B.J. Zhang, Kinetics of Fischer-Tropsch synthesis over precipitated iron catalyst, J. Nat. Gas Chem. 4 (1994) 385–400.
[106] S. Ledakowicz, H. Nettelhoff, R. Kokuun, W.D. Deckwer, Kinetics of the Fischer-Tropsch synthesis in the slurry phase on a potassium promoted iron catalyst, Ind. Eng. Chem. Process Des. Dev. 24 (1985) 1043–1049.
[107] G.P. van der Laan, A.A.C.M. Beenackers, Hydrocarbon selectivity model for the gas-solid Fischer – Tropsch synthesis on precipitated iron catalysts, Ind. Eng. Chem. Res. 38 (1999) 1277–1290.
[108] G.A. Huff Jr, C.N. Satterfield, Intrinsic kinetics of the Fischer-Tropsch synthesis on a reduced fused-magnetite catalyst, Ind. Eng. Chem. Process Des. Dev. 23 (1984) 696–705.
[109] H. Nettelhoff, R. Kokuun, S. Ledakowicz, W.D. Deckwer, Studies on the kinetics of Fischer-Tropsch synthesis in slurry phase, Ger. Chem. Eng. 8 (1985) 177–185.
[110] B. Sarup, B.W. Wojciechowski, Studies of the Fischer-Tropsch synthesis on a cobalt catalyst II. Kinetics of carbon monoxide conversion to methane and to higher hydrocarbons, Can. J. Chem. Eng. 67 (1989) 62–74.
[111] C.S. Kellner, A.T. Bell, The kinetics and mechanism of carbon monoxide hydrogenation over alumina-supported ruthenium, J. Catal. 70 (1981) 418–432.

[112] C.G. Takoudis, Power rate law studies in heterogeneously catalyzed reactions, Ind. Eng. Chem. Prod. Res. Dev. 23 (1984) 149–153.
[113] H.E. Atwood, C.O. Bennett, Kinetics of the Fischer-Tropsch reaction over iron, Ind. Eng. Chem. Process Des. Dev. 18 (1979) 163–170.
[114] W.D. Deckwer, R. Kokuun, E. Sanders, S. Ledakowicz, Kinetic studies of Fischer-Tropsch synthesis on suspended iron/potassium catalyst-rate inhibition by carbon dioxide and water, Ind. Eng. Chem. Process Des. Dev. 25 (1986) 643–649.
[115] R.S. Dixit, L.L. Tavlarides, Kinetics of the Fischer-Tropsch synthesis, Ind. Eng. Chem. Process Des. Dev. 22 (1983) 1–9.
[116] I.C. Yates, C.N. Satterfield, Intrinsic kinetics of the Fischer-Tropsch synthesis on a cobalt catalyst, Energy Fuels 5 (1991) 168–173.
[117] C.N. Satterfield, R.T. Hanlon, S.E. Tung, Z.M. Zou, G.C. Papaefthymiou, Effect of water on the iron-catalyzed Fischer-Tropsch synthesis, Ind. Eng. Chem. Prod. Res. Dev. 25 (1986) 407–414.
[118] M.A. Vannice, The catalytic synthesis of hydrocarbons from carbon monoxide and hydrogen, Catal. Rev. 14 (1976) 153–191.
[119] W. Zimmerman, D. Bukur, S. Ledakowicz, Kinetic model of Fischer–Tropsch synthesis selectivity in the slurry phase, Chem. Eng. Sci. 47 (1992) 2707–2712.
[120] Y.-N. Wang, W.-P. Ma, Y.-J. Lu, J. Yang, Y.-Y. Xu, H.-W. Xiang, et al., Kinetics modelling of Fischer–Tropsch synthesis over an industrial Fe–Cu–K catalyst, Fuel 82 (2003) 195–213.
[121] B. Sarup, B.W. Wojciechowski, Studies of the Fischer-Tropsch synthesis on a cobalt catalyst. III. Mechanistic formulation of the kinetics of selectivity for higher hydrocarbon formation, Can. J. Chem. Eng. 67 (1989) 620–627.
[122] J.L. Feimer, P.L. Silveston, R.R. Hudgins, Steady-state study of the Fischer-Tropsch reaction, Ind. Eng. Chem. Prod. Res. Dev. 20 (1981) 609–615.
[123] J. Na, K.S. Kshetrimayum, U. Lee, C. Han, Multi-objective optimization of microchannel reactor for Fischer-Tropsch synthesis using computational fluid dynamics and genetic algorithm, Chem. Eng. J. 313 (2017) 1521–1534.
[124] X. Wei, H. Wang, Z. Yin, S. Qaseem, J. Li, Tubular electrocatalytic membrane reactor for alcohol oxidation: CFD simulation and experiment, Chin. J. Chem. Eng. 25 (2017) 18–25.
[125] V.S. Ermolaev, K.O. Gryaznov, E.B. Mitberg, V.Z. Mordkovich, V.F. Tretyakov, Laboratory and pilot plant fixed-bed reactors for Fischer–Tropsch synthesis: mathematical modeling and experimental investigation, Chem. Eng. Sci. 138 (2015) 1–8.
[126] G. Chabot, R. Guilet, P. Cognet, C. Gourdon, A mathematical modeling of catalytic milli-fixed bed reactor for Fischer–Tropsch synthesis: influence of tube diameter on Fischer Tropsch selectivity and thermal behavior, Chem. Eng. Sci. 127 (2015) 72–83.
[127] Y. Sun, J. Wei, J.P. Zhang, G. Yang, Optimization using response surface methodology and kinetic study of Fischer–Tropsch synthesis using SiO_2 supported bimetallic Co–Ni catalyst, J. Nat. Gas Sci. Eng. 28 (2016) 173–183.
[128] D.G. Rethwisch, J.A. Dumesic, Adsorptive and catalytic properties of supported metal oxides: III. Water-gas shift over supported iron and zinc oxides, J. Catal. 101 (1986) 35–42.
[129] M.S. Challiwala, B.A. Wilhite, M.M. Ghouri, N.O. Elbashir, Multidimensional modeling of a microfibrous entrapped cobalt catalyst Fischer-Tropsch reactor bed, AIChE J. (2017) 1–9.
[130] N. Park, J.R. Kim, Y. Yoo, J. Lee, M.J. Park, Modeling of a pilot-scale fixed-bed reactor for iron-based Fischer–Tropsch synthesis: two-dimensional approach for optimal tube diameter, Fuel 122 (2014) 229–235.

[131] N. Moazami, M.L. Wyszynski, K. Rahbar, A. Tsolakis, H. Mahmoudi, A comprehensive study of kinetics mechanism of Fischer-Tropsch synthesis over cobalt-based catalyst, Chem. Eng. Sci. 171 (2017) 32–60.
[132] A. Jess, R. Popp, K. Hedden, Fischer–Tropsch-synthesis with nitrogen-rich syngas: fundamentals and reactor design aspects, Appl. Catal., A: Gen. 186 (1999) 321–342.
[133] Y.N. Wang, Y.Y. Xu, Y.W. Li, Y.L. Zhao, B.J. Zhang, Heterogeneous modeling for fixed-bed Fischer–Tropsch synthesis: reactor model and its applications, Chem. Eng. Sci. 58 (2003) 867–875.
[134] J.W.A. De Swart, R. Krishna, S.T. Sie, Selection, design and scale up of the Fischer-Tropsch reactor, in: M. de Pontes, R.L. Espinoza, C.P. Nicolaides, J.H. Scholtz, and M.S. Scurrell (Eds.), Nat. Gas Convers. IV, Elsevier, Amsterdam, 1997, pp. 213–218.
[135] F.A. Fernandes, S.J. Cartaxo, An object-oriented graphics interface simulator for analysis of Fischer-Tropsch synthesis, Chem. Prod. Process Model. 5 (2010) 9.
[136] K.S. Kshetrimayum, I. Jung, J. Na, S. Park, Y. Lee, S. Park, et al., CFD simulation of microchannel reactor block for Fischer–Tropsch synthesis: effect of coolant type and wall boiling condition on reactor temperature, Ind. Eng. Chem. Res. 55 (2016) 543–554.
[137] G. Gumuslu, A.K. Avci, Parametric analysis of Fischer-Tropsch synthesis in a catalytic microchannel reactor, AIChE J. 58 (2012) 227–235.
[138] J.A. Díaz, H. Akhavan, A. Romero, A.M. Garcia-Minguillan, R. Romero, A. Giroir-Fendler, et al., Cobalt and iron supported on carbon nanofibers as catalysts for Fischer–Tropsch synthesis, Fuel Process. Technol. 128 (2014) 417–424.
[139] S. Buller, J. Strunk, Nanostructure in energy conversion, J. Energy Chem. 25 (2016) 171–190.
[140] J. Lu, L. Yang, B. Xu, Q. Wu, D. Zhang, S. Yuan, et al., Promotion effects of nitrogen doping into carbon nanotubes on supported iron Fischer–Tropsch catalysts for lower olefins, ACS Catal. 4 (2014) 613–621.
[141] C.G. Visconti, E. Tronconi, G. Groppi, L. Lietti, M. Iovane, S. Rossini, et al., Monolithic catalysts with high thermal conductivity for the Fischer–Tropsch synthesis in tubular reactors, Chem. Eng. J. 171 (2011) 1294–1307.
[142] L.C. Almeida, O. Sanz, D. Merino, G. Arzamendi, L.M. Gandía, M. Montes, Kinetic analysis and microstructured reactors modeling for the Fischer–Tropsch synthesis over a Co–Re/Al_2O_3 catalyst, Catal. Today 215 (2013) 103–111.
[143] P. Serp, E. Castillejos, Catalysis in carbon nanotubes, ChemCatChem 2 (2010) 41–47.
[144] C. Pham-Huu, N. Keller, V.V. Roddatis, G. Mestl, R. Schlogl, M.J. Ledoux, Large scale synthesis of carbon nanofibers by catalytic decomposition of ethane on nickel nanoclusters decorating carbon nanotubes, Phys. Chem. Chem. Phys. 4 (2002) 514–521.
[145] E. Rytter, N.E. Tsakoumis, A. Holmen, On the selectivity to higher hydrocarbons in Co-based Fischer–Tropsch synthesis, Catal. Today 261 (2016) 3–16.
[146] V. Vosoughi, S. Badoga, A.K. Dalai, N. Abatzoglou, Modification of mesoporous alumina as a support for cobalt-based catalyst in Fischer-Tropsch synthesis, Fuel Process. Technol. 162 (2017) 55–65.
[147] A.P. Savost'yanov, R.E. Yakovenko, S.I. Sulima, V.G. Bakun, G.B. Narochnyi, V. M. Chernyshev, et al., The impact of Al_2O_3 promoter on an efficiency of C_{5+} hydrocarbons formation over Co/SiO_2 catalysts via Fischer-Tropsch synthesis, Catal. Today 279 (2017) 107–114.
[148] V.R.R. Pendyala, G. Jacobs, W. Ma, W.D. Shafer, D.E. Sparks, A. MacLennan, et al., Fischer-Tropsch synthesis: effect of carbonyl sulfide poison over a Pt promoted Co/alumina catalyst, Catal. Today 299 (2018) 14–19.

[149] E. Rytter, A. Holmen, On the support in cobalt Fischer–Tropsch synthesis—emphasis on alumina and aluminates, Catal. Today 275 (2016) 11–19.
[150] M. Mansouri, H. Atashi, A.A. Mirzaei, R. Jangi, Kinetics of the Fischer-Tropsch synthesis on silica-supported cobalt-cerium catalyst, Int. J. Ind. Chem. 4 (2013) 1–10.
[151] M.K. Gnanamani, G. Jacobs, V.R.R. Pendyala, U.M. Graham, S.D. Hopps, G.A. Thomas, et al., Fischer-Tropsch synthesis: anchoring of cobalt particles in phosphorus modified cobalt/silica catalysts, Appl. Catal., A: Gen. 523 (2016) 146–158.
[152] K. Shimura, T. Miyazawa, T. Hanaoka, S. Hirata, Fischer–Tropsch synthesis over alumina supported bimetallic Co–Ni catalyst: effect of impregnation sequence and solution, J. Mol. Catal. A: Chem. 407 (2015) 15–24.
[153] Z. Yu, Ø. Borg, D. Chen, B.C. Enger, V. Frøseth, E. Rytter, et al., Carbon nanofiber supported cobalt catalysts for Fischer-Tropsch synthesis with high activity and selectivity, Catal. Lett. 109 (2006) 43–47.
[154] Z. Yu, Ø. Borg, D. Chen, E. Rytter, A. Holmen, Role of surface oxygen in the preparation and deactivation of carbon nanofiber supported cobalt Fischer-Tropsch catalysts, Top. Catal. 45 (2007) 69–74.
[155] G.L. Bezemer, P.B. Radstake, U. Falke, H. Oosterbeek, H.P.C.E. Kuipers, A.J. van Dillen, et al., Investigation of promoter effects of manganese oxide on carbon nanofiber-supported cobalt catalysts for Fischer–Tropsch synthesis, J. Catal. 237 (2006) 152–161.
[156] G.L. Bezemer, U. Falke, A.J. van Dillen, K.P. de Jong, Cobalt on carbon nanofiber catalysts: auspicious system for study of manganese promotion in Fischer-Tropsch catalysis, Chem. Commun. (2005) 731–733.
[157] S.H. Kang, H.M. Koo, A.R. Kim, D.H. Lee, J.H. Ryu, Y.D. Yoo, et al., Correlation of the amount of carbonaceous species with catalytic performance on iron-based Fischer–Tropsch catalysts, Fuel Process. Technol. 109 (2013) 141–149.
[158] M.C. Bahome, L.L. Jewell, D. Hildebrandt, D. Glasser, N.J. Coville, Fischer–Tropsch synthesis over iron catalysts supported on carbon nanotubes, Appl. Catal., A: Gen. 287 (2005) 60–67.
[159] M.C. Bahome, L.L. Jewell, K. Padayachy, D. Hildebrandt, D. Glasser, A.K. Datye, et al., Fe-Ru small particle bimetallic catalysts supported on carbon nanotubes for use in Fischer–Tröpsch synthesis, Appl. Catal., A: Gen. 328 (2007) 243–251.
[160] J.L. Tu, M.Y. Ding, Q. Zhang, Y.L. Zhang, C.G. Wang, T.J. Wang, et al., Design of carbon-encapsulated Fe_3O_4 nanocatalyst with enhanced performance for Fischer–Tropsch synthesis, ChemCatChem 7 (2015) 2323–2327.
[161] G.L. Bezemer, A. van Laak, A.J. van Dillen, K.P. de Jong, Cobalt supported on carbon nanofibers—a promising novel Fischer-Tropsch catalyst, in: X. Bao, and Y. Xu (Eds.), Nat. Gas Convers. VII, Elsevier, Amsterdam, 2004, pp. 259–264.

CHAPTER 8

Constraints, impacts and benefits of lignocellulose conversion pathways to liquid biofuels and biochemicals

Amalia Zucaro[1], Gabriella Fiorentino[1] and Sergio Ulgiati[1,2]
[1]Department of Science and Technology, Parthenope University of Naples, Centro Direzionale – Isola C4, Naples, Italy
[2]School of Environment, Beijing Normal University, Beijing, P.R. China

Contents

8.1 Introduction

The urgent need to create new conversion pathways to reduce the dependency on fossil resources, due to the increasing environmental concerns related to the large use of fossil-based fuels and carriers in the industrial and transport sectors worldwide, call for effective actions to identify

Lignocellulosic Biomass to Liquid Biofuels
DOI: https://doi.org/10.1016/B978-0-12-815936-1.00008-3

innovative and more sustainable bio-based supply chains. Over the last decades, within the transition to a decarbonized energy system, bioenergy production increased at a remarkable extent and is still expected to grow, in particular, in the form of biofuels, such as bioethanol (EtOH) and biodiesel, which are projected to boost from the present 2% of market share up to 27% in 2050 [1,2]. In particular, bioethanol production has been strongly endorsed, United States and Brazil being the leader producing countries with 14.8 billion gallons of EtOH produced only in the United States in 2015 [3,4]. The European Union (EU) energy policies [5] encourage the replacement of fuels and electricity from fossil sources with bio-based alternatives. A substitution of 50 billion liters of fossil fuels in the EU transportation sector is promoted. Furthermore, based on the national energy action plans of the member states, the expectation is that in 2020, 56% of the renewable energy will be produced from biomass [6]. The contribution of biomass to the global primary energy consumption amounted to 58.5 EJ in 2014 [7], representing by far the largest renewable energy use. The International Renewable Energy Agency estimates an almost doubled amount of biomass to be used by 2030, namely, 108 EJ/year to meet the Sustainable Energy for all (SE4All) target [8]. The Global Energy Assessment predicts an extensive use of lignocellulosic material and agricultural residues leading to a significant growth in bioenergy, up to 80–140 EJ by 2050 [9]. Likewise, the Intergovernmental Panel on Climate Change outlined a comparable increase in bioenergy share, depending on climate ambitions and chosen policy instruments [10]. More recently, the EU issued a new directive [11] to encourage the use of energy from renewable sources, highlighting that fuels produced from biomass need to comply with strict environmental requirements. From 2021 the greenhouse gas (GHG) emissions from the use of biomass-derived fuels have to be at least 70% less than their fossil-based counterparts. In the cases of biomass-based electricity, heating and cooling the threshold for the reduction of GHG emissions is increased up to 80%. In general, bio-based products, such as biofuels, biocomposites, and biochemicals, will follow voluntary standards on sustainability according to the international association of Roundtable on Sustainable Biomaterials [12].

Since the biomass conversion into biofuels has been proven to be technically feasible and considering that the economic convenience depends on strategic and market factors, that are not easily predictable (political equilibrium, fossil fuel price, diffusion of other energy sources, competition with electric vehicles), in this chapter, special focus will be placed on the investigation of the physical and environmental constraints for the large-scale implementation of lignocellulosic biofuels chains. The aim is to explore the potential improvements, in order to remove the existing infrastructural, energy, and environmental barriers. This may pave the way to the penetration of lignocellulosic biofuels into the energy market as a complement to other renewable and nonrenewable options.

8.2 Constraints placed by logistics of biomass production, transport, and processing

The dynamics of the use of biomass as renewable source of energy have evolved in the last few years. In a first stage the feedstocks generally used for EtOH production were sugar and starch crops, such as corn, maize, wheat, sugar beet, cassava, sweet sorghum, sugar cane, among others [13], whereas biodiesel was traditionally produced from edible vegetable oils (e.g., soybean and palm oil). These so-called first generation (1G) crops have grown worldwide mainly driven by dedicated subsidization policies supporting the production and use of transport biofuels [14]. Nevertheless, 1G energy carriers have been charged with the rise of the competition between food and fuel production. Moreover, the energy return on energy investment (EROEI) in biofuels results actually low, as a consequence of both the relatively small yields per hectare of dedicated crops and the high energy investment required in the whole supply chain [15,16]. Further integrated assessments on energy efficiency and societal metabolism highlighted the limitations of 1G biofuels as a possible sustainable energy carrier for industrial societies [17]. As a consequence, it seems that the proliferation of subsidy policies even worsened the global picture, as they incentivized the increased cropping of a single culture, thus promoting a major agroecological risk [18]. In this perspective the employment of second generation (2G) and third generation (3G) feedstocks is toughly encouraged, in order to reduce the land use change (LUC), to ensure the food security and to offset the GHGs emissions, in the framework of the EU and worldwide policies [5,19]. 2G feedstocks, recently object of growing interest, consist of nonfood sources, such as inedible oilseed crops and lignocellulosic biomass, including (1) agricultural waste, such as straw, corn stover, corn cobs, bagasse, and molasses; (2) forestry residues; (3) fraction of municipal and industrial waste; and (4) dedicated energy crops (e.g., giant reed and cardoon). Moreover, waste or recycled oil as well as animal fats have recently gained significant attention for the production of 2G biodiesel [20].

There are two main conversion pathways for biomass feedstocks, namely, thermochemical and biochemical (including both biological and chemical-conversion processes).

The thermochemical conversions include four alternative processes: (1) combustion to produce heat and/or electricity, (2) pyrolysis to obtain bio-oil (as main output), (3) gasification to generate synthesis gas (syngas), and (4) high-pressure liquefaction [21].

The biochemical conversions of lignocellulosic biomass into bioethanol, through hydrolysis and fermentation, or into biodiesel, through oleaginous microorganisms [22,23], seem very promising, both in qualitative and quantitative terms. From a qualitative point of view the nonedible nature of this kind of biomass allows to overcome concerns about food prices and quantity as when starchy feedstocks from edible food crops are utilized. Quantitatively, the present abundant availability at low price is considered a strength [24]. Agricultural biomass residues included in the waste streams from commercial crop processing plants represent an abundant, inexpensive, and readily available source of renewable energy. In addition, their conversion into biofuels contributes to solve a disposal problem. Although the large potential of cellulosic biomass as bio-based feedstock, several practical hurdles in the efficient processing and handling of biomass are still encountered at each stage of the implementation of a bio-based value chain, that require to be properly addressed for a large-scale and cost-effective production of cellulosic biofuels. Such a production involves a series of different steps: first, biomass has to be grown and harvested (agricultural phase), then transported to a conversion plant (biorefinery), where it is converted to bioenergy and bioproducts, and finally end products have to be delivered to distribution centers or markets.

8.2.1 Constraints placed by biomass and land availability

The main criticalities involved in the agricultural phase concern the production of the required biomass and, as a tight consequence, the availability of a sufficient amount of land. Major points of concern are therefore (1) the amount, cost, and availability of biomass needed for the implementation of a biorefinery producing bio-based fuels and materials and (2) the area of agricultural land or forest needed to produce such a biomass.

The amount of biomass needed to sustain bio-based fuel-production pathways, within the EU and worldwide, has been carefully estimated [25]. These evaluations, however, depend on the assumed conversion efficiency and are often reported in different units, thus hampering a direct comparison of data. For instance the amount of biomass energy required to meet the targets of the EU energy strategy was estimated to increase from 3.8 EJ in 2005 to 10.0 EJ in 2020 by Bentsen and Felby [25], whereas Bos and Sanders [26] calculated a biomass requirement of 700 Mt/year in Europe, in accordance with EU policy then in force [5], achievable without significant changes in the agricultural land use [27]. In the analyses from these authors the main part of the increasing biomass demand to supply energy is addressed by means of dedicated energy crops, namely, crops grown dedicatedly on liberated agricultural lands or marginal lands. Of course, specific cases of energy cropping need to be carefully evaluated concerning their water, nutrients, and labor demand, according to the site specific pedo-climatic conditions. An additional contribution is expected to come from agricultural and forestry residues, with a technical potential estimated around 30 and 35 EJ/year, respectively [28]. Consistently, Kajaste [29] concluded that biomass production (including lignocellulosic matter plus algae and microalgae) conservatively overcomes the world sugar production, ensuring the long-term sustainability of available feedstocks.

Of course, biomass production may be even increased through further mobilization of existing resources, intensification of current production or expansion into "new" land. Nonetheless, an expanded production may even generate nonnegligible environmental and social impacts, so that the alternative between a sustainable use of biomass and an extensive use of land likely to cause irreversible environmental and social impacts needs to be thoroughly defined. The supply chain of biomass feedstock requires the use of energy, pesticides, and fertilizers and, above all, it may have significant implications on land availability in general, affecting land use, changes in land use and competition in land use resources, and the related GHGs dynamics. When switching from one cropping system to another (for instance, from current use of land as forestry to the cultivation of biomass as feedstock for food, feed, bioenergy or bio-based materials), a direct LUC is implied, and it may affect the state of agricultural lands, in terms of biogenic carbon emissions, carbon loss from soils, soil erosion, nutrient depletion, water consumption, and loss of biodiversity. By contrast an indirect LUC (iLUC) is an unintentional LUC that is induced by the production of biomass feedstock, even if occurring outside the bio-based material's feedstock production area (for instance, when land used for food and feed production becomes rededicated to biomass production for bioenergy, then additional land is going to be used elsewhere for food and feed production). As a consequence, iLUCs may imply the unintended impact of exacerbating GHGs emissions through land intensification, thus offsetting one of the potential benefits of using bioenergy. In addition to the potential negative effects on climate change from losses/emissions of terrestrial carbon, another issue to ponder is the land use competition between food, feed, and fuel, distrusting that bioenergy developments can occur without a reduction of food availability [30]. Therefore growing energy crops on surplus or poor quality agricultural land can be potentially regarded as an advantageous solution to go beyond the land use competition. Soils characterized by poor physical and chemical properties, aridity, and susceptibility to erosion may result unsuitable for uses other than bioenergy, although their suitability for the latter option also needs to be carefully checked. These lands are often referred to as marginal or degraded lands. In general, marginal lands can be defined as *lands that are not suitable for food-based agriculture and have limited economic potential for fulfilling other ecosystem services* [31]. The use of such lands as a base for feedstock production is expected to entail a beneficial impact in comparison to the use of primary land, that could rather be used for food production. The bioenergy production potential on marginal and degraded lands has been widely evaluated [32]. However, threats to biodiversity and conservation areas as well as the lack of economic incentives, low biomass productivity, and high production costs are currently delaying improvements of biomass production on marginal lands [33]. Furthermore, some estimates of "eligible areas" for cultivation of energy crops, not overlapping with arable lands, irrigated crops, protected area (e.g., parks and reserves) and readily available for LUC, are not presently encouraging. Therefore the actual availability of larger areas to be converted into lands, supporting the possible longer term penetration targets of biofuels and biomaterials supply chains, is a recognized critical issue [34].

On the other hand, if agricultural residues are supposed to be used as feedstock for biofuels production instead of dedicated crops, the role of protecting soils from the erosive forces of wind and rain and of providing the building blocks for soil organic matter is also endangered. In fact the removal of crop residues from soil surface can cause both direct and indirect adverse impacts, over short and long terms. The amount of residues that can be sustainably removed has to be determined by means of an integrated approach, taking into consideration multiple agronomic and environmental factors simultaneously, new technologies and approaches to soil conservation, crop rotation, and crop management [35].

8.2.2 Constraints placed by transport: feedstock biomass transportation and biofuel distribution

The switching to supply chains based on 2G lignocellulosic feedstocks calls for logistic changes in feedstock biomass transportation, biofuel distribution, and input flows over the supply chains. Once biomass has been selected and produced for bioenergy and biomaterials purposes, logistic issues such as collection, storage, and transport of cellulosic biomass to the biorefinery (feedstock conversion facility) are of critical relevance (mainly due to its low density, large volumes, large water content), since logistic costs were estimated to account for 35% or more of the total production costs of cellulosic biofuels [36]. Lignocellulosic biomass is spatially distributed and its collection is limited to certain times of the year, whereas biomass supply should necessarily be continuous throughout the whole year to guarantee a suitable production system. If focus is placed only on the production of feedstock, the costs associated with collecting and transporting cellulosic biomass can represent up to 50% of the total feedstock costs, in economic terms [36]. From an environmental point of view a difference has to be made when the lignocellulosic feedstocks are provided to biofuel supply—use chain on a regional scale (covering distances less than 100 km) or when they serve large-scale supply and distribution systems (i.e., national or transnational). In the first case the transport phase, including both feedstock transport to facility plant and bioethanol transport to service station (bioethanol distribution), raises not more than 10% of the total environmental impacts [37]. In the second case of a wider scale the transport required for collection, storage, and distribution of lignocellulosic feedstocks and processed biofuels, is supposed to generate increased impacts proportionally to longer distances traveled. Therefore to reduce biomass transport cost, optimal biorefinery locations should be within the regions of higher biomass yield density. Since there are several variables that impact biomass transport cost and feedstock availability, strategic decisions on specific location should be based on geospatial tools [38].

8.2.3 Constraints placed by water consumption in the bio-based supply chain

The World Economic Forum has considered the water crisis as one of the largest global risks [39]. The need to investigate the water-energy nexus has become of paramount importance in order to achieve sustainable bio-based production processes. The overall environmental impact could be substantially influenced by the increased water demand, specifically in areas that are already water stressed. Freshwater is essential for biomass production and worldwide about 70% is used within the agricultural sector [40]. Most crops require heavy-duty amount of water during the growing season, with a ratio that in average is above a ton of water per kg of produced crop [41]. Moreover, the agricultural production contributes to the highest risk of water pollution, due to the excess use of nutrients, pesticides and other pollutants [40]. In the biofuels production chains, freshwater is needed not only in the biomass production phase, but also in the downstream processing of biomass, when feedstock is converted into bio-based products. A large impact on the demand for water resources has been ascertained for 1G biofuels, especially due to the requested maximization of agricultural biomass production yield [42]. The alternative use of 2G biomass may apparently solve the problem, but the additional amount of water requested in the feedstock conversion facility plant (biorefinery) has to be accounted for. Therefore a proper evaluation of the efficiency, in terms of output per unit of natural resource use (e.g., water use), is mandatory. Moreover, water consumption is affected by both direct use (water for irrigation or makeup water in the biorefinery) and indirect use from upstream water inputs (such as water for the production of fertilizers and machineries). Several studies were carried out on the water footprint of 2G bioethanol, highlighting smaller water consumption rates in comparison with 1G biofuel. In particular the employment of crop residues instead of dedicated lignocellulosic crops (like *Miscanthus* or woody crops) was confirmed to be relatively more water-efficient [42]. In addition to the chosen biomass (dedicated energy crops vs residual biomass) the total water footprint of bio-based production chains also depends on the conversion pathway selected for processing the biomass feedstocks (thermochemical vs biochemical conversion). In fact, available 2G conversion techniques for biofuels production are differently efficient in terms of water demand per unit of bio-based energy carrier generated [42,43]. Lower water usage may be possible by means of new technologies, which include the use of recycled wastewater. In addition a scrupulous water footprint cannot disregard that an area can be characterized by water abundance, but there may also be an area where water is scarce and where minimum environmental flow requirements are not met because of overdraft. The collection and evaluation of site specific characteristics are required in order to achieve a water sustainability of the whole bio-based production chain.

8.2.4 Constraints placed by net energy rate

The impact of the biomass production system on net energy performance needs to be considered on both local and global scale. The local scale evaluates all input and output flows related to the investigated process and further optimization techniques. The global scale accounts for the compatibility of the production system with the ecosystem in which the bio-based production takes place, in terms of environmental performance. As discussed in Sections 8.2.1 and 8.2.3, the performance of biofuel systems needs to be explored considering the biomass and the marginal land availability, the fresh water requirements, but also considering the net energy actually delivered.

Gross energy production has to be distinguished from net energy production. In the first case, only the total output of energy (namely, the energy content of produced biofuel) is considered. In the second case, all the energy inputs (energy costs of biomass production, collection, and conversion) are subtracted from the gross value [44]. Another parameter that is most often used to assess the performance of biofuel system is the "EROEI." The latter represents the capacity of biofuel to deliver more useful energy than the amount required for its production (energy output/energy input) [45]. The EROEI is considered a suitable energy-efficiency index for the large-scale biofuel supply chain, since it represents the ratio between the energy content of the fuel (measured in MJ/kg or as oil equivalent/kg) and the nonrenewable primary energy consumed to produce 1 kg of that fuel [46]. This indicator mainly depends on the selected 2G conversion facility and on the type of feedstock. For each lignocellulosic biomass cultivation, the performance of the whole production chain is highly influenced by the amount of fossil fuel required, considering its variability in composition (e.g., electricity mix) and the use of both direct energy (e.g., for machinery used in the harvest operations or for irrigation systems) and indirect energy (e.g., for the production of fertilizers and pesticides).

As highlighted by Giampietro and Ulgiati [47], the biomass production does generate a net energy in most cases, whilst the feedstock conversion into biofuel erodes a large fraction of the initial energy saving. According to these authors, biomass is storing solar energy at the rate determined by ecological processes and thermodynamics, and it will likely be unable to supply energy flows above this natural rate. They point out the difficult management and feasibility of the whole biofuels production chain in that the energy efficiency of biomass-production ranges approximately from 2 to 4, whereas the energy efficiencies of biofuel conversion processes drop to values very close to 1 (and lower) [47]. Increasing the energy efficiency of the production step, that is, higher biomass yield per hectare, with reference to 2G pathways, might provide a nonnegligible improvement. A recent review on the environmental profitability of bioethanol supply chain [46] pointed out that the energy ratios are higher for lignocellulosic bioethanol than for 1G bioethanol, thanks to the possible combustion of lignin residues in a cogeneration unit system (combined heating and power plant, CHP) that generates process heat and electricity on-site, thus avoiding the withdrawal of fossil-based energy from the grid. Consequently, cogeneration based on the use of by-products is a key issue in order to achieve the sustainability of bioethanol supply chains. If the lignocellulosic feedstock conversion plant is not self-sufficient from the energetic point of view, a too low output/input energy ratio is unavoidable for the biofuel system.

The abovementioned limits and related considerations lead to the necessity of a thorough evaluation of a large-scale prospective production of biofuels and bio-based materials as an environmental friendly solution for world energy security. Biofuel production chains requires a comprehensive performance monitoring by means of suitable evaluation tools.

8.3 Life cycle assessment as environmental evaluation tool

Innovative environmental strategies and suitable alternatives to fossil energy are increasingly required, with the aim to address the needs and expectations of the present and future generations in a sustainable way. Such challenge concerns the management of a complex agroindustrial and social system. For this reason it is important to adopt a system-wide assessment framework capable to provide a reliable, comprehensive, and quantitative evaluation of the environmental sustainability of the investigated supply chains.

The life cycle assessment (LCA) tool, widely used worldwide for industrial processes evaluation, can be suggested for such a goal. A system perspective is the core of the life cycle approach to assess the environmental aspects and potential impacts associated with a product or process throughout the whole lifecycle, from raw material acquisition, through production and use stages, up to the disposal phase (cradle to grave or cradle to cradle approach). The LCA is a standardized internationally accepted methodology providing qualitative, quantitative, confirmable, and manageable information on the environmental performance of the analyzed system, as defined by ISO standards and ILCD Handbook guidelines [48–51]. According to the standard procedures [48,49], the LCA stages include goal and scope definition, life cycle inventory (LCI), life cycle impact assessment (LCIA), and life cycle interpretation of results, as illustrated in Fig. 8.1.

The aims of the goal and scope definition are (1) defining objective and limits of the investigated system, including information on the quality of data, method issues and value choices; (2) describing the product/process to be analyzed; (3) establishing the context in which the assessment is made; (4) identifying both the functional unit (reference to which all other data in the investigated system are normalized) and the system

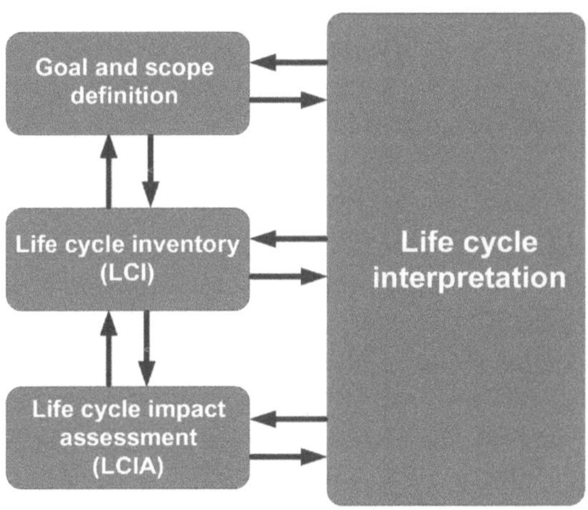

Figure 8.1 Framework for Life cycle assessment (LCA). *Adapted from European Commission, International Reference Life Cycle Data System (ILCD) Handbook – General Guide for Life Cycle Assessment – Detailed Guidance, 2010 [50].*

boundaries (physical, geographical, and time horizon delimitations); and (5) recognizing the environmental effects to be reviewed for the assessment.

Regarding the system boundaries, a cradle to grave LCA considers all the steps of the life cycle of a product, from the raw material extraction, through the transportation, manufacturing and use, up to the product's disposal. A cradle to gate LCA, instead, is limited at the production stage, whereas a cradle-to-cradle LCA also takes into account reuse, recovery, or recycle of product or its parts.

The second LCA stage is the LCI, in which all data are quantified (specifying their units) for each investigated phase of the selected system boundary (e.g., biofuel production chain), namely, inputs (e.g., energy, water, and materials usage), outputs (e.g., products and coproducts), and environmental releases (e.g., air emissions, solid waste disposal, and wastewater discharges). The main criticality of this stage is the availability of data, both primary (data related to the foreground system) and secondary (data from background system). The reliability of LCI depends on the amount of primary data (site specific data) and the accuracy of databases used for secondary data (updated and geographically contextualized).

The LCIA (third phase) is the technical process through which the potential human and ecological effects of energy, water and material usage, and the environmental local releases are assessed. All the foreground and background data collected during the inventory analysis are assigned to an impact category, and the number and type of estimated environmental impacts depend on the impact assessment method chosen. The LCIA methods include specific impact factors for each substance. These factors are calculated through established formulas that convert inventory data into a common unit defined for each impact category, for example, "CO_2 equivalent" for climate change or "SO_2 equivalent" for terrestrial acidification (TA). The potential environmental impacts reflecting the pressure on environment and human health as well as the potential or actual resource scarcity are computed based on the selected functional unit and are calculated by means of free or commercial software (e.g., SimaPro, Gabi, and Open LCA). Instead, the LCIA methods can be distinguished in endpoint and midpoint. Methods linked to the cause—effect chain (environmental mechanism) of an impact category are defined as midpoint [52]. Common examples of midpoint impact categories are climate change, ozone depletion, TA, freshwater eutrophication (FE), marine eutrophication (ME), photochemical oxidant formation, particulate matter formation (PMF), water depletion (WD), fossil depletion, human toxicity, terrestrial ecotoxicity, land use, among others. The endpoint methods are focused on the damage. They quantify the effects of the emissions on the objective to be protected: ecosystem diversity (in terms of potentially disappeared fractions of species), human health (in terms of disability adjusted life years), and resource availability (resource depletion to be preserved for future generations [52].

The final stage, once the inventory has been completed and the impacts have been quantified, is the Interpretation, in which all results related to the preferred product or process system are evaluated with a clear understanding of the uncertainty and the assumptions used to generate the results. The interpretation, according to the first LCA phase, quantifies the bottlenecks and identifies input flows, processes, and stages which mostly affect a given impact category or the total environmental performance to gain conclusions and recommendations.

Performing an LCA is an iterative process: once the goal of the study is defined, the initial scope settings, that define the requirements on the subsequent work, are derived. However, as more information becomes available during the LCI step of data collection and the subsequent impact assessment and interpretation steps, the initial goal and scope setting will typically need to be refined and sometimes also revised. The double arrows between phases in Fig. 8.1 indicate the typical interactive nature of LCA, which contributes to the comprehensiveness and consistency of the study and the reported results. These main characteristics of the LCA tool can be recognized in which there is not a single way to make a LCA and also that the procedure can be applied at different in-depth levels. Irrespective of the chosen level of sophistication, there are basic requirements to be addressed in an LCA, that is, clear and explicit statement of study purpose and goal and reference to the methodology used (e.g., definition of functional units, system boundaries, and allocation criteria). These requirements can be summarized in the LCA report, for reliability and replicability purposes.

Therefore the LCA approach differs from other environmental accounting methods since it analyzes the whole life cycle under the perspective of multiple impact dimensions, instead of focusing only on one single stage of the process or impact category. The environmental burden is actually associated to the very step where it belongs, thus allowing to identify the main hotspots where improvements are needed for a better performance of the overall process or product.

When applying LCA to bio-based production chains, an important issue to be addressed is the possibility to avoid the allocation procedure (namely, assigning input flows and environmental burdens to more than one product output, based on physical or economic criteria). The problem occurs when the production process generates multiple marketable coproducts (e.g., biofuels, electricity and bio-based chemicals/materials are coproduced within a biorefinery system; for details see Section 8.5). In this case the inputs and the environmental impacts from the production process have to be allocated among the various products (allocation LCA procedure). ISO 14040 [48] recommends avoiding allocation, wherever possible, through the system expansion adding other related product systems (consequential LCA procedure) or by dividing the unit process into two or more subprocesses. The attributional LCA procedure is considered as a valuable method to assess the emissions and resource use associated with different product systems and it might be considered a suitable tool to implement individual consumption choices.

The consequential LCA procedure needs to go beyond product systems, and for this reason, it is considered more appropriate to support robust choices in policy making. Nevertheless, when applying this procedure, weakness problems can occur due to the absence of an adequate standardization procedure [53]. Therefore the main challenge of an LCA, dealing with a bio-based system, is to overcome the dichotomy between the attributional and consequential approach, with a clear statement on the LCA procedure chosen by the analyst. Moreover, when it is possible, the LCA results should be presented together with the results obtained from other models that analyze, for example, the rebound effect or revenue recycling.

The following Section will report examples of the application of the LCA procedure to the production and use chain of 2G lignocellulosic biofuels, demonstrating the validity of the LCA methodology as an evaluation tool for the environmental sustainability of bio-based systems.

8.4 Environmental assessment of biofuels: a selection of life cycle assessment studies

Biofuels are generally considered synonymous with "green" and "environmental friendly." Actually, this is not necessarily true. As highlighted by pertinent scientific literature, biofuels entail tradeoffs among positive and negative environmental effects [43,46,54]. Fuels produced from renewable resources are acknowledged to show an improved performance in terms of reduction of GHG emissions as well as of depletion of fossil sources [11], but at the same time, they can produce worsened impacts on soil, forests, and natural resources [18]. Most of the studies based on the LCA approach highlighted large differences in the environmental performance not only of different biofuels (bioethanol or biodiesel, for example), but also of the same biofuel produced in different ways (for instance, bioethanol produced from different lignocellulosic feedstocks or by using different 2G conversion technologies) (Fig. 8.2).

In the context of 2G biofuel production the most widely used feedstock is the lignocellulosic material that represents the majority of the cheap and abundant nonfood materials, readily available, through (1) the cultivation of dedicated energy crops on marginal lands like short and medium rotation forestry crops (e.g., poplar, willow, and eucalyptus), perennial grasses (e.g., *Miscanthus*, switchgrass, giant reed, and canary grass), and annual grasses (e.g., sorghum and Napier grass); or (2) the use of residues from agriculture, forestry, and wood industry. Advanced biofuels can be obtained from lignocellulosic materials either through hydrolysis and fermentation (i.e., bioethanol) or through gasification (i.e., Fischer–Tropsch biodiesel, biodimethyl ether, and biosynthetic natural gas) [55].

Regardless of the relevant expectations associated with 2G advanced biofuels, in the EU, the developed technologies have not reached yet the commercial maturity for a large-scale production, whereas in the United States, the biofuels production and consumption are more extensively established, reaching 51% and 52% of the world production and consumption, respectively, in terms of thousand barrels per day [56]. From now on the focus will be placed on the environmental assessment of lignocellulosic biofuels, obtained by means of biochemical conversion processes, due to the more extensive literature on LCA evaluation of these conversion routes [3,46]. In this context the aim of this study is to identify the benefits and the weaknesses of biofuel production/use chain in a circular economy (CE) perspective, so that specific efforts can be placed to minimize the generated environmental impacts shifting the observed constraints into added values to the whole bio-based chain.

To the best of our knowledge, LCA studies on biodiesel production from lignocellulosic materials have never been addressed. Contrariwise, the environmental performance of lignocellulosic 2G bioethanol fuel has been widely investigated. The LCA studies, referring to the bioethanol supply chain, were recently reviewed by Morales et al. [46] and the

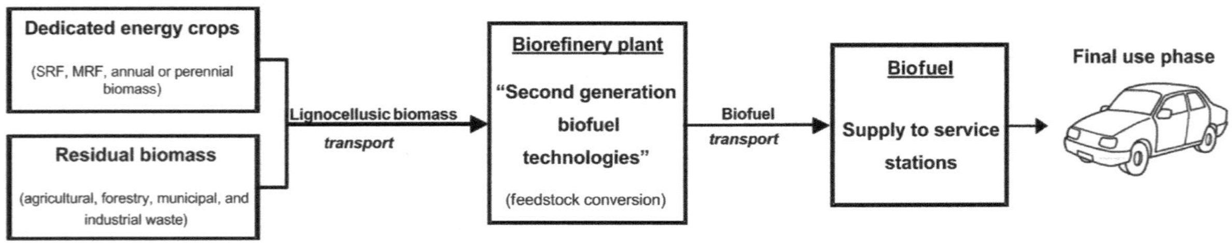

Figure 8.2 Schematic flowchart of the investigated biofuel supply-use chain. *MRF*, Medium rotation forestry; *SRF*, short rotation forestry.

generated impacts resulted to be highly influenced by two factors: (1) the source of raw material (different dedicated crops or residual biomass) and (2) the feedstock conversion technologies used at industrial scale. Moreover, there are several methodological issues to overcome in order to address the environmental profitability of the production of bioethanol and its use as a fuel. The first issue is the lack of a detailed description of the direct field emissions (DFE) and how to calculate them. Many studies provide almost no information on the methods and data used for estimating field emissions [57]. Although the lignocellulosic feedstock production is identified as an hotspot in bio-based supply/use chains [37,57–59], site specific data on climate and field conditions are still missing and the direct filed emissions of agricultural phase can be hardly assessed. Special attention should be placed on the appropriate calculation of the DFE related to (1) the nitrogen fertilizer application, like volatilized ammonia (NH_3), nitrogen oxides (NO_x), biogenic dinitrogen monoxide (N_2O), nitrate leaching (NO_3), fossil carbon dioxide (CO_2); (2) the application of phosphorous fertilizers, like soluble phosphate to groundwater and surface and the particulate phosphorus emitted to river; (3) the application of pesticides; and (4) the use of agricultural machinery (heavy metals and tailpipe emissions).

Moreover, the issue of the soil carbon storage (SCS) dynamics is still debated in many LCA studies, due to the spatial variability and time-horizon definition [34]. Therefore a proper harmonization and standardization of soil carbon dynamics and their inclusion in GHGs inventory should be considered as a prerequisite for the environmental studies on perennial lignocellulosic feedstocks [60]. As a consequence of the lack of DFE or miscalculations of SCS, the bioethanol LCA results obtained for some impact categories, like climate change, eutrophication potential, acidification potential, human and ecosystem toxicity, and particular matter formation, are often not comparable. In addition the choice of the LCIA method appears to be not always consistent among the bioethanol LCA studies. The LCIA methods account for different impact categories with different in-depth examination models (Tier 1, Tier 2, Tier 3, [61]) and consider different units of measurement. Other different accounting procedures in works based on LCA are (1) the lack of standardized functional unit (e.g., kilogram of produced EtOH, MJ of biofuel energy content as well as kilometer traveled), (2) the request of primary data or clear assumption on yield pattern along the crop lifespan for perennial lignocellulosic crops, (3) the different choice related to the selected system boundary (e.g., from cradle to plant gate or from cradle to wheel system), (4) the variation in the choice of allocation methods and the consequent issue of attributional versus consequential approach (see Section 8.2 for further details), and (5) the effect of biomass removal from soils (if lignocellulosic agricultural residues are used). This wide range of variables is further worsened by a large use of lab experiments that are generally difficult to be modeled into real industrial situations. This calls for a wider standardization of the application of LCA to bio-based systems, as previously done for other processes and products [62].

The main question to be answered is "Can the lignocellulosic bioethanol option be addressed as a sustainable logistic and technical alternative to decarbonize the transport sector?" The lignocellulosic bioethanol can be used in different blending without engineers' modification in E10 vehicles (10% of EtOH and 90% of gasoline). A mixture of 85% of bioethanol and 15% gasoline is possible in the so-called flexi-fuel vehicles (E85). Instead, specifically designed engines are required for E100 vehicles (100% bioethanol). The environmental advantages and disadvantages of using bioethanol-gasoline blends as alternative to fossil-based fuels have been analyzed by numerous LCA studies [46]. If on the one hand, the bioethanol chain shows GHGs mitigation and fossil energy savings, impacts on eutrophication and acidification categories are worsened [37,43,46,63,64]. In deeper details the LCA of E10 and E85 passenger cars entailed, in comparison with gasoline, reduced GHG emissions by about 5%–10% and 30–60%, respectively, due to the biogenic origin of carbon [64]. Focusing on the environmental profile of E85 flexi-fuel vehicles (given their better environmental performance compared to bioethanol-gasoline mixtures with lower ethanol percentage), fossil energy savings range from 30% up to 70% [43,46]. The inclusion for lignocellulosic perennial crops of the SCS into the GHG inventory would turn the crop phase from a net emitter into a net sink [37,46]. Moreover, for the E85 bioethanol mixture, the production and conversion of lignocellulosic feedstock resulted to highly affect the following impact categories: TA, FE, ME, PMF, and WD. In particular, for TA, ME, and PMF impacts, a key role was played by the DFE linked to the volatilized ammonia in the lignocellulosic biomass production [43,63]. For the FE impact, the feedstock cultivation and the conversion facility stages resulted the main hotspots, whilst the feedstock conversion plant often was a prevailing contributor for WD impact category, due to the large direct water demand in the industrial phase [37,43].

A cradle to wheel LCA analysis of 2G bioethanol has highlighted the relevance of the biomass production (agricultural phase) in almost all impact categories for the E85 vehicles, whilst in the case of E10 passenger cars, gasoline production and the related tailpipe emissions resulted to be the main drawbacks [37,43]. These outcomes pointed out that (1) the diverse proportion of gasoline in the blends strongly affected the final results [43,46,63]; (2) for the blends with high bioethanol content, the crop stage was the main hotspot [58,59]; and (3) generally, the energy efficiency of the system depends on the achieved fossil energy savings [37,46].

Therefore the LCA identified the main hotspots of the biofuel production/use chain, showing how different biofuel production yields, as a function of the different conversion pathways, could highly influence the environmental profile of the whole supply chain [65]. In this regard the environmental performance of the bioethanol systems can be improved through (1) the optimization of pretreatment and other conversion technologies; (2) the recycling of chemicals, water and coproducts; and (3) the minimization of the energy demand also through the energetic valorization of by-products [46,66].

Different results can be achieved depending on selected crops and pedo-climatic site specific conditions. Therefore a thorough evaluation of benefits and constrains has to be performed for each bioethanol supply chain. Moreover, the biofuels are a product made by a complex and dynamic system interconnected with ecological and socio-economic systems (the biophysical capacity of the territory, the societal energy demand and structure, the environmental performance of the production chain, the expected social and monetary benefits). The sustainability of the investigated bio-based system depends on the complicated interlinkages between biofuels commercialization on the large scale and the agronomic supply chains locally developed [64]. The biofuel system may be considered economically competitive with its fossil counterpart, only reducing the cost of feedstock conversion technologies, which have not yet reached maturity on the market, and decreasing the selling price of lignocellulosic biomass which can account for 35%–50% of the total cost of bioethanol production [67].

8.5 Toward multiproduct biorefinery processes

The controversial performance of biofuels as a unique product of biomass conversion and the difficulties met in evidencing their environmental sustainability led to broaden the boundaries of the exploitation of bio-based resources, throughout the upgrade of biomass feedstocks into a large array of products, beyond the hotspot of biofuels. Therefore focus was recently shifted towards the evaluation of platform chemicals that can be produced from biomass as end products themselves or as key intermediates between raw materials and a multitude of high-value added marketable products, coupled to biofuel production, in order to optimize conversion processes and remove constraints and barriers.

8.5.1 The added value of biochemicals and biomaterials chains

As already highlighted for biofuels, the employment of renewable feedstock does not necessarily guarantee an environmental positive performance, and therefore the sustainability of a bio-based product should not be taken for granted. Hence, for bio-based chemicals and materials as for biofuels, the necessity of a thorough assessment of the environmental sustainability is strongly required on a case-by-case basis [68].

There are a few LCA evaluations on bio-based platform chemicals: Cok et al. [69] highlighted a better environmental performance of biosuccinic acid from corn-derived dextrose in comparison with the production of maleic anhydride, succinic acid, and adipic acid from fossil sources. Similarly, Zucaro et al. [70] confirmed the relevant GHG and the nonrenewable energy benefits for the biosuccinic acid from lignocellulosic giant reed versus the conventional fossil adipic acid. However, if compared with the fossil maleic anhydride and fossil succinic acid counterparts, the lignocellulosic biosuccinic acid might provide relevant but not renewable energy saving, but comparable or even higher gross GHG emissions [69]. The issue on the accounting of gross GHG emissions (total GHG emissions) or net GHG emissions (subtracting the credits from biogenic carbon storage to the total GHG emissions) remains an open discussion in the LCA literature. Moussa et al. [71] pointed out that, by using the Myriant technologies and production routes for the sorghum grain-based biosuccinic acid, the electricity and heat generation were the main environmental hotspots along the production chain. Nevertheless, the biosuccinic acid from sorghum grain showed lower GHG values in comparison to petroleum-based succinic acid [71].

Potential environmental benefits were retrieved from the comparison between the production of fossil 1,4-butanediol and the bio-based 1,4-butanediol, produced via direct fermentation of sugars from wheat straw feedstock [72]. A net energy consumption and wastes reduction was achieved throughout the energetic valorization of lignin and other residual unconverted solids in a CHP unit within the integrated biorefinery. Likewise, Urban and Bakshi [73] concluded that the bio-based 1,3 propanediol from corn was 46%–71% less GHG-intensive than its fossil counterpart, depending on the different system boundaries encompassed in the life cycle evaluation.

A wider bibliography can be referred to bio-based materials as end products. Polyethylene terephthalate from fossil sources was checked against polyethylene furandicarboxylate from corn starch accounting for both nonrenewable energy use (NREU) and GHG emissions [74]. Direct and iLUC was also included in the LCA of low density polyethylene from sugarcane and polyvinyl chloride from bio-based ethylene [75–77], whereas Hottle et al. [78] underlined the relevance of the end of life phase in the overall performance assessment, when reviewing the sustainability of polylactic acid, polyhydroxyalkanoate, and thermoplastic starch. High density polyethylene from biomass was evaluated from the environmental viewpoint and compared with its fossil counterpart by different authors [79,80].

Generally, the published LCA studies on bio-based chemicals and materials are referred only to GHG emissions and nonrenewable energy consumption, possibly missing potential unintended consequences on other impact categories. Moreover, the observed environmental benefits are not always sufficient to justify a production shift from fossil to bio-based sources. These findings evidence the unsustainability of a single bio-based production process and call for integrated production pathways.

8.5.2 Evaluation of biorefinery concept: a review

The simultaneous production of several bio-based products (bioenergy, in the form of biofuels, electricity, and heat, as well as biochemicals and biomaterials) increases the value of biomass feedstocks and their efficient conversion in line with the concept of an integrated biorefinery, as illustrated in Fig. 8.3. Integrated biorefineries may be regarded as technological means for the transition to a bio-based economy, endorsed as a sustainable alternative for replacing fossil resources in the production of energy, chemicals, and materials.

The term "biorefinery" was established in the 1990s referring to a facility (or a network of facilities) that integrates the biomass-conversion processes and equipment to yield useful products, using renewable

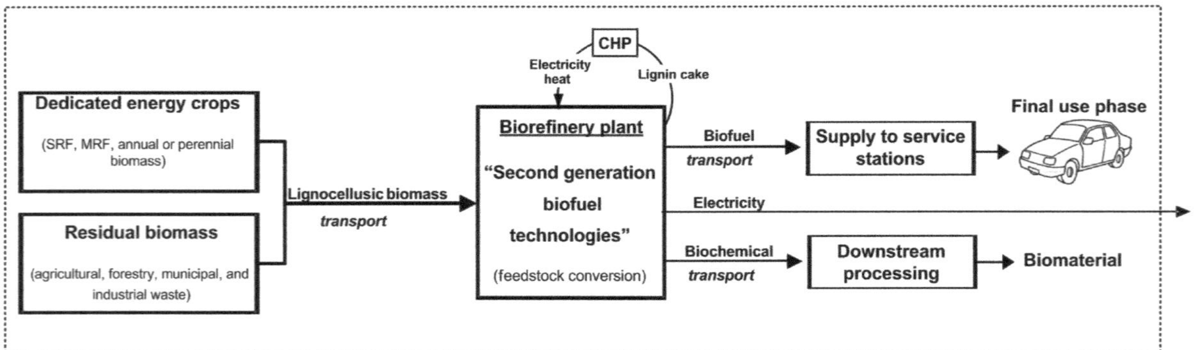

Figure 8.3 Integrated biorefinery-production processes for the coproduction of bioenergy and other biobased products. *CHP*, Combined heating and power plant; *MRF*, medium rotation forestry; *SRF*, short rotation forestry.

resources like biomass as input, in analogy with the "oil refinery" where a fossil fuel is used to generate power, fuels, and chemicals (www.nrel.gov/biomass/biorefinery.html). Namely, the biorefinery was firstly defined as "the sustainable processing of biomass into a spectrum of value-added products (chemicals, materials, food and feed) and energy (biofuels, power and heat)" [72]. A more recent definition from the International Energy Agency states that "biorefining is the sustainable synergetic processing of biomass into a spectrum of marketable food and feed ingredients, products (chemicals, materials) and energy (fuels, power, heat)" (http://www.iea.org). Thus the concept of biorefinery is referred to as a process, a facility, or a cluster of facilities integrally covering the upstream, midstream, and downstream processing of biomass. In any case the main goal is the use of all biomass fractions in order to maximize the product yield per biomass input; in the same way, conventional refineries have been optimized to produce a multitude of products by exploiting all the crude oil component [81].

Beyond the renewable nature of used raw material, a biorefinery differs from an oil refinery due to the application of a wide range of different existing and emerging technologies to separate biomass into its principal constituents (C5 and C6 sugars, lignin, proteins, triglycerides, among others), which can subsequently be transformed into value-added products. Expected products from biorefineries include energy in the form of heat or biofuels produced at large volumes and lower sale prices, and molecules for fine chemistry, cosmetics, or medicinal applications with low volumes and higher sale prices. In addition, products that cannot be obtained from crude oil can be gained at the same time (namely, food products and animal feed).

The classification of biorefineries can be based on features such as (1) platforms (including sugars, oil and, syngas); (2) product groups, both energy and material, like bioethanol, glycerol, or lactic acid; (3) feedstock groups, that is, dedicated sugar and oil crops or residues; (4) conversion processes, such as fermentation, gasification, and pyrolysis; and (5) sustainability and flexibility [82,83]. A distinction can be made according to the feedstock used as raw material, in terms of 1G or 2G. 1G biorefineries use food crop resources and represent the most established type of biorefineries. 2G biorefineries, instead, are based on agricultural residues, wood, and lignocellulosic energy crops, and started commercial operations only recently [84]. Finally, biorefineries processing algae biomass are considered 3G biorefineries and are still facing technical or economic challenges

[85,86]. An up-to-date classification system of biorefineries, based on the feedstock groups, includes (1) "yellow" biorefineries that utilize dry lignocellulosic materials; (2) "green" biorefineries that utilize nature's wet grasses and immature crops; (3) "blue" biorefineries that use algae; and (4) "gray" biorefineries that utilize food waste [81].

Regardless of the type of biorefinery system, the efficient use of biomass components is not only enabled by using a mix of biomass feedstocks and employing a combination of mechanical, thermal, chemical, and biological technologies to produce different products simultaneously, but it is also fostered against increasing the field production of crops. The full use of biomass as well as the minimization of transportation costs, together with appropriate use of energy, water, materials, and cost-effective application of renewable energy sources (solar, wind, biomass energy) or energy conservation practices and indirect savings, are essential prerequisites for a successful biorefinery to be implemented preferably at regional scale rather than larger scale. The implementation of sustainable biorefineries cannot be based on a single perspective or a single design approach but requires a deep understanding, across different disciplines and research areas, of production technology, chemistry, conversion technology of biomass, economics- and environmental-related issues, to allow an improved effectiveness of biomass processing for the coproduction of biofuels and bioproducts in a typical biorefinery pathway.

The opportunity of coproducing 2G bio-based chemicals together with biofuels, if adequately implemented, can be particularly advantageous in terms of (1) valorization of marginal lands, through a proper accounting of the possible LUC; (2) improvement of waste-management strategies and valorization of waste, through a reduction of additional energy and economic investments for waste disposal, on the one hand, and for production of raw materials, on the other hand; (3) additional sources of income to the agricultural and manufacturing sectors, thus generating a wide social consensus through socio-economic benefits; and, last but not least, (4) a possible reduction of environmental impacts and GHGs emissions.

As illustrated in Section 8.4, a widespread scientific literature concerns the environmental evaluation of 1G and 2G biofuels, whereas a few environmental evaluations were performed for bio-based chemicals and materials (Section 8.5.1). On the other hand, biorefineries were widely investigated by means of techno-economic and environmental analyses [87–89]. Also in this case, the environmental perspective is usually accomplished by using the LCA methodology to identify the most sustainable biorefining route, and several LCA studies were published concerning 2G biorefinery systems with multiple output products [54,68,86,90]. In a biorefinery framework the first attempts to evaluate the environmental performance of the coproduction of bioethanol, energy, and biochemicals from lignocellulosic raw materials, namely, switchgrass and corn stover and wheat straw, were realized by Cherubini and Jungmeier [90] and Cherubini and Ulgiati [91] respectively. Common findings from these studies were that the investigated biorefinery systems provide an effective option for (1) markedly increasing the GHG savings, due to the on set-aside cultivation of perennial switchgrass (thanks to SCS) or due to the use of agricultural residues (with a reduction of GHG emissions of about 50%); (2) reducing dependence on imported fossil fuels (reaching more than 80% of nonrenewable energy savings); and (3) enhancing cleaner production chains based on local and renewable resources. However, in comparison with the reference fossil-based systems, additional environmental impacts were generated in other impact categories (eutrophication, terrestrial and fresh water ecotoxicity, acidification, human toxicity) that cannot be disregarded for a comprehensive environmental assessment. Fiorentino et al. [54] evaluated the coproduction of biodiesel and bio-based chemicals, respectively, from seeds and agricultural and processing wastes from *Brassica carinata*, a non-food oil crop supposed to be grown on nonfertile soil in the Campania Region (Italy). The assessment of such integrated biorefinery system confirmed that the production of the only biodiesel was not favorable either from the energetic (very low net energy ratio) or from the environmental point of view (with many impact categories affected more than from the fossil counterparts). If the coproduction of ethyl levulinate, a derivative of levulinic acid, and glycerol, via an innovative conversion route (Biofine process), were associated to the production of the energy carrier, then the whole process resulted feasible. An optimization of the environmental and economic performance was achieved in terms of energy consumption and global warming as well as in terms of human toxicity, acidification, eutrophication, and photochemical oxidation potentials.

Integrated configurations of sugarcane-based biorefineries producing 1G and 2G ethanol, sugar, molasses (for animal feed), and electricity were analyzed and compared using techno-economic value-based approach and LCA methodology [87]. Benefits in terms of climate change and fossil depletion were gained as compared to the fossil reference systems.

However, when considering only the production of fuels, less efficiency was retrieved for human toxicity, freshwater ecotoxicity, and eutrophication impact categories. If ethanol production was boosted, the best economic performance was attained, while if coproduction (1G and 2G) was favored, the lowest environmental impact was recorded, although not corresponding to the cheapest solution [87]. Similar results were also achieved in the assessment of the biorefining of palm empty fruit bunches, abundant lignocellulosic residues from the palm oil/biodiesel industry, leading to the production of fuel ethanol, heat, and power and C5 sirup (as feed for cattle) [89].

Recently, Parajuli et al. [92] evaluated the environmental impacts of two standalone biorefinery plants, separately producing bioethanol from wheat straw and bio-based lactic acid from alfalfa, and of an integrated system producing the both stated products. The LCA evaluation showed that both bioethanol and bio-based lactic acid had net environmental gains compared to petrol and conventional lactic acid, respectively. Moreover, the integrated system again achieved a better environmental performance in most of the impact categories (GHG emissions, NREU, and eutrophication potential) than both standalone systems.

Alternative lignocellulosic biorefineries, producing ethanol, lactic acid, or methanol and coproducing ethanol and lactic acid from lignocellulosic residues of a typical sugar mill (sugarcane bagasse and trash), were investigated by Mandegari et al. [88]. Two options of bioenergy systems were considered: a self-sufficient biorefinery scenario, in which a portion of lignocellulosic feedstock was burned into a centralized CHP unit, and another one in which coal is cocombusted with biomass residues. The main outcomes of the environmental assessment were that, even with cocombustion of coal, environmental benefits were gained for all the investigated scenarios.

In conclusion, according to the definitions for "process integration" and "feedstock and product integration" [93], the integration of biorefinery systems was proven to deliver additional synergies to optimally utilize the resources and minimize the related burdens. The implementation of locally integrated biorefinery systems may thus overcome the main barriers and constraints linked to the production of a single bioproduct chain.

8.6 Concluding remarks: the perspective of circular economy pathways

The implementation of integrated biorefinery systems requires the development of highly innovative approaches for a deeper understanding and improved management of water-energy-materials-environment nexus towards new CE pathways.

The flowchart in Fig. 8.4 displays an example of an integrated system in a CE prospective.

Different agricultural production chains are networked to provide not only the traditional main products but also a set of coproducts and substrates to be used as raw resources for downstream conversion activities. Bulk chemicals, animal feedstocks, pharmaceuticals, fertilizers, biomaterials, and bioenergy, in the form of biogas, liquid fuels, electricity, and heat, can be cogenerated. For this to happen, a number of local manufacturing activities must be identified, developed, and integrated, capable to process each substrate into products. Each small or medium enterprise becomes a valuable component of the network, develops technology and innovation, yields added value products, and contributes to new jobs and income.

Integration between agricultural productions and industrial activities is an optimal solution to valorize every residue and maximize recycling and exchange of secondary materials. CE is a new way of deriving resources from a territory, respecting its specific vocation and tradition without competing with food production.

In this context the CE becomes the next framework and business model, in which a breakthrough innovation is needed to address the shift from a linear production pattern (where waste and pollution are the rule) to complex networks (where waste from a process is the raw input to another process and emissions are minimized). The key issues addressed by this new model of production are

- the concept of "zero waste;"
- the necessity to implement the already existing activities;
- the employment and valorization of different available substrates (e.g., agricultural residues), suppliers and final users; and
- the optimization of input resources as well as intermediate recycling and output flows.

Such complex patterns also require complex and flexible technologies and comprehensive evaluation tools. Understanding and properly addressing network, substrates, technology, and methodological complexities are therefore important goals to develop a theoretical integrated framework and operational model. The implementation of a CE, the appropriate resource management, the shift from providing services instead of products (e.g., good mobility instead of cars), and the better image that such

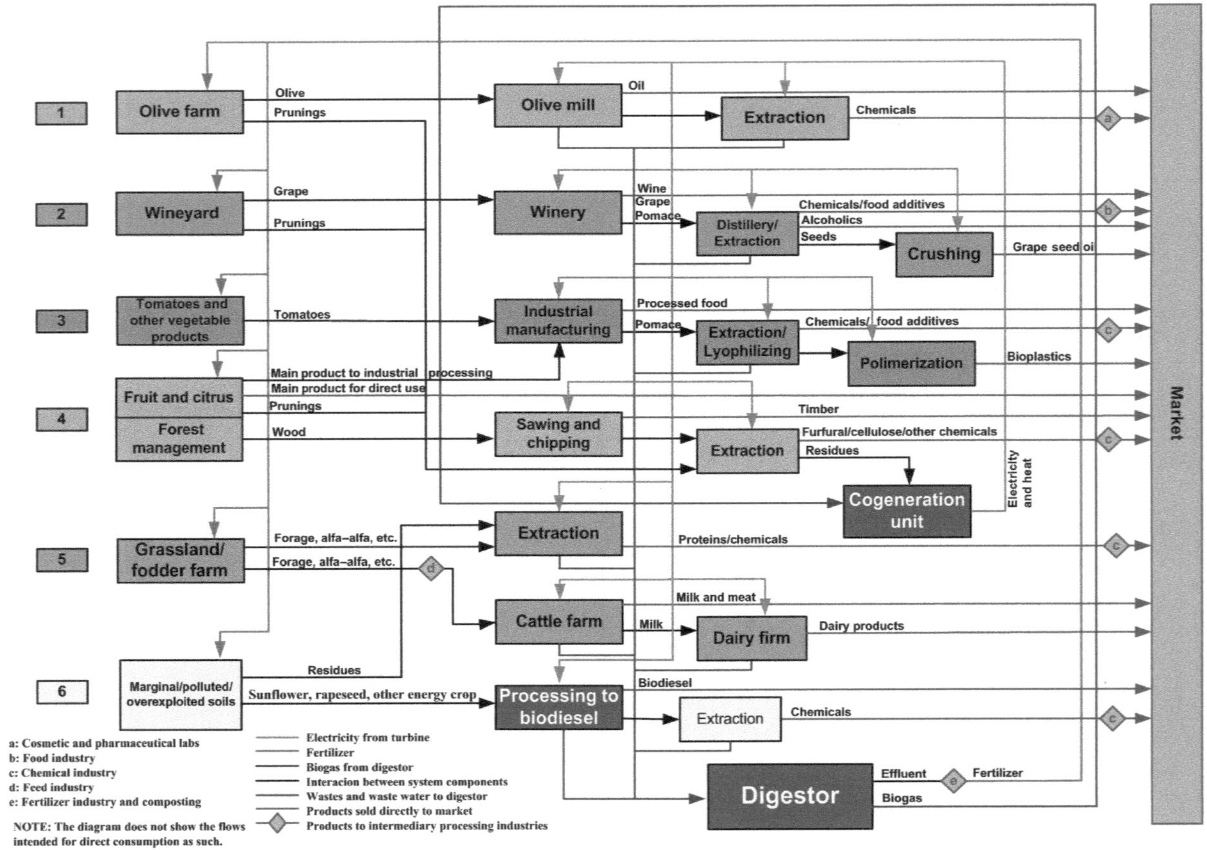

Figure 8.4 An example of integrated system in a circular economy pathway.

improvements provide are likely to attract new companies and generate innovative businesses, thus generating a new set of jobs over the logistic chain, that replace old, no longer sustainable jobs, while also decreasing environmental impacts.

Circular model where products and materials continue to circulate seems to be the only solution to overcome the current consumption and production trends. According to Ellen MacArthur Foundation [94] in the CE, products and materials should continue to circulate through maintenance, reuse and redistribution, refurbishment and remanufacturing, and finally recycling. The CE aims at keeping products, components, and materials at their highest utility and value at all times preserving and enhancing natural capital, optimizing resource yields, and minimizing system risks by managing finite stocks and renewable flows. Nonetheless, the profitability of integrated biorefinery in a CE prospective has to be carefully assessed, in view of recent legislative regulations and of the abovementioned controversial scientific results. In fact the pertinent scientific literature of different bio-based production pathways highlighted the relevance of (1) feedstock cultivation or used raw materials; (2) selected transformation routes for each final bioproduct; and (3) the biotechnological processes involved in their transformation [37,43]. In this regard the choice among different conversion routes as well as the selected technological processes might amplify or reduce the impact of the whole supply chain and establish the economic competitiveness with the fossil counterpart. The environmental loads generated through the whole chain of bio-based products has to be carefully quantified and assessed to ensure adequate decisions and to avoid too optimistic claims as with 1G biofuels. The expected advantages of bio-based production routes consist of significantly reducing GHG emissions and fossil fuel use, with net benefits on climate change. However important, climate change cannot be considered the only gauge for the environmental sustainability. Other issues have to be given due weight. LUC, water consumption, biodiversity loss, and variations of biogeochemical cycles are an absolute priority. Therefore the evaluation of bio-based supply chains cannot disregard a comprehensive picture of all the impacts on water, air, and soil affecting both human and ecosystems health.

References

[1] G. Alberts, M. Ayuso, A. Bauen, F. Boshel, C. Chudziak, J. Peer Gebauer, et al., Innovation Outlook: Advanced Liquid Biofuels, 2016. <www.irena.org/Publications>.

[2] W. Stafford, A. Lotter, A. Brent, G. von Maltitz, Biofuels Technology. A Look Forward, 2017. <https://www.wider.unu.edu/sites/default/files/wp2017-87.pdf>.

[3] E.I. Wiloso, R. Heijungs, G.R. de Snoo, LCA of second generation bioethanol: a review and some issues to be resolved for good LCA practice, Renew. Sustain. Energy Rev. 16 (2012) 5295–5308.
[4] H. Zabed, J.N. Sahu, A. Suely, A.N. Boyce, G. Faruq, Bioethanol production from renewable sources: current perspectives and technological progress, Renew. Sustain. Energy Rev. 71 (2017) 475–501.
[5] European Commission, Directive 2009/28/EC of the European Parliament and of the Council of 23 April 2009 on the Promotion of the Use of Energy From Renewable Sources and Amending and Subsequently Repealing Directives 2001/77/EC and 2003/30/EC, 2009.
[6] L.W.M. Beurskens, M. Hekkenberg, Renewable Energy Projections as Published in the National Renewable Energy Action Plans of the European Member States, 2011. <www.ecn.nl/docs/library/report/2010/e10069.pdf>.
[7] REN21, Renewables 2015 Global Status Report, 2015. <http://www.ren21.net/wp-content/uploads/2015/07/REN12-GSR2015_Onlinebook_low1.pdf>.
[8] S. Nakada, D. Saygin, D. Gielen, Global bioenergy: supply and demand projections, in: A Working Paper for REmap 2030, 2014. <http://www.globalbioenergy.org/uploads/media/1409_IRENA_-_REmap_2030_Biomass_paper_2014.pdf>.
[9] GEA (Global Energy Assessment), Toward a Sustainable Future, Cambridge University Press, Cambridge, UK and New York, NY, USA and the International Institute for Applied Systems Analysis, Laxenburg, Austria, 2012. <http://www.globalenergyassessment.org/>.
[10] T. Bruckner, I.A. Bashmakov, Y. Mulugetta, H. Chum, A. de la Vega Navarro, J. Edmonds, et al., Energy systems, in: O. Edenhofer, R. Pichs-Madruga, Y. Sokona, E. Farahani, S. Kadner, K. Seyboth, et al. (Eds.), Climate Change 2014: Mitigation of Climate Change. Contribution of Working Group III to the Fifth Assessment Report of the Intergovernmental Panel on Climate Change, Cambridge University Press, Cambridge and New York, 2014, pp. 511–597. Available from: https://ar5-syr.ipcc.ch/resources/htmlpdf/ipcc_wg3_ar5_chapter7/.
[11] European Commission, Proposal for a Directive of the European Parliament and the Council on the promotion of the use of energy from renewable sources (recast), in: COM (2016) 767 Final/2, 2017.
[12] E. Gnansounou, Coproducts performances in biorefineries: development of claiming-based allocation models for environmental policy, Bioresour. Technol. 254 (2018) 31–39.
[13] S.C. de Vries, G.W.J. van de Ven, M.K. van Ittersum, K.E. Giller, Resource use efficiency and environmental performance of nine major biofuel crops, processed by first-generation conversion techniques, Biomass Bioenergy 34 (2010) 588–601.
[14] P. Lamers, C. Hamelinck, M. Junginger, A. Faaij, International bioenergy trade—a review of past developments in the liquid biofuel market, Renew. Sustain. Energy Rev. 15 (2011) 2655–2676.
[15] M. Giampietro, S. Ulgiati, D. Pimentel, Feasibility of large-scale biofuel production. Does an enlargement of scale change the picture? Bioscience 47 (1997) 587–600.
[16] I.B. Banković-Ilić, O.S. Stamenković, V.B. Veljković, Biodiesel production from non-edible plant oils, Renew. Sustain. Energy Rev. 16 (2012) 3621–3647.
[17] M. Giampietro, K. Mayumi, A.H. Sorman (Eds.), Energy Analysis for a Sustainable Future, Taylor & Francis Group, Routledge, London, 2013.
[18] T. Gomiero, Large-scale biofuels production: a possible threat to soil conservation and environmental services, Appl. Soil Ecol. 123 (2018) 729–736.
[19] COM 15 final, A Policy Framework for Climate and Energy in the Period From 2020 to 2030, Brussels, 2014.

[20] M. Ripa, C. Buonaurio, S. Mellino, G. Fiorentino, S. Ulgiati, Recycling waste cooking oil into biodiesel: a life cycle assessment, Int. J. Performability Eng. 10 (2014) 347−356.
[21] T. Kan, V. Strezov, T.J. Evans, Lignocellulosic biomass pyrolysis: a review of product properties and effects of pyrolysis parameters, Renew. Sustain. Energy Rev. 57 (2016) 1126−1140.
[22] A. Yousuf, Biodiesel from lignocellulosic biomass − prospects and challenges, Waste Manage. 32 (2012) 2061−2067.
[23] D. Kumar, B. Singh, J. Korstad, Utilization of lignocellulosic biomass by oleaginous yeast and bacteria for production of biodiesel and renewable diesel, Renew. Sustain. Energy Rev. 73 (2017) 654−671.
[24] R.A. Sheldon, Utilisation of biomass for sustainable fuels and chemicals: molecules, methods and metrics, Catal. Today 167 (2011) 3−13.
[25] N. Bentsen, C. Felby, Biomass for energy in the European Union − a review of bioenergy resource assessments, Biotechnol. Biofuels 5 (2012) 25.
[26] H.L. Bos, J.P.M. Sanders, Raw material demand and sourcing options for the development of a bio-based chemical industry in Europe. Part 1: Estimation of maximum demand, Biofuels Bioprod. Biorefin. 7 (2013) 246−259.
[27] J.P.M. Sanders, H.L. Bos, Raw material demand and sourcing options for the development of a bio-based chemical industry in Europe. Part 2: Sourcing options, Biofuels Bioprod. Biorefin. 7 (2013) 260−272.
[28] N.S. Bentsen, I.M. Møller, Solar energy conserved in biomass: sustainable bioenergy use and reduction of land use change, Renew. Sustain. Energy Rev. 71 (2017) 954−958.
[29] R. Kajaste, Chemicals from biomass − managing greenhouse gas emissions in biorefinery production chains − a review, J. Clean Prod. 75 (2014) 1−10.
[30] T.D. Searchinger, T. Beringer, A. Strong, Does the world have low-carbon bioenergy potential from the dedicated use of land? Energy Policy 110 (2017) 434−446.
[31] F. Kraxner, K. Aoki, G. Kindermann, S. Leduc, F. Albrecht, J. Liu, et al., Bioenergy and the city − what can urban forests contribute? Appl. Energy 165 (2016) 990−1003.
[32] M. Saha, M.J. Eckelman, Geospatial assessment of regional scale bioenergy production potential on marginal and degraded land, Resour. Conserv. Recycl. 128 (2018) 90−97.
[33] A. Monti, W. Zegada-Lizarazu, Sixteen-year biomass yield and soil carbon storage of giant reed (*Arundo donax* L.) grown under variable nitrogen fertilization rates, BioEnergy Res. 9 (2016) 248−256.
[34] A. Zucaro, A. Forte, S. Faugno, A. Impagliazzo, A. Fierro, Effects of urea-fertilization rates on the environmental performance of giant reed lignocellulosic feedstock produced for biorefinery purpose, J. Clean. Prod. 172 (2018) 4200−4211.
[35] D.J. Muth, K.M. Bryden, An integrated model for assessment of sustainable agricultural residue removal limits for bioenergy systems, Environ. Model. Softw. 39 (2013) 50−69.
[36] K.-Q. Fan, P.-F. Zhang, Z.J. Pei, An assessment model for collecting and transporting cellulosic biomass, Renew. Energy 50 (2013) 786−794.
[37] A. Forte, A. Zucaro, M. Fagnano, A. Fierro, Potential environmental impact of bioethanol production chain from fiber sorghum to be used in passenger cars, Sci. Total Environ. 598 (2017) 365−376.
[38] R. Golecha, J. Gan, Biomass transport cost from field to conversion facility when biomass yield density and road network vary with transport radius, Appl. Energy 164 (2016) 321−331.
[39] WEF (World Economic Forum), Global Risks Report 2017, 12th ed., Geneva, 2017. <http://www3.weforum.org/docs/GRR17_Report_web.pdf>.

[40] OECD (Organisation for Economic Co-operation and Development), OECD Council Recommendation on Water, 2016. <http://www.oecd.org/environment/resources/Council-Recommendation-on-water.pdf>.
[41] D. Pimentel, B. Berger, D. Filiberto, M. Newton, B. Wolfe, E. Karabinakis, et al., Water resources: agricultural and environmental issues, Bioscience 54 (2004) 909–918.
[42] V. Mathioudakis, P.W. Gerbens-Leenes, T.H. Van der Meer, A.Y. Hoekstra, The water footprint of second-generation bioenergy: a comparison of biomass feedstocks and conversion techniques, J. Clean. Prod. 148 (2017) 571–582.
[43] A. Zucaro, A. Forte, R. Basosi, M. Fagnano, A. Fierro, Life cycle assessment of second generation bioethanol produced from low-input dedicated crops of *Arundo donax* L., Bioresour. Technol. 219 (2016) 589–599.
[44] D. Pimentel, Ethanol fuels: energy balance, economics, and environmental impacts are negative, Nat. Resour. Res. 12 (2003) 127–134.
[45] C.J. Cleveland, R. Costanza, Energy return on investment (EROI), in: C.J. Cleveland (Ed.), Encyclopedia of Earth. Environmental Information Coalition, National Council for Science and the Environment, Washington, DC, 2013, pp. 329–354. <https://editors.eol.org/eoearth/wiki/Energy_return_on_investment_(EROI)>.
[46] M. Morales, J. Quintero, R. Conejeros, G. Aroca, Life cycle assessment of lignocellulosic bioethanol: environmental impacts and energy balance, Renew. Sustain. Energy Rev. 42 (2015) 1349–1361.
[47] M. Giampietro, S. Ulgiati, Integrated assessment of large-scale biofuel production, CRC Crit. Rev. Plant. Sci. 24 (2005) 365–384.
[48] ISO 14040, Environmental Management – Life Cycle Assessment – Principles and Framework, Geneva, 2006.
[49] ISO 14044, Environmental Management – Life Cycle Assessment – Requirements and Guidelines, Geneva, 2006.
[50] European Commission, International Reference Life Cycle Data System (ILCD) Handbook – General Guide for Life Cycle Assessment – Detailed Guidance, 2010.
[51] European Commission, Supporting Environmentally Sound Decisions for Waste Management – A Technical Guide to Life Cycle Thinking (LCT) and Life Cycle Assessment (LCA) for Waste Experts and LCA Practitioners, 2011.
[52] Goedkoop, M., Heijungs, R., Huijbregts, M., De Schryver, A., Struijs, J., Description of the ReCiPe Methodology for Life Assessment Impact Assessment, 2013. <http://www.lcia-recipe.net>.
[53] B.E. Dale, S. Kim, Can the predictions of consequential life cycle assessment be tested in the real world? Comment on "Using attributional life cycle assessment to estimate climate-change mitigation...", J. Ind. Ecol. 18 (2014) 466–467.
[54] G. Fiorentino, M. Ripa, S. Mellino, S. Fahd, S. Ulgiati, Life cycle assessment of *Brassica carinata* biomass conversion to bioenergy and platform chemicals, J. Clean. Prod. 66 (2014) 174–187.
[55] S.N. Naik, V.V. Goud, P.K. Rout, A.K. Dalai, Production of first and second generation biofuels: a comprehensive review, Renew. Sustain. Energy Rev. 14 (2010) 578–597.
[56] F. Bilgili, E. Koçak, Ü. Bulut, S. Kuşkaya, Can biomass energy be an efficient policy tool for sustainable development? Renew. Sustain. Energy Rev. 71 (2017) 830–845.
[57] C. Bessou, C. Basset-Mens, T. Tran, A. Benoist, LCA applied to perennial cropping systems: a review focused on the farm stage, Int. J. Life Cycle Assess. 18 (2013) 340–361.
[58] C. Godard, J. Boissy, B. Gabrielle, Life-cycle assessment of local feedstock supply scenarios to compare candidate biomass sources, GCB Bioenergy 5 (2013) 16–29.

[59] S. Cadoux, F. Ferchaud, C. Demay, H. Boizard, J.-M. Machet, E. Fourdinier, et al., Implications of productivity and nutrient requirements on greenhouse gas balance of annual and perennial bioenergy crops, GCB Bioenergy 6 (2014) 425–438.
[60] P. Pawelzik, M. Carus, J. Hotchkiss, R. Narayan, S. Selke, M. Wellisch, et al., Critical aspects in the life cycle assessment (LCA) of bio-based materials – reviewing methodologies and deriving recommendations, Resour. Conserv. Recycl. 73 (2013) 211–228.
[61] IPCC, Climate change 2013: the physical science basis, in: Contribution of Working Group I to the Fifth Assessment Report of the Intergovernmental Panel on Climate Change, Cambridge, 2013.
[62] A. Zucaro, G. Fiorentino, A. Zamagni, S. Bargigli, P. Masoni, A. Moreno, et al., How can life cycle assessment foster environmentally sound fuel cell production and use? Int. J. Hydrogen Energy 38 (2013).
[63] Y. Bai, L. Luo, E. van der Voet, Life cycle assessment of switchgrass-derived ethanol as transport fuel, Int. J. Life Cycle Assess. 15 (2010) 468–477.
[64] A. Forte, A. Zucaro, S. Faugno, R. Basosi, A. Fierro, Carbon footprint and fossil energy consumption of bio-ethanol fuel production from *Arundo donax* L. crops on marginal lands of Southern Italy, Energy 150 (2018) 222–235.
[65] Y. Sun, J. Cheng, Hydrolysis of lignocellulosic materials for ethanol production: a review, Bioresour. Technol. 83 (2002) 1–11.
[66] A. Margeot, B. Hahn-Hagerdal, M. Edlund, R. Slade, F. Monot, New improvements for lignocellulosic ethanol, Curr. Opin. Biotechnol. 20 (2009) 372–380.
[67] E. Gnansounou, A. Dauriat, Techno-economic analysis of lignocellulosic ethanol: a review, Bioresour. Technol. 101 (2010) 4980–4991.
[68] G. Fiorentino, M. Ripa, S. Ulgiati, Chemicals from biomass: technological versus environmental feasibility. A review, Biofuels Bioprod. Biorefin. 11 (2017) 195–214.
[69] B. Cok, I. Tsiropoulos, A.L. Roes, M.K. Patel, Succinic acid production derived from carbohydrates: an energy and greenhouse gas assessment of a platform chemical toward a bio-based economy, Biofuels Bioprod. Biorefin. 8 (2014) 16–29.
[70] A. Zucaro, A. Forte, A. Fierro, Greenhouse gas emissions and non-renewable energy use profiles of bio-based succinic acid from *Arundo donax* L. lignocellulosic feedstock, Clean Technol. Environ. Policy 19 (2017) 2129–2143.
[71] H.I. Moussa, A. Elkamel, S.B. Young, Assessing energy performance of bio-based succinic acid production using LCA, J. Clean. Prod. 139 (2016) 761–769.
[72] A. Forte, A. Zucaro, R. Basosi, A. Fierro, LCA of 1,4-butanediol produced via direct fermentation of sugars from wheat straw feedstock within a territorial biorefinery, Materials (Basel) 9 (2016) 563.
[73] R.A. Urban, B.R. Bakshi, 1,3-Propanediol from fossils versus biomass: a life cycle evaluation of emissions and ecological resources, Ind. Eng. Chem. Res. 48 (2009) 8068–8082.
[74] A.J.J.E. Eerhart, A.P.C. Faaij, M.K. Patel, Replacing fossil based PET with biobased PEF; process analysis, energy and GHG balance, Energy Environ. Sci. 5 (2012) 6407.
[75] C. Liptow, A.-M. Tillman, A comparative life cycle assessment study of polyethylene based on sugarcane and crude oil, J. Ind. Ecol. 16 (2012) 420–435.
[76] R.A. Alvarenga, J. Dewulf, S. De Meester, A. Wathelet, J. Villers, R. Thommeret, et al., Life cycle assessment of bioethanol-based PVC. Part 1: Attributional approach, Biofuels Bioprod. Biorefin. 7 (2013) 386–395.
[77] R.A. Alvarenga, J. Dewulf, S. De Meester, A. Wathelet, J. Villers, R. Thommeret, et al., Life cycle assessment of bioethanol-based PVC. Part 2: Consequential approach, Biofuels Bioprod. Biorefin. 7 (2013) 396–405.
[78] T.A. Hottle, M.M. Bilec, A.E. Landis, Sustainability assessments of bio-based polymers, Polym. Degrad. Stab. 98 (2013) 1898–1907.

[79] I. Tsiropoulos, A.P.C. Faaij, L. Lundquist, U. Schenker, J.F. Briois, M.K. Patel, Life cycle impact assessment of bio-based plastics from sugarcane ethanol, J. Clean. Prod. 90 (2015) 114–127.
[80] S. Belboom, A. Léonard, Does biobased polymer achieve better environmental impacts than fossil polymer? Comparison of fossil HDPE and biobased HDPE produced from sugar beet and wheat, Biomass Bioenergy 85 (2016) 159–167.
[81] A. Corona, M. Ambye-Jensen, G.C. Croxatto Vega, M.Z. Hauschild, M. Birkved, Techno-environmental assessment of the green biorefinery concept: combining process simulation and LCA at an early stage of the design process, Sci. Total Environ. 635 (2018) 100–111.
[82] E. Gnansounou, A. Pandey, Classification of biorefineries taking into account sustainability potentials and flexibility, in: E. Gnansounou, A. Pandey (Eds.), Life-Cycle Assessment of Biorefineries, Elsevier, 2017, pp. 1–39.
[83] M. Palmeros Parada, P. Osseweijer, J.A. Posada Duque, Sustainable biorefineries, an analysis of practices for incorporating sustainability in biorefinery design, Ind. Crops Prod. 106 (2017) 105–123.
[84] R.F. Service, Renewable energy. Cellulosic ethanol at last? Science 345 (2014) 1111.
[85] S.J. Gerssen-Gondelach, D. Saygin, B. Wicke, M.K. Patel, A.P.C. Faaij, Competing uses of biomass: assessment and comparison of the performance of bio-based heat, power, fuels and materials, Renew. Sustain. Energy Rev. 40 (2014) 964–998.
[86] R. Parajuli, T. Dalgaard, U. Jørgensen, A.P.S. Adamsen, M.T. Knudsen, M. Birkved, et al., Biorefining in the prevailing energy and materials crisis: a review of sustainable pathways for biorefinery value chains and sustainability assessment methodologies, Renew. Sustain. Energy Rev. 43 (2015) 244–263.
[87] E. Gnansounou, P. Vaskan, E.R. Pachón, Comparative techno-economic assessment and LCA of selected integrated sugarcane-based biorefineries, Bioresour. Technol. 196 (2015) 364–375.
[88] M. Mandegari, S. Farzad, J.F. Görgens, A new insight into sugarcane biorefineries with fossil fuel co-combustion: techno-economic analysis and life cycle assessment, Energy Convers. Manage. 165 (2018) 76–91.
[89] P. Vaskan, E.R. Pachón, E. Gnansounou, Techno-economic and life-cycle assessments of biorefineries based on palm empty fruit bunches in Brazil, J. Clean. Prod. 172 (2018) 3655–3668.
[90] F. Cherubini, G. Jungmeier, LCA of a biorefinery concept producing bioethanol, bioenergy, and chemicals from switchgrass, Int. J. Life Cycle Assess. 15 (2010) 53–66.
[91] F. Cherubini, S. Ulgiati, Crop residues as raw materials for biorefinery systems – a LCA case study, Appl. Energy 87 (2010) 47–57.
[92] R. Parajuli, M.T. Knudsen, M. Birkved, S.N. Djomo, A. Corona, T. Dalgaard, Environmental impacts of producing bioethanol and biobased lactic acid from standalone and integrated biorefineries using a consequential and an attributional life cycle assessment approach, Sci. Total Environ. 598 (2017) 497–512.
[93] P.R. Stuart, M.M. El-Halwagi (Eds.), Integrated Biorefineries: Design, Analysis, and Optimization, Taylor and Francis/CRC Press, 2012.
[94] Ellen Macarthur Foundation, Towards the Circular Economy, 2012. <https://www.ellenmacarthurfoundation.org/>.

Further reading

B. Kamm, M. Kamm, Principles of biorefineries, Appl. Microbiol. Biotechnol. 64 (2004) 137–145.

CHAPTER 9

Environmental and socioeconomic impact assessment of biofuels from lignocellulosic biomass

Naveenji Arun and Ajay K. Dalai
Catalysis and Chemical Reaction Engineering Laboratories, Department of Chemical and Biological Engineering, University of Saskatchewan, Saskatoon, SK, Canada

Contents

9.1 Introduction

The production of third-generation biofuels processes employ lignocellulosic biomass, nonedible crops, and nonfood parts of plant as feedstock. As the source for biodiesel shifts from first to third generation, it indicates that the characteristics of the processes also differ [1]. Environmentally benign biofuel-production process is expected to possess low energy inputs in all steps from crop cultivation to processing of plants and production of biofuels. Moreover, it should possess better energy balance, enhanced engine performance, and lower greenhouse gas (GHG) emissions [2]. These characteristics of biofuels can be quantitatively analyzed using life cycle analysis (LCA) tools.

Lignocellulosic Biomass to Liquid Biofuels
DOI: https://doi.org/10.1016/B978-0-12-815936-1.00009-5

Three major energy-related concerns that are influencing the society are growth in population, increase in global emissions, and depletion in fossil fuel. The increase in population indicates that usage of land for cultivation, timber production, and other agricultural activities will increase. The Renewable Transport Fuel Obligation (RTFO) policy defined "metastandard" that includes five environmental and two socioeconomic principles for analysis. Currently, Renewable Fuels Agency administers RTFO and it monitors the report on carbon balance and sustainability from different parties [3].

In the United States, current rate of ethanol production from corn is adequate to commercialize E10 blend (maximum blend of 10% of ethanol in gasoline) throughout the country. In such a scenario, it is challenging to convince farmers to cultivate crops such as switchgrass, miscanthus, or other feedstocks for biomass production without providing financial security [4]. Globally, most of the arable lands are being employed to produce food crops. Usage of marginal or degraded lands for biomass production can be promising to deal with concerns related to indirect land use change. Scale of operation of biofuels production industry will be of major concern as biofuels production directly depends on factors, such as biomass growth and its productivity rate, choice of processing technology, and logistics-related issues. The objectives of this chapter are to discuss the trends in the growth of biofuels sector, importance of by-products in life cycle analysis and the immediate need for socioeconomic and environmental assessment to ensure the credibility of novel biofuels-production processes. The chemical processes involved in the chemical processing of canola oil to produce third-generation biofuels are discussed in this chapter.

9.2 Feedstocks for sustainable biofuels production

Regional preference toward a particular feedstock is an important parameter for the successful development and commercialization of biomass-based bioenergy-production process. Attention toward palm oil, sugarcane, and jatropha is comparatively less and this could be attributed to the regional preference toward particular feedstock based on its commercial value and availability.

Brazil and other South American countries primarily focus on the usage of sugarcane for biofuels (especially bioethanol) production. However, Asian countries, such as Malaysia and India, are presently

Table 9.1 Vegetable oil availability in different parts of the world [5].

Continent	Common available vegetable oil
Asia	Palm, corn, soybean, karanja, rubber seed, peanut, and coconut
America	Canola, soybean, sunflower, jojoba, castor, olive, safflower, corn, and peanut
Australia	Sunflower, canola, almond, soybean, palm, linseed, and olive
Africa	Palm, silk cotton (*Ceiba pentandra*), elais olenfera, and castor
Europe	Canola, sunflower, olive, soybean, sesame, corn, and peanut

focusing on palm and karanja oil, respectively. Table 9.1 indicates the different feed sources that are potential feedstocks for biofuel production [5]. In North America, most promising feedstocks are corn, soybean, and canola oil. However, debate over food versus fuel on the choice of feedstocks is well known. In Canada, green-seed canola oil, the nonedible form of canola oil, is gaining attention as promising feedstock for biofuel production. It is projected that the overall biofuels share in Canadian fuel scenario will increase in comparison to its past contribution (Table 9.2).

Review published by Cherubini and Strømman [7] provides comprehensive information about the global approach toward LCA of bioenergy systems. Souza et al. [8] compared between traditional sugarcane ethanol production system (TSES) and joint production system (JSEB) using ISO 14040:2006 and ISO 14044:2006. In South America, sugarcane (*Saccharum* spp.) is established as a potential feedstock for ethanol production due to its high yield (7.6 m^3/ha/year). It was concluded that the emission levels for JSEB was 23% lesser than the emission levels for TSES.

According to Soratana [9] and Kadam [10], the usage of recyclable materials from industrial resources as feedstocks is recommended to minimize the by-product generation from the bioprocess. Krohn and Fripp [4] performed LCA on biodiesel production using *Camelina sativa* (L.), and it was concluded that using *C. sativa* (L.) crop overcomes challenges, such as food versus fuel issues and land use change (LUC), and also results in the reduction of GHG emissions by 40%–60% annually in comparison to the traditional fossil fuels. However, it should be noted that the availability of these crops for sustained and continuous production of alternate fuels is essential and it is required to have the crop yield of at least 1000–2000 kg/ha on continuous basis. Another important challenge with respect to the environmental credibility is the usage of fertilizers for the growth of these crops. *C. sativa* (L.) has great potential as agricultural crop

Table 9.2 Canadian canola industry goals for 2025 [6].

Element	2015 Results	2025 Target
Exported seed	9.3 MMT[1]	12 MMT
Domestic processing	7.7 MMT	14 MMT
Acres	20 million	22 million
Yield	38.0 bu/acre	52 bu/acre
Production	17.2 MMT	26 MMT
High oleic and specialty Oil	11% of canola acres[2]	33% of canola acres
Oil content (average of no. 1 grade)	44.3%[3]	Maintain global competitiveness in oil content (10 years average = 44.2%)
Saturated fat content	6.8%	Global leadership position in oil saturated fat content
Meal crude protein content (oil-free, 12% moisture basis)	38.9%	Increase protein availability by target species (10 years average = 38.2%)

MMT, Million metric tons. All statistics are for 2015 Calendar year. (1) Based on Canadian Grain Commission data; (2) industry estimate; and (3) Canadian Grain Commission preliminary quality report.

for biofuels production in North America, as it is tolerant to harsh winter conditions that prevail from October to April in most of the European and North American nations. It is also efficient in terms of its agricultural requirements and water usage.

Grau et al. [3] analyzed the environmental impact of conventional crop-rotation system in Spanish northeastern area using rapeseed oil as a potential feedstock. It was concluded that the introduction of rapeseed oil into the crop rotation reduced the environmental impact. For the vegetable oil the crop energy ratio was 21.6% superior over the conventional diesel fuels. During LCA, inventory of all emissions associated with the process, consumption of resources and energy are carried out based on material and energy balance. In the case of corn the following two types of feedstocks for alternate fuels production are available: (1) corn grain that can be used for the production of bioethanol, and (2) corn stover from which biochar and bio-oil can be obtained by the fast pyrolysis of process and the bio -oil can be further upgraded by hydrodeoxygenation route to remove oxygen to enhance its combustion properties. LCAs of both the pathways were conducted, and it was concluded that a reduction in GHG emission of 52.1% was observed for both the processes [11].

9.3 Hydroprocessing of biomass—a promising route for biofuels production

Hsu [12] compared gasoline and diesel production through fast pyrolysis and hydroprocessing of biomass on factors, such as GHG emissions and net energy value. Production of diesel through pyrolysis process is reported to be favorable as it gave lesser GHG emissions (98 g/km CO_2 eq.) in comparison to the production of gasoline (117 g/km CO_2 eq.). Table 9.3 compares the engine and chemical properties of the biofuels to commercial fossil fuels [5]. As seen in Table 9.3, the sulfur levels of biofuels are less (<0.09%) in comparison to fossil fuels (>0.15%). Flash points of biofuels are greater than those of fossil fuels making its storage economical and safe. However, major drawback is the high water content (>0.04%) in biofuels. During hydrotreatment of plant oils, oxygen content in the fuel can be completely removed yielding straight chain paraffins, alkenes, and isomerized products.

Table 9.3 Comparison of fuel properties [5].

Fuel property	Waste cooking oil	Biodiesel from waste cooking oil	Commercial diesel fuel	Commercial gasoline
Kinematic viscosity (mm^2/s, at 313K)	36.4	5.3	1.9–4.1	–
Density (kg/L, at 288K)	0.924	0.897	0.075–0.840	0.718–0.778
Flash point (K)	485	469	340–358	265.8
Pour point (K)	284	262	254–260	
Cetane number	49	54	40–46	(Octane) 86–92
Ash content (%)	0.006	0.004	0.008–0.010	–
Sulfur content (%)	0.09	0.06	0.35–0.55	0.15 (max)
Carbon residue (%)	0.46	0.33	0.35–0.40	–
Water content (%)	0.42	0.04	0.02–0.05	0.01 (max)
Higher heating value (MJ/kg)	41.4	42.65	45.62–46.48	47.8
Free fatty acid (mg KOH/g oil)	1.32	0.1	–	–
Saponification value	188.2	–	–	–
Iodine value	141.5	–	–	–

Renewable diesel is produced by the hydrotreatment of vegetable oil producing diesel fuel substitutes. In soybean to biodiesel conversion, soy meal and glycerin are obtained as coproducts. During the hydrotreatment pathway, soy meal, fuel gas, and heavy oils are obtained as coproducts. From well-to-wheel stage, biodiesel and renewable diesel pathways were compared for their total energy usage. It was concluded that the choice of method for analysis plays a crucial role in the determination of well-to-wheel results. Moreover, maintaining the transparency of the selected method plays a vital role in the accuracy of the LCA. Sobrino et al. [13,14] developed a LCA model for the comparison of biofuels- and fossil fuels–production processes and they provided model tools for management and optimization of production resources. For the growth of biomass, usage of fertilizers and other equipment are essential. Hence, the growth of biomass may contribute to GHG emissions and energy spent for the growth of biomass is debatable [15].

Hydrotreatment of vegetable oil yields CO_2 and CO as the by-products as shown in Fig. 9.1. It should be noted that decarboxylation or decarbonylation reaction results in product with reduced carbon number (original carbon number—1). Preservation of carbon number is essential to assure desirable combustion and engine properties of biofuels. Most desired reaction route will be the direct deoxygenation of feedstock resulting in product with same carbon number and water as by-product. It should be noted that the reaction mechanism shown in Fig. 9.1 is generalized and there will be several intermediate reactions involved in this process [16].

Bio-oil obtained through pyrolysis of lignocellulosic biomass has considerable proportion of solid mass. Physical and chemical treatments are employed for the removal of solid mass and other impurities. Purified bio-oil is hydrotreated in the presence of excess hydrogen to produce renewable gasoline, LPG and diesel fuel as products and water are obtained as byproduct during direct hydrodeoxygenation (Fig. 9.2).

As stated earlier, hydrotreatment is an energy-intensive process and globally, researchers are working on the development of novel catalyst that can favor out hydrotreatment reactions in less severe conditions on a commercial scale making the process energy efficient and environment friendly.

9.4 Woodchips—a socio-environmentally benign feedstock for biofuels production

Terrestrial or aquatic biomass, such as woodchips, agricultural crops, and aquatic plants, and their waste have gained attention as potential

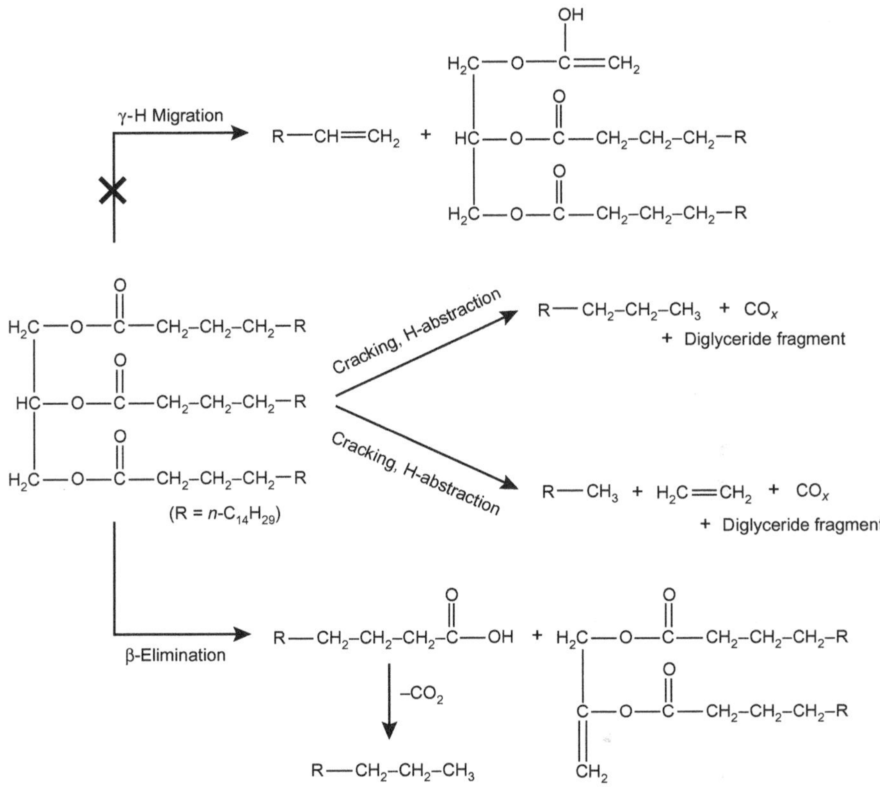

Figure 9.1 Simplified reaction scheme for tristearin deoxygenation [16].

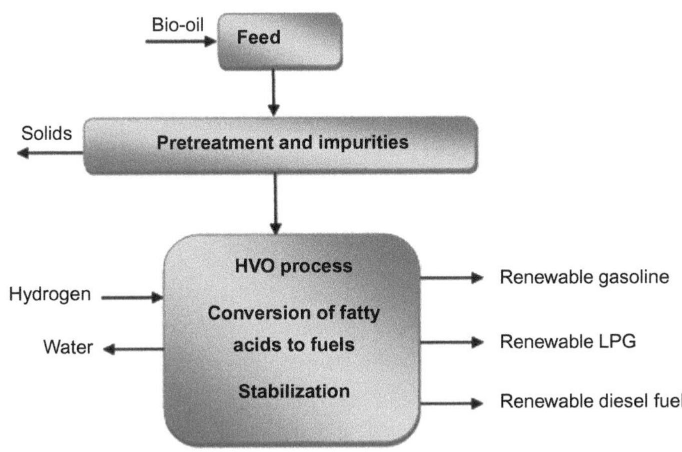

Figure 9.2 Hydrotreatment concept—from waste oils to valuable products [17]. *HVO*-Hydrogenated Vegetable Oil.

renewable energy source for biofuels production. By and large, LCA is the only means to justify the fact that alternate fuels from bio-based resources are comparatively efficient than fossil fuels in terms of energy usage and cost. US Environmental Protection Agency (EPA) carried out LCA of biofuels production pathway to find the methodology that can satisfy the GHG emission conditions put forth by RFS2 and it released a final rule that was effective from July 2010.

Neupane et al. [18] carried out LCA of bioethanol production from woodchips. During the cradle-to-grave analysis of this process, it was concluded that the environmental emissions were higher during harvesting, processing of woodchips, and during their transportation to the facility for biofuel production. Moreover, second-generation fuels pose concerns related to feedstock collection networks and technology commercialization. To conduct LCA, EPA mandates the measurement of generation of carbon dioxide equivalent per mega joule of biofuels. Usually, calculations are based on both direct and indirect emissions. However, this approach has a major pitfall, as it doesn't provide allowance for emission measurements in biofuels production pathway from a single feedstock. This case is particularly important if there is scarce agricultural land source.

9.5 Analysis on food versus fuel debate on third-generation biofuels production

As explained by Kauffman et al. [19], a case study can be considered where a hectare of land is used for cultivating corn and another hectare used for the cultivation of switchgrass. Hectare of land with corn yields corn seeds and the stover. Corn seeds can be used as a source for producing oil that acts as a feedstock for human consumption and for the production of alternate fuels. Nonedible corn stover can be used as a feedstock for the production of bio-oil and biochar. Melamu and Blottnitz [20] analyzed the carbon debt and the energy efficiency of a South African industry that produced bagasse. They considered seven scenarios related to the production and usage of bagasse. In one scenario the manufacture of bagasse was carried out without any diversion with another scenario involving 100% diversions. Factors, such as global warming potential (GWP), usage of nonrenewable energy sources (coal for heat and natural gas), eutrophication of aquatic systems, and acidification of land and terrestrial components, were considered during LCA. It was concluded that the scenario which involved 100% diversion of bagasse to bioethanol proved to be the worst in terms of energy consumption and expenditures. The base scenario involving no diversion of bagasse was concluded to outperform the other scenarios with respect to the various environmental factors. García et al. [21] analyzed one current and four possible future modalities for the production of sugarcane ethanol fuel in Mexico. GHG emissions and the energy balance for the five modalities were estimated. Usually CO_2 release from biofuel combustion is given less importance as CO_2 is sequestered by plants during their growth and respiration, and this concern is still in debate. The GWP due to CO_2 from biofuels is taken as zero. In Mexico, sugarcane is widely used as feedstock for ethanol production like Brazil. To avoid food versus fuel conflict, certain assumptions were accounted for the LCA. Some of the important assumptions are as follows: (1) sufficient rainfall ranges, (2) long frost-free period, (3) considerable potential yields, and (4) usage of present lands [21].

Guinee et al. [2] predicted the trend in growth of LCA of biofuel production systems. Future LCA studies will involve sustainability analysis (SA) for deepening the scope of LCA and promoting its interdisciplinary relationship. As stated earlier, traditional LCA (based on ISO 14040 framework) involves definition of scope and goals, inventory analysis, and impact assessment followed by interpretation. Results from the LCA usually have impact on strategic planning, drafting of public policy, marketing, and product development and improvement.

Lin et al. [17] have performed sensitivity analysis for the production of bio-based fuels from corn seeds and corn stover. Biomass (corn stover) was employed for the production of bio-oil which can be further hydrotreated to produce diesel fuel substitutes, such as higher carbon paraffins and olefins. Requena et al. [1] compared the environmental impact of three different processes that employ rapeseed, sunflower, and soybean oil as feedstocks for alternate fuels production. Table 9.4 illustrates the environmental impact of biodiesel-production process from different feedstocks [22]. It can be observed that the usage of algae for biofuels production is energy intensive with higher GWP (2,756,614 t CO_2 eq.) in comparison to vegetable oils. The higher energy requirements can be attributed to the area usage and incorporation of new technologies for algae growth.

9.6 Influence of by-products/coproducts on socioeconomic benefits

During LCA, it is very important to find commercial value for all the by-products formed during the process (Table 9.5). In the case of transesterification of plant-based oils to produce biofuels, glycerol, and seed cake

Table 9.4 Life cycle environmental impacts of biodiesel production [22].

Items	Units	Soybean	Jatropha fruits	Vegetable seeds	Castor seeds	Algae	Waste cooking oil	Waste extraction oil
Fruits	Ton	214,354	267,682	235,787	219,509	234,184	–	–
Biomass oil	Ton	204,293	217,397	208,340	219,290	234,184	–	–
Energy	Terajoule	8305	5481	5260	8093	40,276	7485	8142
Freshwater	10,000 t	138,207	102,386	71,679	118,059	4791	1866	9223
GWP	Ton CO_2 eq.	566,384	430,219	417,447	570,883	2,756,614	602,539	647,602
HTP	Ton 1,4-dichlorobenzene eq.	11,584,047	5,426,562	5,920,674	8,441,050	14,222,412	24,740,478	8,925,107
FAETP	Ton 1,4-dichlorobenzene eq.	1,205,481	558,859	615,000	871,386	1,437,457	2,459,580	911,559
MAETP	Ton 1,4-dichlorobenzene eq.	233,000	108,780	119,084	169,432	284,414	496,285	281,919
TETP	Ton 1,4-dichlorobenzene eq.	0.226	0.085	0.106	0.141	0.171	0.240	0.142
POCP	Ton ethylene eq.	152	136	128	163	852	195	204
AP	Ton SO_2-eq.	4393	3971	3721	4713	24,836	5664	5952
EP	Ton PO_4-eq.	385	361	353	398	1177	640	859
Solid wastes	10,000 t	36.3	8.5	9.3	11.8	45.1	13.9	35.3

AP, Aquatic potential; *EP*, eutrophication potential; *FAETP*, freshwater aquatic ecotoxicity potential; *GWP*, global warming potential; *HTP*, human toxicity potential; *MAETP*, marine aquatic ecotoxicity potential; *POCP*, photochemical oxidation potential; *TETP*, terrestrial ecotoxicity potential.

Table 9.5 Coproducts and their uses in biofuels industry [23].

Feedstock	Main product	Coproducts	Coproduct uses
Soybean	Biodiesel	Soy meals and glycerine	Soy meal—animal feed; glycerine—specialty chemical
Soybean	Renewable diesel	Fuel gas and heavy oils	Energy sources for plant internal use or energy products for sale
Rapeseed	Biodiesel	Rapeseed meal and glycerine	Rapeseed meal—animal feed; glycerine—specialty chemical
Palm oil	Biodiesel	Residual fertilizer and glycerine	Fertilizer—farming; glycerine—specialty chemical
Cellulosic biomass	Ethanol	Lignin	Steam and electricity production in cellulosic ethanol plants

after oil extraction are obtained as by-products. Usually, seed cakes are used as organic fertilizer and glycerol is used to produce cosmetics and biolubricants. Flash points of biofuels are greater than those of fossil fuels making its storage economical and safe. However, major drawback is the high water content ($>0.04\%$) in biofuels. During hydrotreatment of plant oils, oxygen content in the fuel can be completely removed yielding straight chain paraffins, alkenes, and isomerized products.

Fig. 9.3 indicates the different processes and products that are involved during the production of biofuels and value-added products from biomass [24]. Till date, comparative LCA coupled with SA on all these processes is not available in literature and it can be a potential area to focus on.

The route that involved the production of electricity from plants (1-7-8) is not economically and environmentally beneficial owing to the enormous consumption of energy and increased GHG emissions. Sugars produced from plants (route 1-2-3) can also be used to produce fermentative alcohols that have high value in market. Integration of biorefinery is very essential to balance the CO_2 emissions from biorefineries. At present, it is tough to avoid the usage of fossil fuels for generation of biomass and their transportation. Hence, it is recommended to focus on alternate ways to neutralize the overall GHG emissions from bioenergy process.

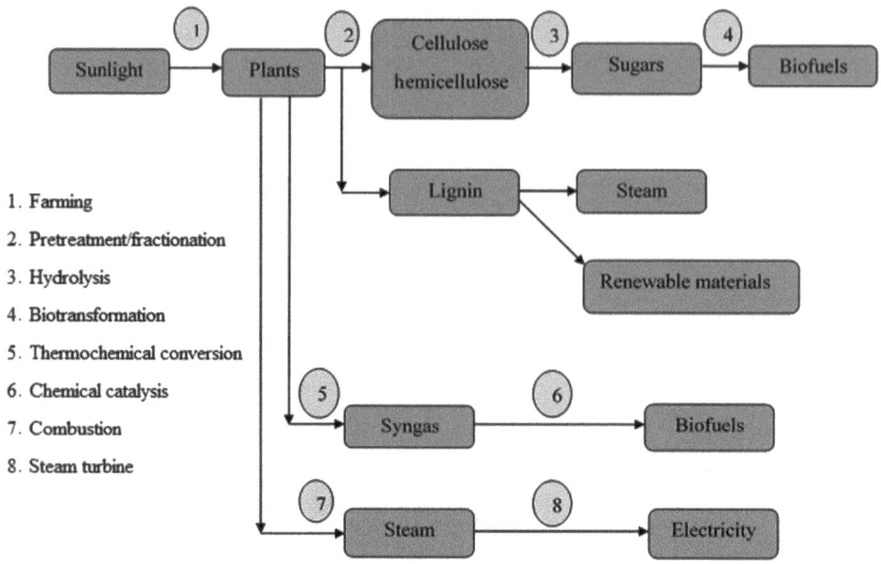

Figure 9.3 Different pathways for biofuels production from lignocellulosic biomass [24].

9.7 Future of biofuels sector

By 2050 the global aviation industry planned to reduce its emissions by 50% through improvement in air traffic management, improved design of aircrafts, and promoting the commercial usage of biomass-based fuels [25]. Renewable jet fuel production (farnesene) from sugarcane base had a life cycle emission of 21 g CO_2 eq./MJ (approx.) while its fossil fuel counterpart gave an emission of 80–95 g CO_2 eq./MJ range [26]. At present, it can be stated that biogas production is more energy favorable in comparison to the process which involves the combination of biogas and bioethanol production. Sensitivity analysis is crucial to test the robustness of the results obtained using LCA. LUC defines the release of sequestrated carbon into the atmosphere when carbon-rich lands, such as forest lands, are converted to low carbon content lands, such as agricultural areas, for the growth of oilseed crops, and this release of carbon creates carbon debt and increases the annual energy usage in a biofuel industry. It is evident that the development of biofuels from biomass is still in rudimentary stages and considerable research work is essential for the optimization of biofuel production cost and energy utilization.

Figure 9.4 LCA of canola oil production. *LCA*, Life cycle assessment.

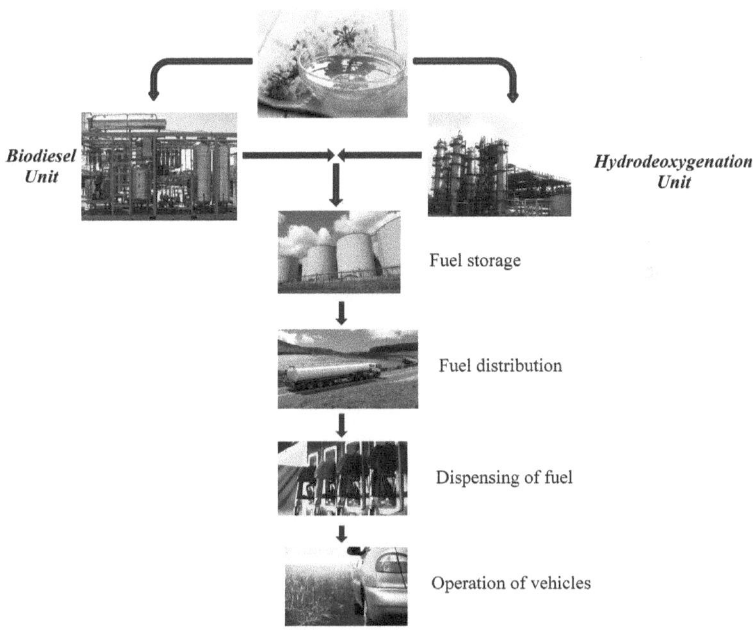

Figure 9.5 LCA of biofuels production from canola (hydrotreatment and transesterification). *LCA*, Life cycle assessment.

Figs. 9.4 and 9.5 indicate the stages involved during cradle-to-grave LCA of canola oil-based biofuel-production process. During the production of feedstock (canola oil), major factors that impact environment are the usage of fertilizers and transportation of canola oil. Usage of fertilizers

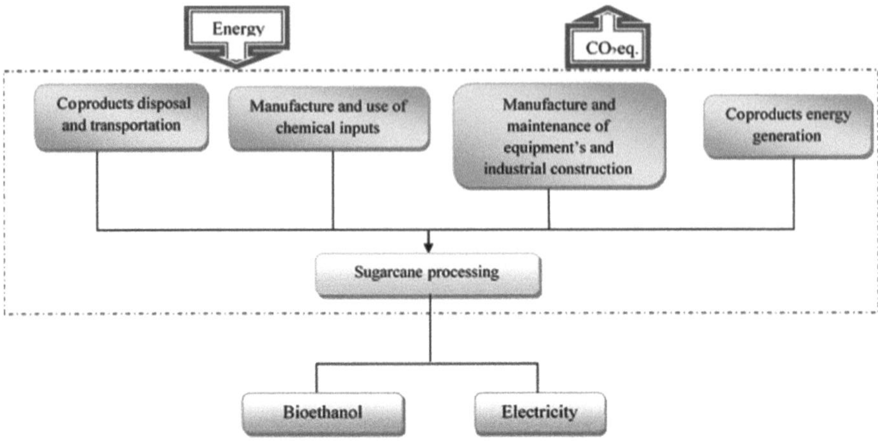

Figure 9.6 Industrial subsystems considered for traditional system sugarcane ethanol life cycle [8].

affects the acidification and human toxicity. Transportation of canola oil and canola crushing unit contribute to GHG emissions.

Canola oil can be used to produce biodiesel through transesterification process or it can be used to produce green diesel using catalytic hydrotreatment process (Fig. 9.5). During the transesterification process, methanol is employed as a reactant with acid/base catalyst. Usage of these chemicals will contribute to human toxicity and acidification.

During hydrotreatment process, hydrogen is employed and the reaction is carried out at high temperature and pressure (>300°C and 1200 Psi). Mostly, CO_2 and CO are obtained as by-products. Severe operating conditions can contribute to the release of energy and to overall GHG emissions. One of the major factors that contribute to increased GHG emissions is the transportation of biomass (feedstock) from its production area to the processing unit (Fig. 9.6).

Holma et al. [27] compared the LCA of processes that use forest based and microalgae as feedstocks for biodiesel production. They are both contrasting system as one feedstock is totally natural (forest-based feedstocks) and another one (microalgae) is based on artificial infrastructure and external growth factors. The two processes were compared on their impact on environmental factors, such as depletion of feed resources, land usage and their impact on soil, and air quality.

Fig. 9.7 indicates the different processing units that can be employed depending on the nature (lignocellulosic, starch rich, and sugar rich) of

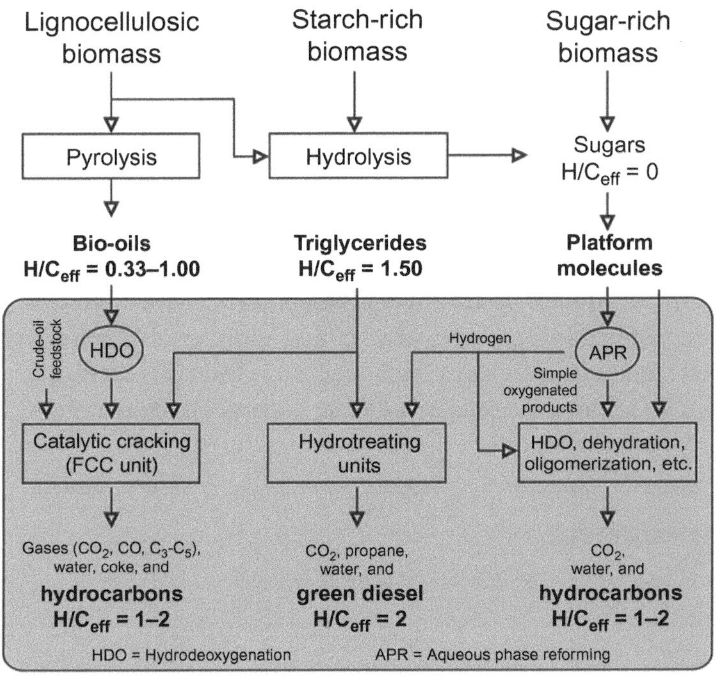

Figure 9.7 Integration of biomass feedstock in conventional refinery processes [28].

biomass [28]. In refinery units, it is possible to integrate catalytic cracking unit and hydrotreating units as they both involve high-temperature operations using catalysts. In most cases the products produced from these units are also similar (CO_2, water, and paraffins).

9.8 Conclusions

Production of third-generation biofuels from agricultural and forest-based feedstocks is a complex process and hence, a holistic approach toward LCA is mandatory to analyze their environmental benefits and capital costs. Most of the LCA incorporates quantity of energy into analysis. Quality of energy is important and the concept of exergy is proven to be a powerful technique to assess the sustainability of bioenergy technology. Renewability or CO_2 neutrality of the futuristic fuels cannot be 100% as many nonrenewable sources of energy are employed for the production of biomass feedstocks.

LCA studies are usually limited to the analysis of balance in GHGs and energy transfer between system and its surrounding. Choice of feedstock for biofuels production is highly dependent on the agricultural subsidiaries available for the growth of crops for biofuels production. Countries with multiple feed sources, such as corn, canola, rapeseed, and soybeans, should concentrate on the minimization of arable land usage for the cultivation of bioenergy crops. Renewability or CO_2 neutrality of these futuristic fuels cannot be 100% as many nonrenewable sources of energy are employed for the production of feedstocks (biomass). Economic and environmental credibility of coproducts are essential to completely justify that investments in biofuels sector are profitable and it will meet the present energy demands and also curb greenhouse emissions. Most of the LCA incorporate quantity of energy into analysis. However, quality of energy (exergy) is an important concept and is proven to be a powerful technique to assess the sustainability of bioenergy technology.

Acknowledgment

The financial support from BioFuelNet Canada is greatly acknowledged.

References

[1] J.F.S. Requena, C. Guimaraes, S.Q. Alpera, E.R. Gangas, S. Hernandez-Navarro, L. M.N. Gracia, et al., Life cycle assessment (LCA) of the biofuel production process from sunflower oil, rapeseed oil and soybean oil, Fuel Process. Technol. 92 (2011) 190–199.

[2] J.B. Guinee, R. Heijungs, G. Huppes, Life cycle assessment: past, present and future, Environ. Sci. Technol. 45 (2011) 90–96.

[3] B. Grau, E. Bernat, P. Rita, R. Jordi-Roger, R. Antoni, Environmental life cycle assessment of rapeseed straight vegetable oil as self-supply agricultural biofuel, Renew. Energy 50 (2013) 142–149.

[4] B.J. Krohn, M. Fripp, A life cycle assessment of biodiesel derived from the "niche filling" energy crop camelina in the USA, Appl. Energy 92 (2012) 92–98.

[5] N. Taufiqurrahmi, S. Bhatia, Catalytic cracking of edible and non-edible oils for the production of biofuels, Energy Environ. Sci. 4 (2011) 1087–1112.

[6] Canola Council of Canada, An Industry Inspired, 2016 Annual Report (accessed 11.18).

[7] F. Cherubini, A.H. Strømman, Life cycle assessment of bioenergy systems: state of the art and future challenges, Bioresour. Technol. 102 (2011) 437–445.

[8] S.P. Souza, M.T. De Ávila, S. Pacca, Life cycle assessment of sugarcane ethanol and palm oil biodiesel joint production, Biomass Bioenergy 44 (2012) 70–79.

[9] K. Soratana, A.E. Landis, Evaluating industrial symbiosis and algae cultivation from a life cycle perspective, Bioresour. Technol. 102 (2011) 6892–6901.

[10] K.L. Kadam, Environmental implications of power generation via coal microalgae cofiring, Energy 27 (2002) 905–922.

[11] V. Vasudevan, R.W. Stratton, M.N. Pearlson, G.R. Jersey, A.G. Beyene, J.C. Weissman, et al., Environmental performance of algal biofuel technology options, Environ. Sci. Technol. 46 (2012) 2451–2459.

[12] D.D. Hsu, Life cycle assessment of gasoline and diesel produced via fast pyrolysis and hydroprocessing, Biomass Bioenergy 45 (2012) 41–47.
[13] F.H. Sobrino, C.R. Monroy, Critical analysis of the European Union directive which regulates the use of biofuels: an approach to the Spanish case, Renew. Sustain. Energy Rev. 13 (2009) 2675–2681.
[14] F.H. Sobrino, C.R. Monroy, J.L. Hernandez, Critical analysis on hydrogen as an alternative to fossil fuels and biofuels for vehicles in Europe, Renew. Sustainable Energy Rev. 14 (2010) 772–780.
[15] T.E. McKone, W.W. Nazaroff, P. Berck, M. Auffhammer, T. Lipman, M.S. Torn, et al., Grand challenges for life-cycle assessment of biofuels, Environ. Sci. Technol. 45 (2011) 1751–1756.
[16] T. Morgan, D. Grubb, E. Santillan-Jimenez, M. Crocker, Conversion of triglycerides to hydrocarbons over supported metal catalysts, Top. Catal. 53 (2010) 820–829.
[17] C.S.K. Lin, L. Pfaltzgraff, L. Herrero-Davila, E.B. Mubofu, S. Abderrahim, J.H. Clark, et al., Food waste as a valuable resource for the production of chemicals, materials and fuels. Current situation and global perspective, Energy Environ. Sci. 6 (2013) 426–464.
[18] B. Neupane, A. Halog, S. Dhungel, Attributional life cycle assessment of woodchips for bioethanol production, J. Clean. Prod. 19 (2011) 733–741.
[19] N. Kauffman, D. Hayes, R. Brown, A life cycle assessment of advanced biofuel production from a hectare of corn, Fuel 90 (2011) 3306–3314.
[20] R. Melamu, H.V. Blottnitz, 2nd Generation biofuels a sure bet? A life cycle assessment of how things could go wrong, J. Clean. Prod. 19 (2011) 138–144.
[21] C.A. García, A. Fuentes, A. Hennecke, E. Riegelhaupt, F. Manzini, O. Masera, Life-cycle greenhouse gas emissions and energy balances of sugarcane ethanol production in Mexico, Appl. Energy 88 (2011) 2088–2097.
[22] S. Liang, M. Xu, T. Zhang, Life cycle assessment of biodiesel production in China, Bioresour. Technol. 129 (2013) 72–77.
[23] M. Wang, H. Huo, S. Arora, Methods of dealing with co-products of biofuels in life-cycle analysis and consequent results within the U.S. context, Energy Policy 39 (2011) 5726–5736.
[24] W.D. Huang, Y.H.P. Zhang, Analysis of biofuels production from sugar based on three criteria: Thermodynamics, bioenergetics, and product separation, Energy Environ. Sci. 4 (2011) 784–792.
[25] D. Mu, M. Min, B. Krohn, K.A. Mullins, R. Ruan, J. Hill, Life cycle environmental impacts of wastewater-based algal biofuels, Environ. Sci. Technol. 48 (2014) 11696–11704.
[26] M. Moreira, A.C. Gurgel, J.E.A. Seabra, Life cycle greenhouse gas emissions of sugar cane renewable jet fuel, Environ. Sci. Technol. 48 (2014) 14756–14763.
[27] A. Holma, K. Koponen, R. Antikainen, L. Lardon, P. Leskinen, P. Roux, Current limits of life cycle assessment framework in evaluating environmental sustainability – case of two evolving biofuel technologies, J. Clean. Prod. 54 (2013) 215–222.
[28] J.A. Melero, J. Iglesias, A. Garcia, Biomass as renewable feedstock in standard refinery units. Feasibility, opportunities and challenges, Energy Environ. Sci. 5 (2012) 7393–7420.

CHAPTER 10

Pretreatment of lignocellulosic sugarcane leaves and tops for bioethanol production

S. Niju, M. Swathika and M. Balajii
Department of Biotechnology, PSG College of Technology, Coimbatore, India

Contents

10.1 Introduction

Energy has been a backbone for many sectors of the economy, which mainly includes industries, agriculture, and transportation, for its

Lignocellulosic Biomass to Liquid Biofuels
DOI: https://doi.org/10.1016/B978-0-12-815936-1.00010-1

sustainable growth and development. The intensified use of energy has been both a result of and a prerequisite for industrialization, which has led to the depletion of energy resources. Meanwhile, energy requirement is also increasing tremendously due to high population growth, industrialization, and urbanization. Currently, fossil fuels are largely used to meet most of the world's energy demand [1], but on the other hand, world is facing reduction of fossil fuels, which are nonrenewable and major contributors of environmental pollution. Dwindling fossil fuels and inability to replenish these limited sources increases the world's energy demand. So it is mandatory to replace the fossil fuels, at least partially by renewable energy resources. The sustainable renewable bioenergy resources offer an attractive solution to produce liquid biofuels by which world's primary energy demand can be accomplished [2]. In the meantime, many countries are promoting several actions for the development of renewable energy, which mainly includes biofuel development [3]. The remarkable features of all the biofuel policies include the less dependence on fossil fuels, minimizing emission of greenhouse gases and maximum utilization of renewable waste [4]. Moreover to produce biofuels, the availability of raw materials and its potentials are to be analyzed in the economic, environmental, and social conditions. Biofuels in turn decrease the dependency, provide energy security, strengthen rural and agricultural economies, and also increase the sustainability of the world's transportation system [2].

At present, one of the most promising alternatives for convectional petrol is bioethanol, the largest produced liquid biofuel in world. As a transportation fuel, it can be either used in blended form along with gasoline or as pure ethanol [5]. With global production crossing 100 billion liters in 2016 as reported in renewable fuel association, ethanol is expected to remain the most prominent and cost-effective biofuel. Bioethanol is considered the prime biofuel in world and its market is continuously growing. The replacement for fossil fuels with bioethanol has already been established and its use is increasing in Brazil (sugarcane) and the United States (maize). The United States and Brazil together are the major producers of the world's ethanol [6]. In addition, using bioethanol safeguards the environment, helps in the development of rural economy, and ensures fuel security. It is the growing renewable transport biofuel that can be produced from a diverse range of agricultural residues [1].

Liquid fuel produced from food crops containing sucrose, such as sugarcane, sugar beet, and sweet sorghum, and starch-rich feedstocks, such as corn, wheat, and cassava, are termed first-generation bioethanol [5]. However, there are several limitations in producing energy from these substrates because of dispute between food and feed production and it also raised the competition between fuel and food. To overcome all these conflicts, the technologies involved in renewable energy generation target the use of lignocellulosic biomass (LCB) as feedstocks for bioethanol production. These processes are produced with high efficiency and also have the flexibility to supply both electricity and biofuels [1]. Interestingly, world's attention is now on cellulosic or plant biomass as raw material for the production of bioethanol rather than corn and sugarcane.

Biomass is a bioresidue that is available in marine vegetation, forest or organic waste, crop production, and agro or food industries waste. Various forms of biomass resources, which can be used for the production of biofuels, are mostly in the form of grasses, woody plants, fruits, vegetables, manures, and aquatic plants. The biomass energy sources can be classified as residue of agricultural crop, energy plantation, and municipal and industrial waste [7]. Biomass is the major contributor to the world-level energy needs. Among both developed and developing countries, it has the wide area of scope. On the cumulative use of biomass, its main influence depends on the land usages for energy purposes. It has some ecological development concerns such as the social development and environmental impacts that are associated with land usages and its alteration [7].

Currently, much focus is given on LCB [8], nonfood crops, industrial and municipal wastes, because it overcomes the limitations of first-generation biofuels and also results in greater greenhouse gas reduction. Among the available feedstock, the LCB has been identified as the cheap substrate and sustainable feedstock for producing biofuels and other valuable products [5].

Sugarcane crops are potent renewable energy resources that are abundantly available and largely grown in countries such as Brazil, China, India, Thailand, and Australia. The main residues obtained in agricultural lands from the sugarcane crops are sugarcane leaves (SCL) and sugarcane tops (SCT). SCL and SCT are rich in cellulose content, which can be exploited for bioethanol production [9]. This chapter summarizes the attributes and agronomy of SCT, different pretreatment methods employed for their bioconversion and hydrolysis techniques that can be used for producing reducing sugars from SCT for fermentation into bioethanol.

10.2 Lignocellulosic biomass: diversity and traits

10.2.1 Lignocellulosic biomass: resource for renewable bioenergy

LCB is the plant biomass, available in large amount on earth, and can be used as raw material for biofuel production especially bioethanol [7]. Nearly 50% of the world's biomass are LCB that do not enter the human food chain and have energy content higher than world's basic energy requirements. These traits establish it an efficient feedstock for bioethanol production to meet the international energy demand. Moreover, conversion of a plant-derived biomass into bioethanol is one of the main sources of renewable energy because it absorbs solar energy and converts it into the chemical energy of organic compounds producing simple and complex carbohydrates with the help of carbon dioxide and water, which also aids in mitigating pollution. Bioethanol produced from the LCB is called second-generation bioethanol, which does not compete with human food and animal feed industry [8]. Apart from bioethanol production, many value-added products, such as lactic acid, acetic acid, furfural, methanol, and hydrogen, can be produced from LCB. In the biomass, lignin can be used for producing certain polymers and aromatic aldehydes [10].

10.2.2 Chemistry of lignocellulosic biomass

LCB is composed of carbohydrate polymers of cellulose and hemicellulose, aromatic polymers of lignin and small amounts of acids, salts, and minerals [7]. The carbohydrate polymers that include cellulose and hemicellulose comprise two-thirds of the biomass, which can be hydrolyzed into reducing sugars and further fermented into bioethanol, while lignin cannot be used for the production of bioethanol.

Cellulose (40%–60% of the dry biomass) is found as structural material in plants formed from common six-carbon sugar glucose, although glucose ($C_6H_{12}O_6$) is the smallest monomer found as a result of cellulose degradation. The basic building block of cellulose is actually cellobiose, which is a two-glucose anhydride unit (dimer). The orientation of the linkages and hydrogen bonds makes cellulose a rigid polymer so cannot be broken down easily. The polysaccharides in cellulose are reduced to simple sugars (glucose) on hydrolysis, which is called saccharification [11].

Hemicellulose (20%–40% of the dry biomass) is the macromolecular polysaccharides and the most soluble fraction of native plant biomass. It consists of highly branched short chains of various sugars such as glucose (six carbon), galactose (six carbon), mannose (six carbon) but mainly xylose (five carbon) and arabinose (five carbon). Hemicellulose binds with pectin to cellulose to form a network of cross-linked fibers. Hemicellulose can be easily hydrolyzed as they are highly branched and amorphous in nature [11].

Lignin (10%–25% of the dry biomass) is a very complex molecule made from phenyl propane and methoxy units and noncarbohydrate polyphenolic substance linked in a three-dimensional structure [7]. It is like glue that holds the cellulose fibrils, making removal of lignin difficult. But it is mandatory to remove lignin for effective ethanol production. This lignin can be degraded only by few organisms and by using certain physical and chemical treatments. For any lignocellulosic ethanol production to be cost-effective, the challenge lies in the removal of lignin and its by-products.

Hemicellulose and lignin form a protective covering surrounding cellulose, which must be removed or modified for the hydrolysis to occur. Therefore advanced pretreatments technologies are required for maximal removal of these compounds [7].

10.2.3 Sources of lignocellulosic biomass

LCB is categorized based on their origin as follows: wood such as softwoods and hardwoods; shrubs; nonfood agricultural crops such as kenaf, rapeseed; residues, such as wheat straw, corncobs, rice husk, sugarcane, and municipal solid wastes related to thinning, gardening, road maintenance, etc. LCB involved in bioethanol production mainly includes agricultural residues [5]. Most of the organic materials are produced while processing the by-products during agricultural crop harvesting. It is also classified as primary and secondary residue. The primary residues are the residual components obtained in the field during harvesting the crops (e.g., SCT and wheat straw) while the secondary residues are assembled during the processes are defined (e.g., sugarcane bagasse) [5]. Though the primary residues are used as animal feed and as fertilizers, their application in energy production is very low and limited. Meanwhile, the secondary residues, gained in the large quantity from its yielding site can also be confined as energy resources. Tons of agricultural residues are not utilized and burnt directly in open field, which has severe impact on the environment with huge loss of energy. Agricultural waste, such as rice husks, wheat straw, rice straw, energy grass, corncobs, sugarcane bagasse, and dry SCL and SCT [6], are abundantly available in the field and believed to be a huge potential resource for bioethanol production.

10.2.4 Lignocellulosic biomass: a biorefinery approach

A biorefinery integrates biomass conversion techniques and processes for the generation of energy such as biofuels, power, and heat. Many valuable chemicals can also be produced from biomass either as a main product or as a by-product during biofuel production. Lignocellulosic feedstocks are an economical resource that are found abundantly and have the capability to support the sustainable production of liquid and gaseous biofuels. Different conversion technologies, which mainly include thermal and mechanical treatments, are employed to exploit maximum energy generation and to produce value-added products. Various treatments employed for the conversion of LCB for energy generation are summarized and produced as a flowchart in Fig. 10.1.

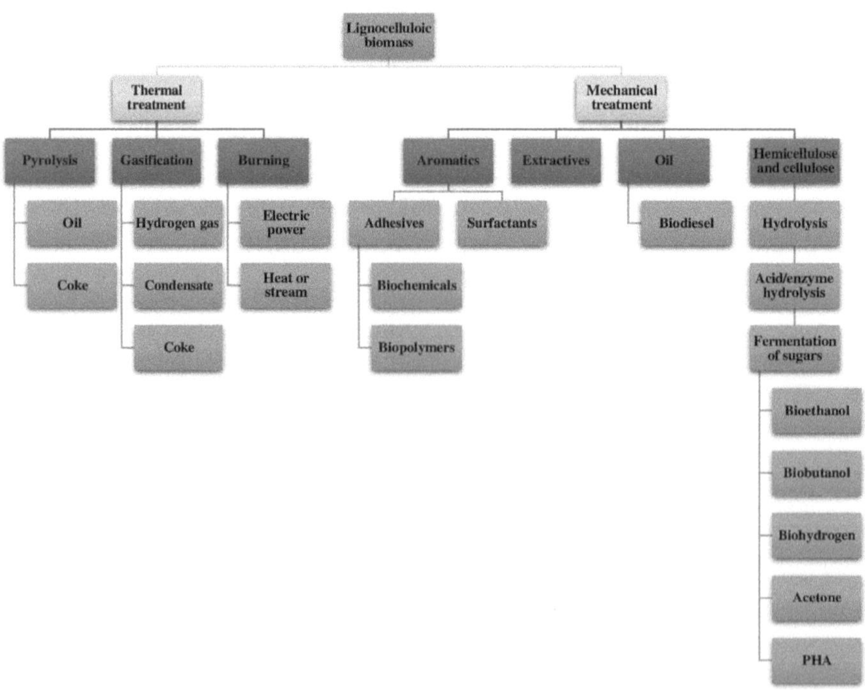

Figure 10.1 Biorefinery approach developed from lignocellulosic biomass.

10.3 Sugarcane tops as potential feedstock for bioethanol production

10.3.1 Sugarcane tops agronomy and production

Sugarcane, *Saccharum officinarum*, is a species of tall perennial true grasses of the genus *Saccharum*, tribe Andropogoneae [12]. The main product of sugarcane is sucrose, which gets accumulated in the stalk internodes and can be extracted and purified in specialized sugar factories, and then it can be used in food industries or can be fermented to produce ethanol. A fully grown sugarcane plant will have about 75% sugarcane stalks, and the leaves and tops will be around 25% [13]. SCT including leaves contain sugar in the form of cellulose and are rich cellulosic materials. Thus ethanol can be extracted from SCT by taking advantage of its rich cellulosic matter and the large biomass availability. Extraction of ethanol from SCT and leaves neither will affect food supply nor will interfere negatively with sugar-based food or juice extracted from stalks of sugarcane plant. SCT are available in surplus during harvest, but they are burnt in the field itself [14], and usually used as animal fodder before the leaves start rotting.

Polyaromatic hydrocarbons are released during the burning process, and some of the compounds released in the air can be carcinogenic or mutagenic. Nearly 1700 million tons of sugarcane are produced annually worldwide as reported by the Food and Agricultural Organization in 2010. This generates huge amount of postharvest residues, mainly SCT, which are inexpensive and readily available source of LCB. Usually, 0.25–0.30 MT of SCT are produced while harvesting 1 MT of sugarcane [9]. Efforts can be taken to utilize SCT as a substrate for bioethanol production. The sugarcane crop is a strong renewable energy resource, which is abundantly available and also largely produced in the countries such as Brazil, China, India, Thailand, and Australia.

10.3.2 Sugarcane tops attributes

The SCT have three major components with an average composition of 29.85% cellulose, 18.85% hemicellulose and 25.80 ± 0.1% lignin, 2.56 ± 0.1% ash [15] while SCL have a composition of 36% cellulose, 21% hemicelluloses, and 16% lignin [16]. Furthermore, the energy of dry SCL is equivalent to the energy in 10 t of coal per hectare. SCT are a rich cellulosic material containing high content of glucan and xylan and found to be an attractive raw material, which can be used for ethanol production. The steps involved in size reduction of SCT were presented in Fig. 10.2.

Figure 10.2 Steps involved in size reduction of SCT. *SCT*, Sugarcane tops.

10.4 Pretreatment for delignification

Pretreatment is one of the most crucial steps in biomass conversion as it directly impacts the efficiency of bioethanol production. LCB mainly consists of densely packed cellulose and hemicelluloses along with lignin, which serves plant for performing several functions [17]. The main aim of the pretreatment is to disrupt the recalcitrant lignin structures in the biomass and to make cellulose more accessible to the enzymes for the conversion into fermentable sugars. The lignin removal efficiency depends on the type of pretreatment employed and the optimum conditions maintained during the treatment. Various pretreatment methods used for delignification of SCT reported in different literatures were presented in Table 10.1. Process flow diagram for bioethanol production with different pretreatment methods for lignin removal from SCT is shown in Fig. 10.3.

10.4.1 Physical pretreatment

10.4.1.1 Microwave pretreatment

Microwaves (MWs) are electromagnetic waves with frequency ranging from 300 MHz to 300 GHz. MWs have high heating efficiency, which

Table 10.1 List of pretreatments employed for delignification of sugarcane tops (SCT).

S. no.	Pretreatment technique	Optimized pretreatment conditions	Percentage removal of lignin/hemicellulose	References
1	SCT—dilute acid pretreatment	3% w/w H_2SO_4, 15% w/w solid loading, mixed particle size, and incubation time of 60 min at 121°C	Hemicellulose content decreased from 18.9% to 6.16%	[9]
2	SCT—surfactant-assisted acid pretreatment	2.5% w/w Triton X-100, 1.5% of H_2SO_4, 30% of biomass loading, and incubation time of 45 min at 121°C	Lignin removed—50% Hemicellulose removed—53%	[18]
3	SCT—surfactant-assisted ultrasound pretreatment	20% w/w of biomass, 3% w/w surfactant concentration (Tween 40) and incubated for 60 min at 121°C followed by sonication for 60 s	Lignin removed—40% Hemicellulose removed—30.2%	[15]
4	SCT—alkali pretreatment	3% NaOH, 15% of biomass loading, and incubated for 60 min at 121°C	Lignin removed—89.80% Hemicellulose removed—45.78%	[19]
5	SCT—enzymatic pretreatment	20% (w/v) substrate concentration, pH 7, 6 h incubation time, and 500 IU/mL enzyme titer at 40°C	Lignin removed—79.1%	[20]
6	SCT—microwave alkali–assisted pretreatment	2% (w/v) NaOH, 10% of biomass loading and microwave at 320 W for 10 min	Lignin removed—67%	[21]
7	SCT—alkali and acid pretreatment	15% w/v biomass loading, 2% w/v of NaOH, autoclave at 121°C, 15 $lb/in.^2$ for 15 min followed by 2% w/v of H_2SO_4, autoclave at 121°C, 15 $lb/in.^2$ for 15 min	–	[22]
8	SCT—waste glycerol–assisted transition metal and alkali pretreatment	3% v/v glycerol in the presence of 1% NaOH and different transition metals at concentration of 1% w/w (manganese sulfate, cadmium acetate, nickel chloride, zinc sulfate, ferrous sulfate, and ferric chloride) at 121°C for 60 min	Lignin removed—21.3% Hemicellulose removed—47.50%	[23]
9	SCT—enzymatic pretreatment	500 IU/mL laccase, pH 7, final substrate concentration of 30% (w/v) incubated at 35°C for 6 h	Lignin removed—70%–80%	[24]

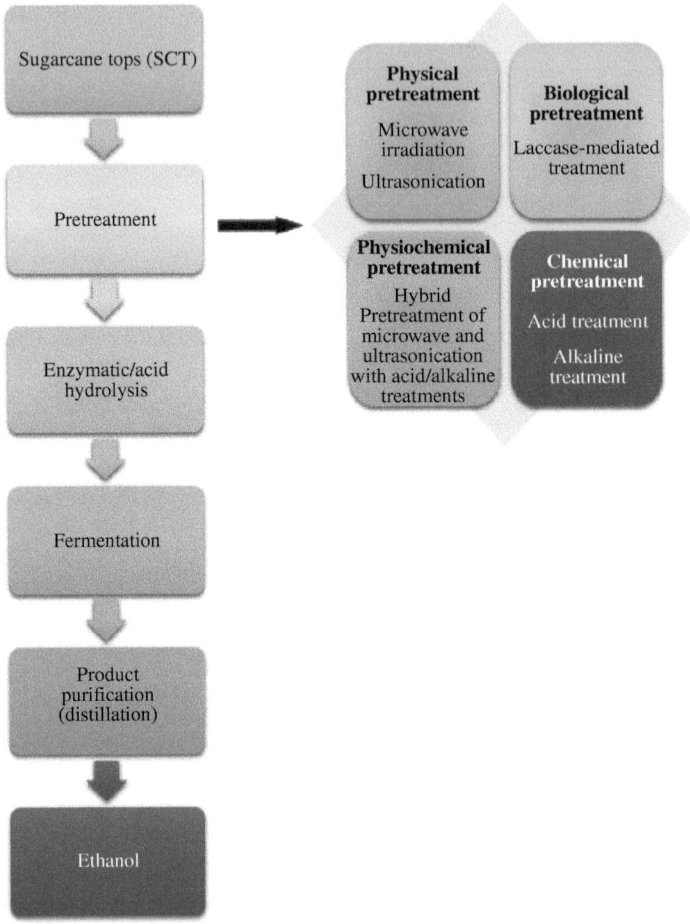

Figure 10.3 Process flow diagram for bioethanol production with different pretreatment methods used for lignin removal from SCT. *SCT*, Sugarcane tops.

increases the reaction rate and reduces the reaction time. MW irradiation employs high-temperature generation that creates thermal hot spots, which accelerates the collision of ions, thus producing many chemical changes in the lignocellulosic matrix [25]. The high temperature produced reduces the cellulose crystallinity and increases delignification. MW pretreatment does not result in any by-product formation but the main drawback is the high cost involved in the energy generation. MW pretreatment is usually combined with other pretreatments for effective delignification.

Maurya et al. [21] performed MW pretreatment and reported 67% of lignin removal from SCT using 2% (w/v) NaOH, which was initially mixed with 10% (w/v of biomass), and then MW pretreatment was carried out at 320 W for 10 min.

10.4.1.2 Ultrasound pretreatment

Ultrasound pretreatment (UP) employs the principle of hydrodynamic cavitation. The application of ultrasound waves results in the formation, growth, and subsequent collapse of microsize cavities [26]. High temperature and pressure gradients are produced by the microsize cavities for few microseconds, which creates high energy densities. These bring alterations in the surface morphology of the biomass thus disrupting the recalcitrant lignocellulosic structures for the effective production of fermentable sugars. By-products are not produced in the pretreatment as it employs only ultrasound waves and also improves the overall yield of sugars [26].

A surfactant-assisted UP of SCT was performed by Sindhu et al. [15] with optimum conditions of 20% (w/w) of biomass loading, 3% (w/w) of Tween 40 (surfactant) and soaked for 60 min followed by sonication (60 s) yielded 0.426 g/g of reducing sugar.

10.4.2 Chemical pretreatment

10.4.2.1 Dilute acid pretreatment

Dilute acid pretreatment (DAP) is the most widely used commercial pretreatment, which breaks the covalent bonds in the composite linkages, thus disrupting the lignocellulosic structure. The most commonly used acids are dilute sulfuric acid, nitric acid, and hydrochloric acid [17]. Among these, dilute sulfuric acid is extensively employed because of its low cost and high efficiency in lignin removal. The DAP is usually performed over a temperature range of 120°C–210°C with an acid concentration of less than 4 wt.% and incubation time varying from several minutes to hours. DAP is simple in operation, cost-effective, and results in maximum lignin removal [17]. The main drawback is the formation of several inhibiting by-products such as hydroxymethyl furfural (HMF), furfural and other by-products such as phenylic compounds and aliphatic carboxylic acids [27]. It also requires pH neutralization for downstream processes. Moreover, it also requires a significant size reduction of biomass, which consume large amounts of energy [17].

Different DAP treatments were carried out in SCT and SCL with varying sulfuric acid and surfactant concentration. Sindhu et al. [9]

employed DAP in SCT with optimum conditions of 3% (w/w) sulfuric acid, 15% (w/w) solid loading with mixed particle size and incubation time of 60 min at 121°C, which yielded 0.437 g/g of reducing sugar.

In another study reported by Sindhu et al. [18], surfactants were used to achieve better dissolution of lignin. The optimum conditions maintained during the study were 25% (w/w) Triton X-100 (surfactant), 1.5% of dilute sulfuric acid, and 30% (w/w) of biomass loading with mixed particle size and incubation time of 45 min at 121°C, which yielded 0.448 g/g of reducing sugar.

Moodley and Kana [16] performed a comparative study by optimizing three acid-based pretreatments (HCl, H_2SO_4, and HNO_3) for recovering reducing sugars from sugarcane leaf with optimum conditions of acid concentration, temperature, solid-to-liquid ratio, and heating time in the range of 0.5%–5.0% (v/v), 60°C–100°C, 30%–50% (w/v), and 60–240 min, respectively, resulted in maximum of 93.15% hemicellulose removal in HCl-based model.

10.4.2.2 Alkaline pretreatment

Alkaline pretreatment is another major chemical pretreatment, which uses sodium or calcium hydroxide for effective cellulose digestibility. It generally involves a delignification process, which influences the solubilization of a significant amount of hemicelluloses. The reaction mechanism is based on the saponification reaction of intermolecular ester bonds cross-linking between hemicellulose and lignin [17]. Saponification causes the cleavage of lignin bonds and makes cellulose more accessible to the enzymes for hydrolysis. The main advantage is that the residual alkali can be recycled and reused and also involves lower reaction temperature and pressure for effective lignin removal. The main drawback is the incorporation of irrecoverable salts in the biomass during the pretreatment [17].

Sindhu et al. [19] employed alkali pretreatment for lignin removal from SCT with optimum conditions of 3% sodium hydroxide, 15% of biomass loading and incubated for 60 min at 121°C in laboratory autoclave and resulted in 89.9% of lignin removal.

Alkaline pretreatment showed high lignin removal of about 89.9% when compared to DAP in SCT [19]. The acid pretreatment, which was carried out using sulfuric acid in the presence of Triton X-100 as surfactant, resulted in 50% of lignin removal [18]. But the biomass loaded in the case of acid was comparatively higher than the alkaline pretreatment, and decrease in solid concentration can increase the lignin removal in the case of acid pretreatment.

10.4.3 Biological pretreatment

10.4.3.1 Enzymatic pretreatment

Biological pretreatment with lignin-degrading enzymes is usually carried out at milder conditions and plays an efficient role in the lignin removal. Peroxidases and oxidases are the commonly available lignocellulosic enzymes among which laccases are the commercially used enzyme. Laccases are copper-containing oxidase enzymes and are involved in degradation of lignin for effective biomass conversion. The main advantage is that they do not require high energy input and do not produce harmful by-products.

Sherpa et al. [13] employed laccase enzyme produced from *Pleurotus djamor* for SCT pretreatment, and the studies showed 71.60 % of lignin removal from SCT. Another study reported by Sherpa et al. [20], using laccase enzyme for the pretreatment of SCT, resulted in maximum delignification of 79.1% with optimum conditions of 21% (w/v) substrate concentration at pH 7 incubated for 6 h with an enzyme concentration of 500 IU/mL at temperature 40°C.

10.4.4 Hybrid pretreatment

Hybrid pretreatment involves the combination of two or more pretreatment techniques for treating the LCB. Maximum lignin removal is possible by combining different pretreatments. Though hybrid pretreatments are costly, they are very effective for maximum lignin removal and also reduce the cost involved in recovery steps after pretreatment as the toxic by-products formed are very less in combined pretreatments.

Srinorakutara et al. [22] employed alkaline treatment followed by acid pretreatment, which yielded the maximum reducing sugar after enzymatic hydrolysis. It was observed that 2% (w/v) NaOH followed by 2% (w/v) sulfuric acid with autoclaving at 121°C for 15 min produced maximum ethanol of 48.17 g/L.

Raghavi et al. [23] employed crude glycerol with ferric chloride and NaOH for the removal of lignin from SCT and observed that it was highly effective in removing hemicelluloses and lignin. Maximum reducing sugar of 0.721 g/g was observed with 10% (w/w) biomass loading with different reactants concentration of crude glycerol (6%, v/v),

NaOH (5%), ferric chloride (1%, w/w), and incubation time (45 min) with autoclaving at 121°C.

Moodley and Kana [16] employed hybrid pretreatment of SCL with hydrochloric acid and moist heat with optimum conditions of 5.28% HCl with 187 min incubation at 94.94°C, which resulted in xylose and glucose yields of 8.92 and 1.68 g/L, respectively.

Pretreatment of sugarcane leaf waste using MW-aided inorganic salt performed by Moodley and Kana [25], with optimum conditions of 5 g of solid residue of mixed particle size immersed in 2 M $Fecl_2$, followed by MW irradiation of 700 W for 3.5 min, yielded 0.406 g/g of reducing sugars.

10.5 Structural characterization of sugarcane tops before and after pretreatment

SCT were subjected to different pretreatment techniques for efficient lignin removal and release of sugars. Structural characterization of native and pretreated SCT was analyzed using Fourier-transform infrared (FTIR) spectra, scanning electron microscope (SEM), and X-ray diffraction (XRD) studies.

10.5.1 Fourier-transform infrared analysis

FTIR spectra were used to analyze the difference in structural modifications between the native and pretreated SCT samples. Different absorption bands are visualized between 4000 and 400 cm^{-1} to examine the functional modifications after pretreatment. FTIR analyses of SCT subjected to different pretreatments are summarized in Table 10.2.

10.5.2 Scanning electron microscope analysis

Structural transformations and the morphology of native and pretreated sugarcane tops were visualized using SEM. The structural modifications observed in SCT subjected to different pretreatments are summarized in Table 10.3.

10.5.3 X-ray diffraction analysis

XRD studies were done by various authors on native and pretreated SCT samples to calculate the crystallinity index with the intensity of diffractions

Table 10.2 Summary of FTIR interpretation of pretreated SCT.

LCB	Pretreatment employed	FTIR analyses		References
		Bands	Inference	
SCT	Dilute acid pretreatment	Band widening at 1316 cm^{-1} Band widening at 3302 cm^{-1} Intensity of the band at 1045 cm^{-1} decreased	CH_2 wagging vibrations in the cellulose and hemicelluloses Stretching of H-bonded OH groups Indication of hemicellulose removal	[9]
SCT	Surfactant-assisted acid pretreatment	Band widening at 2889 cm^{-1} Band widening at 1045 cm^{-1} 1516 cm^{-1} band was absent	CH_2 stretching and this region is distinguished feature of cellulose Vibrational modes of $-CH_2$ OH groups and the C−O stretching vibration coupled with C−O bending of the C−OH groups of carbohydrates Assigned to C = C in guaiacyl ring of lignin of the aromatic ring of lignin, which indicates that the lignin is removed during pretreatment	[18]
SCT	Surfactant-assisted ultrasound pretreatment	1382, 1743, 2862, and 2924 cm^{-1} were absent Peak found near 898 cm^{-1} Peak at 1064 cm^{-1} Carbonyl band at 1743 cm^{-1} weakened	Most prominent bands that are found in native biomass but absent after pretreatment C−O−C stretching of ß-(1,4)-glycosidic linkages Stretching of C−O−C−O−C bonds in cellulose Indication of hemicelluloses removal	[15]
SCT	Alkali pretreatment	Band widening at 1318 cm^{-1} Band at 897 cm^{-1} Characteristic band at 1516 cm^{-1} Band at 1045 cm^{-1}	Reveals CH_2 wagging vibrations in cellulose C−O−C stretching of ß-(1,4)-glycosidic linkages in cellulose Assigned to C = C in guaiacyl ring of lignin of the aromatic ring of lignin, which indicates that the lignin is removed during pretreatment Responsible for the vibrational modes of CH_2OH groups and the C−O stretching vibration coupled with C−O bending of the C−OH groups of carbohydrates	[19]

(Continued)

Table 10.2 (Continued)

LCB	Pretreatment employed	FTIR analyses		References
		Bands	Inference	
SCT	Enzymatic pretreatment	Decrease in absorption band at 3543, 3468, and 3416 cm^{-1}	Removal of cellulose in the form of reducing sugar	[13]
		Intensity band of 1618 cm^{-1}	Corresponds to the carbonyl group	
		Decrease in intensity of band at 1055 cm^{-1}	Corresponds to the xylan region and indicates the removal of hemicelluloses	
SCT	Waste glycerol–assisted transition metal and alkali pretreatment	Peaks at 1000–1200 cm^{-1} region were enhanced	Indication of increase in cellulose content after pretreatment	[23]
		Broad absorption band at 3000–3500 cm^{-1}	Stretching of –OH groups	
		Absorption at 2900 cm^{-1}	Related to $-CH_2$ groups	
		Reduction in band intensity of 1509, 1464, and 1422 cm^{-1} bands	Indication of lignin removal	
		Band widening at 2800 cm^{-1}	CH_2 stretching and this region is distinguished feature of cellulose	
SCT	Enzymatic pretreatment	Decrease in the absorbance of following bands compared to native biomass:		[24]
		3435 cm^{-1}	O–H stretching in hydroxyl group	
		2427 cm^{-1}	C–H stretching of methyl and methylene	
		1635 cm^{-1}	C = O stretching of lignin	
		1384 and 1113 cm^{-1}	C–H, C–O deformation, bending or stretching vibration of lignin, cellulose, and hemicelluloses	
		830 and 527 cm^{-1}	Glycosidic linkage of cellulose and hemicellulose	
SCT	Enzymatic pretreatment	1510 cm^{-1} band absorption decreased	Corresponds the aromatic skeletal vibration of lignin that constituted the conjugated C = C, aryl-substituted C = C, and alkenyl C = C stretch indicating the lignin removal	[20]
		Decrease in the intensity at 1250 cm^{-1}	Characteristic of aromatic lignin indicating lignin removal	

FTIR, Fourier-transform infrared; *LCB*, Lignocellulosic biomass; *SCT*, sugarcane tops.

Table 10.3 Summary of scanning electron microscope inference of native and pretreated SCT.

LCB	Pretreatment employed	Observations		References
		Native	Pretreated	
SCT	Dilute acid pretreatment	Displayed a regular, compact, and smooth surface, indicating a highly ordered surface structure	Displayed a rough surface that exposed some internal areas in the biomass compared with the native biomass	[9]
SCT	Surfactant-assisted acid pretreatment	Native samples have a compact rigid structure	Pretreated samples showed a distorted structure indicating the partial removal of lignin and hemicelluloses	[18]
SCT	Surfactant-assisted ultrasound pretreatment	A highly ordered and compact structure	A highly distorted structure indicating removal of the external fibers by destroying the cellulose–hemicelluloses lignin network	[15]
SCT	Alkali pretreatment	Native samples have a compact rigid	Pretreated SCT have a distorted structure. This structure and increase in surface area of the pretreated SCT improve the hydrolysis efficiency	[19]
SCT	Waste glycerol–assisted transition metal and alkali pretreatment	Showed a compact and highly ordered structure.	Showed a highly distorted structure with removal of some external fibers and with more roughness and surface area	[23]
SCT	Enzymatic pretreatment	Showed compact and smooth surface	Increased surface area after delignification with laccase induces easy depolymerization of holocelluloses by cellulase and xylanase	[24]
SCT	Enzymatic pretreatment	a firm and compact structure	Rough distorted and porous structure	[20]

LCB, Lignocellulosic biomass, *SCT*, sugarcane tops.

Table 10.4 Summary of X-ray diffraction studies presenting crystallinity index of native and pretreated SCT.

LCB	Pretreatment employed	Crystallinity index (%)		References
		Native	Pretreated	
SCT	Dilute acid pretreatment	37.74	43.33	[9]
SCT	Surfactant-assisted acid pretreatment	37.74	52.9	[18]
SCT	Surfactant-assisted ultrasound pretreatment (without[a] and with[b] autoclaving)	37.74	48.85[a] 63.40[b]	[15]
SCT	Microwave alkali–assisted pretreatment	31.46	37.51	[21]
SCT	Alkali pretreatment	37.74	63.40	[19]
SCT	Enzymatic pretreatment	76.05	80.3	[20]
SCT	Waste glycerol–assisted transition metal and alkali pretreatment	45	57	[23]
Sugarcane leaf Waste	Microwave-assisted inorganic salt pretreatment($NaCl^c$, $ZnCl_2{}^d$, and $FeCl_3{}^e$)	20.45	25.70[c] 27.04[d] 31.26[e]	[25]

LCB, Lignocellulosic biomass; *SCT*, sugarcane tops.
[a]Corresponds to surfactant-assisted ultrasound pretreatment without autoclaving [15].
[b]Corresponds to surfactant-assisted ultrasound pretreatment with autoclaving [15].
[c]Corresponds to microwave-assisted inorganic salt pretreatment with NaCl [25].
[d]Corresponds to microwave-assisted inorganic salt pretreatment with $ZnCl_2$ [25].
[e]Corresponds to microwave-assisted inorganic salt pretreatment with $FeCl_3$ [25].

observed in XRD studies. The calculated crystallinity index after different pretreatments is summarized in Table 10.4.

From the structural characterization results of native and pretreated SCT, it is evident that pretreatment is the key process for the removal of lignin and pretreatment improved crystallinity index.

10.6 Saccharification of pretreated sugarcane tops

Hydrolysis plays a major role, and it is important step for bioethanol production, which mainly involves the production of fermentable sugars. During the hydrolysis process, the long chains of glucose molecules present in the cellulose are broken down to release the sugars for fermentation. The two major types of hydrolysis processes are chemical reaction using acids and enzymatic hydrolysis.

10.6.1 Acid hydrolysis

Acid hydrolysis is carried out using dilute or concentrated acids, such as sulfurous, sulfuric, hydrochloric, hydrofluoric, phosphoric, nitric, and formic acid. Among all, sulfuric and hydrochloric acids are most widely used for the hydrolysis of LCB [5]. Dilute acid hydrolysis are usually carried out at higher temperature range of 200°C–240°C, while lower temperature is required for concentrated acid hydrolysis and higher glucose yields of about 90% is achieved using 30%–70% sulfuric acid. But concentrated acid hydrolysis requires large amount of acid causing corrosion to equipment, and the main drawback with concentrated acid is the huge cost of acid, and a recovery step is necessary to lower the cost [28].

However, dilute hydrolysis process uses lower amount of acid, which accounts for 2%–5%. As this process is carried out at higher temperature, formation of toxic compounds, such as furfural and 5-HMF, occurs with the decomposition of hemicelluloses sugars [28]. This leads to the reduction of fermentable sugars with 50% glucose yield as the maximum. Furthermore, HMF produced is fermentation inhibitor, which reduces the ethanol production rate. Detoxification steps are necessary to overcome these drawbacks.

10.6.2 Enzymatic hydrolysis

Enzymatic hydrolysis involves enzymes for the conversion of LCB into sugars. This process provides a greater scope for the advancement of bioethanol production. Cellulases are the widely used enzymes to convert cellulose into simple sugars. Cellulase is a group of enzyme produced by bacterial and fungal species, which are aerobic or anaerobic, and work within a range of temperature [5]. It hydrolysis ß-(1,4)-glycosidic linkages in the cellulose chains by converting them into simple sugars. Steps involved in enzyme hydrolysis are shown in Fig. 10.4. Enzymatic hydrolysis of SCT reported in the literature is summarized in Table 10.5.

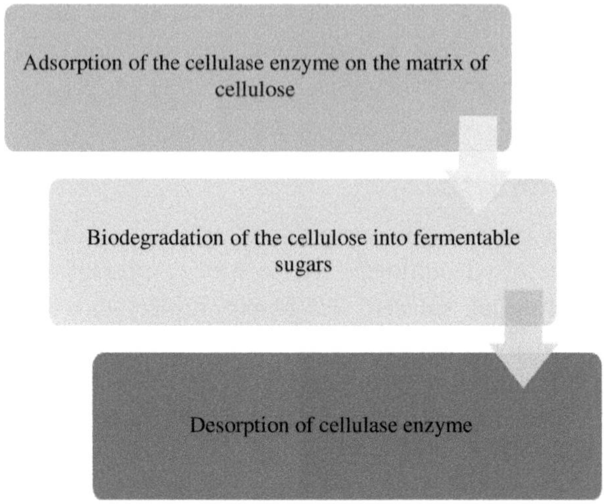

Figure 10.4 Steps involved in enzyme hydrolysis.

Table 10.5 List of enzymatic hydrolysis of SCT and their sugar yield reported in the literature.

S. no.	Hydrolysis/ saccharification technique	Optimized process conditions	Sugar yield	References
1	SCT—enzymatic hydrolysis (cellulase)	Biomass loading: 11.25% (w/w) Enzyme concentration: 50 FPU Surfactant concentration: 0.2% (Tween-80) Incubation time: 60 h Temperature: 50°C	0.685 g/g	[9]
2	SCT—enzymatic hydrolysis (cellulase)	Biomass loading: 11.25% (w/w) Enzyme concentration: 50 FPU Surfactant concentration: 0.2% (Tween-80) Incubation time: 60 h Temperature: 50°C	0.798 g/g	[18]

(*Continued*)

Table 10.5 (Continued)

S. no.	Hydrolysis/ saccharification technique	Optimized process conditions	Sugar yield	References
3	SCT—enzymatic hydrolysis (cellulase)	Biomass loading: 11.25% (w/w) Enzyme concentration: 50 FPU Surfactant concentration: 0.2% (Tween-80) Incubation time: 60 h Temperature: 50°C	0.661 g/g	[15]
4	SCT—enzymatic hydrolysis (cellulase)	Biomass loading: 11.25% (w/w) Enzyme concentration: 50 FPU Surfactant concentration: 0.2% (Tween-80) Incubation time: 42 h	0.775 g/g	[19]
5	SCT—enzymatic hydrolysis (cellulase)	Biomass loading: 14% w/v Incubation time: 7 h Temperature: 45° C–55°C pH 5 Enzyme titer: 19.33 IU/mL	508.92 mg/g	[13]
6	SCT—enzymatic hydrolysis (cellulase)	Biomass loading: 10% (w/w) Enzyme concentration: 100 FPU/g Surfactant concentration: 0.04% Incubation time: 72 h	0.376 g/g	[21]
7	SCT—enzymatic hydrolysis (cellulase)	Biomass loading: 15% (w/w) Enzyme concentration: 50 FPU pH 5 Incubation time: 48 h	0.509 g/g	[22]

(*Continued*)

Table 10.5 (Continued)

S. no.	Hydrolysis/ saccharification technique	Optimized process conditions	Sugar yield	References
8	SCT—enzymatic hydrolysis (cellulase)	Biomass loading: 5% (w/w) Enzyme concentration: 80 FPU/g Surfactant concentration: 0.05% Incubation time: 24 h	0.796 g	[23]

SCT, Sugarcane tops.

10.7 Conclusion

From the overall study, it is clear that most of the pretreatments resulted in maximum lignin removal from SCT. Though many pretreatments are effective, it cannot be used for commercial production as some of them are expensive, time–consuming, and most importantly some chemical treatments end up with the formation of inhibitors, which require further separation and purification that leads to increase in the process cost. To overcome these shortcomings, it could be effective to use hybrid pretreatments with the combination of physical, chemical, and biological pretreatments. From reported findings, it is also evident that enzyme hydrolysis efficiently converts delignified SCT into fermentable sugars. Further improvements can be employed in hybrid pretreatments for sustainable utilization of LCB for energy generation.

References

[1] M. Devarapalli, H.K. Atiyeh, A review of conversion processes for bioethanol production with a focus on syngas fermentation, Biofuel Res. J. 2 (2015) 268–280.

[2] R. Muktham, S.K. Bhargava, S. Bankupalli, A.S. Ball, A review on 1st and 2nd generation bioethanol production-recent progress, J. Sustain. Bioenergy Syst. 6 (2016) 72–92.

[3] Biofuels: Policies, Standards and Technologies, (2010). Available from: https://pdfs.semanticscholar.org/bc30/d69a1a183371f6b2b4df75700cbb22573c61.pdf?_ga = 2.125827146.111865585.1569565645-640486752.1569565645

[4] Government of India, National Policy on Biofuels - 2018 (2018) 1–23. Available from: http://petroleum.nic.in/sites/default/files/biofuelpolicy2018_1.pdf

[5] M. Balat, Production of bioethanol from lignocellulosic materials via the biochemical pathway: a review, Energy Convers. Manage. 52 (2011) 858–875.
[6] G. Jahnavi, G.S. Prashanthi, K. Sravanthi, L.V. Rao, Status of availability of lignocellulosic feed stocks in India: biotechnological strategies involved in the production of bioethanol, Renew. Sustain. Energy Rev. 73 (2017) 798–820.
[7] Z. Anwar, M. Gulfraz, M. Irshad, Agro-industrial lignocellulosic biomass a key to unlock the future bio-energy: a brief review, J. Radiat. Res. Appl. Sci. 7 (2014) 163–173.
[8] L. Rocha-Meneses, M. Raud, K. Orupõld, T. Kikas, Second-generation bioethanol production: a review of strategies for waste valorisation, Agron. Res. 15 (2017) 830–847.
[9] R. Sindhu, M. Kuttiraja, P. Binod, K.U. Janu, R.K. Sukumaran, A. Pandey, Dilute acid pretreatment and enzymatic saccharification of sugarcane tops for bioethanol production, Bioresour. Technol. 102 (2011) 10915–10921.
[10] F.H. Isikgor, C.R. Becer, Lignocellulosic biomass: a sustainable platform for the production of bio-based chemicals and polymers, Polym. Chem. 6 (2015) 4497–4559.
[11] C.N. Hamelinck, G. Van Hooijdonk, A.P.C. Faaij, Ethanol from lignocellulosic biomass: techno-economic performance in short-, middle- and long-term, Biomass Bioenergy 28 (2005) 384–410.
[12] L. Canilha, A.K. Chandel, T. Suzane dos Santos Milessi, F.A. Antunes, W. Luiz da Costa Freitas, M. das Graças Almeida Felipe, et al., Bioconversion of sugarcane biomass into ethanol: an overview about composition, pretreatment methods, detoxification of hydrolysates, enzymatic saccharification, and ethanol fermentation, J. Biomed. Biotechnol. 2012 (2012) 989572.
[13] K.C. Sherpa, M.M. Ghangrekar, R. Banerjee, Optimization of saccharification of enzymatically pretreated sugarcane tops by response surface methodology for ethanol production, Biofuels 7269 (2017) 1–8.
[14] R.K. Sukumaran, V.J. Surender, R. Sindhu, P. Binod, K.U. Janu, K.V. Sajna, et al., Lignocellulosic ethanol in India: prospects, challenges and feedstock availability, Bioresour. Technol. 101 (2010) 4826–4833.
[15] R. Sindhu, M. Kuttiraja, V. Elizabeth Preeti, S. Vani, R.K. Sukumaran, P. Binod, A novel surfactant-assisted ultrasound pretreatment of sugarcane tops for improved enzymatic release of sugars, Bioresour. Technol. 135 (2013) 67–72.
[16] P. Moodley, E.B.G. Kana, Comparative study of three optimized acid-based pretreatments for sugar recovery from sugarcane leaf waste: a sustainable feedstock for biohydrogen production, Eng. Sci. Technol. Int. J. 21 (1) (2017) 107–116.
[17] Z. Xu, F. Huang, Pretreatment methods for bioethanol production, Appl. Biochem. Biotechnol. 174 (2014) 43–62.
[18] R. Sindhu, M. Kuttiraja, P. Binod, V.E. Preeti, S.V. Sandhya, S. Vani, et al., Surfactant-assisted acid pretreatment of sugarcane tops for bioethanol production, Appl. Biochem. Biotechnol. 167 (2012) 1513–1526.
[19] R. Sindhu, M. Kuttiraja, P. Binod, R.K. Sukumaran, A. Pandey, Physicochemical characterization of alkali pretreated sugarcane tops and optimization of enzymatic saccharification using response surface methodology, Renew. Energy 62 (2014) 362–368.
[20] K.C. Sherpa, M.M. Ghangrekar, R. Banerjee, A green and sustainable approach on statistical optimization of laccase mediated delignification of sugarcane tops for enhanced saccharification, J. Environ. Manage. 217 (2018) 700–709.
[21] D.P. Maurya, S. Vats, S. Rai, S. Negi, Optimization of enzymatic saccharification of microwave pretreated sugarcane tops through response surface methodology for biofuel, Indian J. Exp. Biol. 51 (2013) 992–996.

[22] T. Srinorakutara, S. Suttikul, E. Butivate, V. Panphan, N. Boonvitthya, Optimization on pretreatment and enzymatic hydrolysis of sugarcane trash for ethanol production, J. Food Sci. Eng. 4 (2014) 148–154.
[23] S. Raghavi, R. Sindhu, P. Binod, E. Gnansounou, A. Pandey, Development of a novel sequential pretreatment strategy for the production of bioethanol from sugarcane trash, Bioresour. Technol. 199 (2016) 202–210.
[24] A. Althuri, R. Banerjee, Separate and simultaneous saccharification and fermentation of a pretreated mixture of lignocellulosic biomass for ethanol production, Biofuels 7269 (2017) 1–12.
[25] P. Moodley, E.B.G. Kana, Microwave-assisted inorganic salt pretreatment of sugarcane leaf waste: effect on physiochemical structure and enzymatic saccharification, Bioresour. Technol. 235 (2017) 35–42.
[26] P.B. Subhedar, P.R. Gogate, Use of ultrasound for pretreatment of biomass and subsequent hydrolysis and fermentation, Biomass Fractionation Technologies for a Lignocellulosic Feedstock Based Biorefinery, Elsevier, 2016, pp. 127–149.
[27] L.J. Jönsson, C. Martín, Pretreatment of lignocellulose: formation of inhibitory by-products and strategies for minimizing their effects, Bioresour. Technol. 199 (2016) 103–112.
[28] A. Verardi, I. De Bari, E. Ricca, V. Calabrò, Hydrolysis of Lignocellulosic Biomass: Current Status of Processes and Technologies and Future Perspectives, Bioethanol InTech, 2012.

Index

Note: Page numbers followed by "*f*" and "*t*" refer to figures and tables, respectively.

D

E

I

J

K

L

X

Y

FSC
www.fsc.org
MIX
Papier | Fördert gute Waldnutzung
FSC® C083411

Kolibri 360 GmbH
Ferdinand-Jühlke-Straße 7
99095 Erfurt, Deutschland
produktsicherheit@kolibri360.de